T0329089

NEUROTECHNOLOGY AND BRAIN STIMULATION IN PEDIATRIC PSYCHIATRIC AND NEURODEVELOPMENTAL DISORDERS

NEUROTECHNOLOGY AND BRAIN STIMULATION IN PEDIATRIC PSYCHIATRIC AND NEURODEVELOPMENTAL DISORDERS

Edited by

LINDSAY M. OBERMAN

Center for Neuroscience and Regenerative Medicine, Henry M. Jackson Foundation for the Advancement of Military Medicine, Rockville, MD, United States

PETER G. ENTICOTT

Cognitive Neuroscience Unit, School of Psychology, Deakin University, Geelong, VIC, Australia

Academic Press is an imprint of Elsevier
125 London Wall, London EC2Y 5AS, United Kingdom
525 B Street, Suite 1650, San Diego, CA 92101, United States
50 Hampshire Street, 5th Floor, Cambridge, MA 02139, United States
The Boulevard, Langford Lane, Kidlington, Oxford OX5 1GB, United Kingdom

British Library Cataloguing-in-Publication Data

A catalogue record for this book is available from the British Library

Library of Congress Cataloging-in-Publication Data

A catalog record for this book is available from the Library of Congress

ISBN: 978-0-12-812777-3

For Information on all Academic Press publications
visit our website at https://www.elsevier.com/books-and-journals

Publisher: Nikki Levy
Acquisition Editor: Melanie Tucker
Editorial Project Manager: Carlos Rodriguez
Production Project Manager: Sujatha Thirugnana Sambandam
Cover Designer: Greg Harris

Typeset by MPS Limited, Chennai, India

Contents

List of Contributors

Christine A. Conelea Department of Psychiatry, University of Minnesota, Minneapolis, MN, United States

Asimina S. Courelli Bioengineering Department, University of California, San Diego, CA, United States

Hristos S. Courellis Bioengineering Department, University of California, San Diego, CA, United States; Swartz Center of Computational Neuroscience, University of California, San Diego, CA, United States

Paul E. Croarkin Department of Psychiatry and Psychology, Division of Child and Adolescent Psychiatry, Mayo Clinic, Rochester, MN, United States

Ugur Damar Neuromodulation Program, Division of Epilepsy and Clinical Neurophysiology, Department of Neurology, Boston Children's Hospital, Harvard Medical School, Boston, MA, United States; F.M. Kirby Neurobiology Center, Department of Neurology, Boston Children's Hospital, Harvard Medical School, Boston, MA, United States

Nick J. Davis Department of Psychology, Manchester Metropolitan University, Manchester, United Kingdom

Peter G. Enticott Cognitive Neuroscience Unit, School of Psychology, Deakin University, Geelong, VIC, Australia

Faranak Farzan School of Mechatronic Systems Engineering, Simon Fraser University, Surrey, BC, Canada

Elisabeth V.C. Friedrich Cognitive Science Department, University of California, San Diego, CA, United States

Donald L. Gilbert Pediatrics and Neurology, Cincinnati Children's Hospital Medical Center, Cincinnati, OH, United States

Gerald A. Grant Division of Pediatric Neurosurgery, Lucile Packard Children's Hospital, Palo Alto, CA, United States

Michaela Gusman Boston Children's Hospital, Boston, MA, United States

Casey H. Halpern Department of Neurosurgery, Stanford Hospitals and Clinics, Stanford, CA, United States

Mustafa Q. Hameed Neuromodulation Program, Division of Epilepsy and Clinical Neurophysiology, Department of Neurology, Boston Children's Hospital, Harvard Medical School, Boston, MA, United States; Department of Neurosurgery, Boston Children's Hospital, Harvard Medical School, Boston, MA, United States; F.M. Kirby Neurobiology Center, Department of Neurology, Boston Children's Hospital, Harvard Medical School, Boston, MA, United States

Anthony J. Hannan Florey Institute of Neuroscience and Mental Health, Melbourne Brain Centre, University of Melbourne, Parkville, VIC, Australia; Department of Anatomy and Neuroscience, University of Melbourne, Parkville, VIC, Australia

Martin Holtmann LWL-University Hospital for Child and Adolescent Psychiatry and Psychotherapy Hamm, Ruhr University Bochum, Germany

Jason Kahn Department of Psychiatry, Boston Children's Hospital, Boston, MA, United States; Department of Psychiatry, Harvard Medical School, Boston, MA, United States; Neuromotion Labs, Boston, MA, United States

Lily H. Kim Department of Neurosurgery, Stanford Hospitals and Clinics, Stanford, CA, United States; Division of Pediatric Neurosurgery, Lucile Packard Children's Hospital, Palo Alto, CA, United States

Melissa Kirkovski Cognitive Neuroscience Unit, School of Psychology, Deakin University, Geelong, VIC, Australia

Inken Kirschbaum-Lesch LWL-University Hospital for Child and Adolescent Psychiatry and Psychotherapy Hamm, Ruhr University Bochum, Germany

Mari A. Kondo School of Psychiatry, University of New South Wales, Sydney, NSW, Australia; John Curtin School of Medical Research, Australian National University, Canberra, ACT, Australia

Tanja Legenbauer LWL-University Hospital for Child and Adolescent Psychiatry and Psychotherapy Hamm, Ruhr University Bochum, Germany

Charles P. Lewis Department of Psychiatry and Psychology, Division of Child and Adolescent Psychiatry, Mayo Clinic, Rochester, MN, United States

Paul MacMullin Neuromodulation Program, Division of Epilepsy and Clinical Neurophysiology, Department of Neurology, Boston Children's Hospital, Harvard Medical School, Boston, MA, United States; F.M. Kirby Neurobiology Center, Department of Neurology, Boston Children's Hospital, Harvard Medical School, Boston, MA, United States

Nicole C.R. McLaughlin Department of Psychiatry and Human Behavior, Alpert Medical School of Brown University, Providence, RI, United States; Butler Hospital, Providence, RI, United States

Michael A. Nitsche Department Psychology and Neurosciences, Leibniz Research Center for Working Environment and Human Factors, Dortmund, Germany; Department of Neurology, University Medical Hospital Bergmannsheil, Bochum, Germany

Lindsay M. Oberman Center for Neuroscience and Regenerative Medicine, Henry M. Jackson Foundation for the Advancement of Military Medicine, Rockville, MD, United States

Jaime A. Pineda Cognitive Science Department, University of California, San Diego, CA, United States; Group in Neurosciences, University of California, San Diego, CA, United States

Laura M. Prolo Department of Neurosurgery, Stanford Hospitals and Clinics, Stanford, CA, United States; Division of Pediatric Neurosurgery, Lucile Packard Children's Hospital, Palo Alto, CA, United States

Caroline A. Quon The PRIME Center, VA Connecticut Healthcare System, West Haven, CT, United States

Jennifer L. Quon Department of Neurosurgery, Stanford Hospitals and Clinics, Stanford, CA, United States; Division of Pediatric Neurosurgery, Lucile Packard Children's Hospital, Palo Alto, CA, United States

Alexander Rotenberg Neuromodulation Program, Division of Epilepsy and Clinical Neurophysiology, Department of Neurology, Boston Children's Hospital, Harvard Medical School, Boston, MA, United States; F.M. Kirby Neurobiology Center, Department of Neurology, Boston Children's Hospital, Harvard Medical School, Boston, MA, United States; Berenson-Allen Center for Noninvasive Brain Stimulation and Division for Cognitive Neurology, Beth Israel Deaconess Medical Center, Harvard Medical School, Boston, MA, United States

Carmelo M. Vicario Department Psychology and Neurosciences, Leibniz Research Center for Working Environment and Human Factors, Dortmund, Germany; Department of Neurology, University Medical Hospital Bergmannsheil, Bochum, Germany; Department of Cognitive Sciences, Educational and Cultural Studies (CSECS), via della concezione 6, Messina, Italy

Suzanne Wintner Neuromotion Labs, Boston, MA, United States; Simmons College School of Social Work, Boston, MA, United States

CHAPTER

1

Introduction to Device-Based Treatments in Pediatric Psychiatric and Neurodevelopmental Disorders

Lindsay M. Oberman[1] *and Peter G. Enticott*[2]

[1]Center for Neuroscience and Regenerative Medicine, Henry M. Jackson Foundation for the Advancement of Military Medicine, Rockville, MD, United States [2]Cognitive Neuroscience Unit, School of Psychology, Deakin University, Geelong, VIC, Australia

OUTLINE

The human brain is the most complicated biological structure in the known universe. We've only just scratched the surface in understanding how it works – or, unfortunately, doesn't quite work when disorders and disease occur. – ***Francis Collins, MD, PhD, Director of the National Institutes of Health.***

Neurotechnological Pediatric Neuropsych
DOI: https://doi.org/10.1016/B978-0-12-812777-3.00001-5

EPIDEMIOLOGY OF PEDIATRIC PSYCHIATRIC AND NEURODEVELOPMENTAL DISORDERS

Pediatric psychiatric and neurodevelopmental disorders include neurodevelopmental [such as autism spectrum disorder (ASD) and attention deficit hyperactivity disorder (ADHD)], emotional [such as major depressive disorder (MDD) and anxiety disorders], and behavioral disorders (such as conduct disorder and oppositional defiant disorder). These disorders often have deleterious effects on psychological and social well-being, academic performance, and quality of life. Having a psychiatric or neurodevelopmental disorder as a child increases one's risk for both psychiatric and general medical conditions later in life (Rutter et al., 2006; Reef et al., 2011; Costello et al., 2016). There has yet to be developed a "cure" for this category of disorders, thus, treatments aim to reduce the disabling symptoms. Psychiatric and neurodevelopmental disorders not only affect the individual but the family members, caretakers, and community as well. There is also a substantial economic cost associated with medical care, education, welfare support, lost workplace productivity, and reduced workforce participation. Children with psychiatric and neurodevelopmental disorders often require significant additional support from families and educational systems and, due to the limited efficacy of validated treatment options, frequently persist into adulthood (Polanczyk and Rohde 2007; Nevo and Manassis 2009; Shaw et al., 2012).

Recent meta-analyses estimate the worldwide prevalence of pediatric psychiatric and neurodevelopmental disorders at approximately 13.4% of the pediatric population (Polanczyk et al., 2015). Based on the most recent estimate, the worldwide pediatric population is approximately 1.8 billion children. Thus, pediatric psychiatric and neurodevelopmental disorders affect approximately 241 million youths around the world. Compounding this clear public health concern is the staggering statistic that an estimated two-thirds of all young people with psychiatric and neurodevelopmental disorders are not receiving adequate treatment. Thus, the need for novel, innovative treatments for these disorders represents a critical and currently unmet need.

HISTORY OF TREATMENT OF PEDIATRIC PSYCHIATRIC AND NEURODEVELOPMENTAL DISORDERS

The concept of pediatric psychiatric and neurodevelopmental disorders first emerged in the late 1800s; however, it was not until the early part of the 20th century that disorders of childhood were seen as unique or distinguishable from adult psychiatric disorders. The first dedicated pediatric psychiatric clinic in the United States was established in 1909 and the first English-language text on child psychiatry was published in 1935 (Kanner, 1935). Autism and ADHD (then known as hyperkinesis) were recognized as childhood disorders in the 1940s and childhood depression in the 1950s. The first multiaxial coding scheme for clinical syndromes in child psychiatry was developed and evaluated in 1975 (Rutter et al., 1975) and formed the basis for subsequent classification and refinement in the *Diagnostic and Statistical Manual of Mental Disorders (DSM)* of the American Psychiatric Association. It was not until the third edition of the *DSM* (American Psychiatric Association, 1980) that child and adolescent mental disorders were assigned a separate and distinct section within that classification system.

The recognition that children and adolescents suffer from psychiatric disorders is thus a very recent phenomenon. The development of treatments, services, and preventive approaches to risk for these disorders is even more recent. However, in the past two decades, stemming in part from the rapid advances in psychopharmacology and neurotechnology as well as federally funded programs specifically focused on neuroscience and mental health research; the knowledge bases on treatments, services, and prevention programs for pediatric psychiatric and neurodevelopmental disorders have greatly expanded. Yet, despite this progress, the burden of childhood mental health problems is not lessening suggesting that the currently available interventions are not adequately addressing the problem.

In 1990, the then President George H.W. Bush officially declared 1990–99 as the "Decade of the Brain" with the promise of "The dawn of a new era of discovery in brain research" (Jones and Mendell, 1999). As promised, the end of the 20th century and early years of the 21st century did see the advent of functional magnetic resonance imaging, as well as noninvasive brain stimulation techniques including repetitive transcranial magnetic stimulation (TMS), and transcranial direct current stimulation (tDCS). Building upon these discoveries [and rediscoveries in the case of tDCS], 2010–19 is being characterized as the "Decade of Translation." Dr. Thomas Insel, former Director of the National Institute of Mental Health, optimistically expressed, "Clinical Neuroscientists are finally beginning to understand what underlies a few mental disorders. We can look forward to seeing this new understanding translate to improved treatments that will finally reduce the occurrence and death rates of these disabling illnesses." (http://www.dana.org/Publications/Details.aspx?id = 43169). Concurrent with the clinical neuroscience developments are advances in neurotechnology. In 2013, NIH began a 12 year "Brain Research Through Advancing Innovative

Neurotechnologies Initiative" (https://www.braininitiative.nih.gov/). In Europe, there is the Human Brain Project, which is cofunded by the European Union and has a large computational component (https://www.humanbrainproject.eu/). The degree to which these large-scale research programs will lead to novel and innovative therapeutic interventions in the area of pediatric psychiatric and neurodevelopmental disorders is yet to be determined.

NEUROTECHNOLOGY AND BRAIN STIMULATION IN PEDIATRIC PSYCHIATRIC AND NEURODEVELOPMENTAL DISORDERS

The chapters in this volume are meant to highlight some of the groundbreaking research being conducted in the area of device-based interventions for pediatric psychiatric and neurodevelopmental disorders. The brain stimulation techniques discussed in this volume include TMS, tDCS, and deep brain stimulation (DBS). Other chapters discuss the use of neurofeedback, chronotherapy and environmental interventions in pediatric psychiatric and neurodevelopmental disorders. This is an emerging field whose full potential is yet to be realized. Though many of these devices and techniques have been applied in adult psychiatry and neurology for many years, recently many researchers and clinicians have begun applying these treatments to children and adolescents. There is much excitement about the promise of these novel therapeutic interventions and as detailed above, there is a clear and urgent need in the area of child psychiatry. This excitement is supported by preliminary data in many cases; however, equally one needs to acknowledge the infancy of this field and the relatively small number of children who have participated in these initial trials. In addition, as with any clinical intervention, these techniques are not without risk for adverse side effects, some quite serious. Risks include transient headaches, pain, or discomfort near the site of stimulation, and in rare cases induction of a seizure or syncopal episode. The relative safety, tolerability, and efficacy of these techniques in children and adolescents as compared to adults, and how application of these techniques may impact and interact with ongoing neurodevelopment, are largely unknown.

Despite the therapeutic promise of neuromodulation, the translation to the clinic poses a number of important challenges and complexities. Some challenges include (1) considerations in selecting the appropriate dose for a child depending on their age and neurodevelopmental stage, (2) the mechanism of action for the various devices and whether modulation of that mechanism impacts the behavioral symptoms that define the targeted disorder, (3) the synergistic or antagonistic effects of concomitant pharmacological interventions, and (4) whether there is a sensitive period where this type of intervention may be most effective. On the one side, there is an obvious unmet need and billions of children who could benefit from novel interventions, but on the other, children remain a vulnerable population that require additional precautions and considerations. It is for this reason that use of these techniques, and especially TMS and DBS, should be limited to institutionally regulated clinical trials conducted by individuals who have been trained both in safe use of these techniques as well as the relevant neurodevelopmental considerations of applying these techniques to children with psychiatric or neurodevelopmental disorders. Some of these ethical considerations are further discussed in Chapter 4, Ethics of Device-Based Treatments in Pediatric Neuropsychiatric Disorders (Davis) and throughout this volume.

Though there are many daunting challenges facing clinical researchers aiming to develop novel, device-based treatments for pediatric psychiatric and neurodevelopmental disorders, we as a field bring state-of-the-art neurotechnology, public and private foundation support, as well as unwavering dedication to innovative solutions. The prevalence rates of pediatric psychiatric and neurodevelopmental disorders have not decreased in several decades, but children born today may not face the same odds. The accumulation of knowledge of pathophysiology underlying these disorders, as well as the development of device-based treatments specifically designed to target those pathophysiological mechanisms, provide the necessary background and tools to develop more effective treatment options. Finally, it is beginning to be recognized that although psychiatric disorders are based in the brain, the specific etiology in any given individual is likely a result of complex genetic, developmental, social, and environmental factors. Thus, treatments must be both multifaceted and individualized.

The first two chapters of this volume focus on brain development across childhood. Chapter 2, The Developing Brain—Relevance to Pediatric Neurotechnology (Hameed, Damar, MacMullin, and Rotenberg) reviews the cellular, molecular, and neural system changes that occur across childhood and adolescence. This chapter emphasizes that the brain of a child is not and should not be considered to be simply a "small adult brain." Trials aiming to develop therapeutic interventions in pediatric psychiatric and neurodevelopmental disorders must consider the fluid and plastic state of the brain of a child, how neural development may be aberrant in the targeted disorder, and how device-based treatments impact neural development. These considerations will impact both the safety and efficacy of the device-based intervention. This is followed by Chapter 3, Environmental Stimulation Modulating the Pathophysiology of Neurodevelopmental Disorders (Kondo and Hannan) that underscores the important contribution of environmental exposure to brain development and how environments can be manipulated to confer positive benefits in preclinical animal models of pediatric psychiatric and neurodevelopmental disorders. Chapter 4, Ethics of Device-Based Treatments in Pediatric Neuropsychiatric Disorders (Davis) highlights the safety and ethical concerns surrounding the use of device-based interventions on the developing brain of a child.

Chapters 5–8 focus on TMS and its use in specific Pediatric Psychiatric Disorders [see Chapter 5: TMS in Autism Spectrum Disorder (Enticott, Kirkovski, and Oberman), ASD; Chapter 6: Transcranial Magnetic Stimulation in Attention Deficit Hyperactivity Disorder (Gilbert), ADHD; Chapter 7: TMS in Child/Adolescent Major Depression (Lewis, Farzan, and Croarkin), MDD; and Chapter 8: TMS in Tourette Syndrome and OCD (Conelea and McLaughlin), Tourette Syndrome and Obsessive Compulsive Disorder]. TMS involves the application of a strong magnetic pulse through an electromagnetic coil that is placed flush against the scalp. The pulsed magnetic field that is discharged from the coil induces an electrical field inside the brain that in turn induces activity in cortical neurons, cortical interneurons, and associated networks. Repeated trains of these magnetic pulses lead to modulation in cortical excitability that lasts beyond the duration of the pulse train. This plastic modulation of local cortical excitability and resulting changes in network connectivity are thought to underlie the therapeutic effects of such interventions. Thus, disorders where the putative pathophysiology involves aberrant cortical excitability and network connectivity may be amenable to modulation with TMS.

Chapter 9, tDCS in Pediatric Neuropsychiatric Disorders (Nitsche and Vicario) discusses the use of tDCS across multiple pediatric psychiatric and neurodevelopmental disorders. Unlike TMS, tDCS uses steady electrical currents passed across the scalp at subthreshold levels for modulation of brain excitability. The neuromodulation induced by tDCS is thought to be largely a result of depolarization or hyperpolarization of the resting membrane potential of cortical neurons. This shift in the resting membrane potential makes the neuron more (or less) likely to fire in response to a given input. Thus, tDCS is commonly combined with behavioral interventions or specific tasks to engage the appropriate brain networks during or immediately following the stimulation. Chapter 10, Deep Brain Stimulation for Pediatric Neuropsychiatric Disorders (Quon, Quon, Prolo, Grant, and Halpern) focuses on DBS, which is an invasive brain stimulation technique involving neurosurgical implantation of an electrical stimulation device inside the brain. Given the significant risks involved in neurosurgery, this is only used in a limited number of cases, and the efficacy of such interventions is largely unknown.

Chapter 11 (Courellis, Courelli, Friedrich, and Pineda) and Chapter 12 (Kahn, Gusman, and Wintner) cover neurofeedback interventions in neurodevelopmental disorders and emotion regulation, respectively. Neurofeedback involves presenting information regarding the individual's brain activity back to them in real time such that they can voluntarily modulate this activity using behavioral strategies. Because the modulation of brain activity is a result of behavioral strategies, as opposed to electrical or magnetic stimulation, the risk profile for this type of intervention is exceedingly benign. Though this technique has been shown to modulate brain activity during a given session, the durability of modulation beyond the session and to related cognitive and behavioral domains in the "real world" appears to be limited. That being said its ease of use and relatively safe side-effect profile make neurofeedback a technique worthy of further exploration, especially for use in pediatric psychiatric and neurodevelopmental disorders.

The final chapter, that is, Chapter 13, Chronotherapy for Child/Adolescent Major Depression (Lesch, Holtmann, and Legenbauer), covers the use of chronotherapeutic approaches for child and adolescent MDD. Chronotherapy uses devices such as light boxes and other behavioral intervention techniques to manipulate sleep and circadian rhythms. These techniques are well established in the treatment of adult MDD and are now being trialed in pediatric populations.

Overall, this volume aims to introduce the reader to the burgeoning field of neurotechnology and brain stimulation for pediatric psychiatric and neurodevelopmental disorders. Child psychologists, psychiatrists, and other behavioral health practitioners will find evidence-based recommendations for non-pharmacological interventions that may be effective for the treatment of their pediatric patients with psychiatric and neurodevelopmental disorders. Many of these techniques are quite novel such that clinicians may not be aware of these device-based interventions. Likewise, clinical researchers may not be aware of use of these techniques in clinical and basic science research. Herein, the researchers will find data and references for device-based technology that can be used in clinical research designs aimed at studying and modulating brain activity in children and adolescents with psychiatric and neurodevelopmental disorders. Students and fellows who are beginning a career in child psychiatry, clinical neuroscience, or clinical or experimental psychology will find useful information about the state of the science in this fast-moving field of neurotechnology. Finally, children with psychiatric and neurodevelopmental disorders and their families have a reason to be hopeful. In this volume, you will find chapters written by the leaders in their respective fields who have and continue to work diligently to apply these non-pharmacologic techniques to the study and treatment of these debilitating disorders. If the past decade is any indication, we are quite hopeful for a future where safe, well-validated, and device-based treatments are available for children with psychiatric and neurodevelopmental disorders.

References

American Psychiatric Association, 1980. Diagnostic and Statistical Manual of Mental Disorders, third ed. Arlington, VA.

Costello, E.J., Copeland, W., Angold, A., 2016. The Great Smoky Mountains Study: developmental epidemiology in the southeastern United States. Soc. Psychiatry. Psychiatr. Epidemiol. 51 (5), 639–646.

Jones, E.G., Mendell, L.M., 1999. Assessing the decade of the brain. Science 284 (5415), 739.

Kanner, L., 1935. Child Psychiatry. Chas C. Thomas, Springfield, IL.

Nevo, G.A., Manassis, K., 2009. Outcomes for treated anxious children: a critical review of long-term-follow-up studies. Depress. Anxiety 26 (7), 650–660.

Polanczyk, G., Rohde, L.A., 2007. Epidemiology of attention-deficit/hyperactivity disorder across the lifespan. Curr. Opin. Psychiatry 20 (4), 386–392.

Polanczyk, G.V., Salum, G.A., Sugaya, L.S., Caye, A., Rohde, L.A., 2015. Annual research review: a meta-analysis of the worldwide prevalence of mental disorders in children and adolescents. J. Child Psychol. Psychiatry 56 (3), 345–365.

Reef, J., Diamantopoulou, S., van Meurs, I., Verhulst, F.C., van der Ende, J., 2011. Developmental trajectories of child to adolescent externalizing behavior and adult DSM-IV disorder: results of a 24-year longitudinal study. Soc. Psychiatry. Psychiatr. Epidemiol. 46 (12), 1233–1241.

Rutter, M., Kim-Cohen, J., Maughan, B., 2006. Continuities and discontinuities in psychopathology between childhood and adult life. J. Child Psychol. Psychiatry 47 (3–4), 276–295.

Rutter, M., Shaffer, D., Shepherd, M., 1975. A Multi-Axial Classification of Child Psychiatric Disorders: An Evaluation of a Proposal Geneva. World Health Organization, Switzerland.

Shaw, M., Hodgkins, P., Caci, H., Young, S., Kahle, J., Woods, A.G., et al., 2012. A systematic review and analysis of long-term outcomes in attention deficit hyperactivity disorder: effects of treatment and non-treatment. BMC Med. 10, 99.

CHAPTER

2

The Developing Brain—Relevance to Pediatric Neurotechnology

Mustafa Q. Hameed[1,2,3], *Ugur Damar*[1,3], *Paul MacMullin*[1,3] *and Alexander Rotenberg*[1,3,4]

[1]Neuromodulation Program, Division of Epilepsy and Clinical Neurophysiology, Department of Neurology, Boston Children's Hospital, Harvard Medical School, Boston, MA, United States [2]Department of Neurosurgery, Boston Children's Hospital, Harvard Medical School, Boston, MA, United States [3]F.M. Kirby Neurobiology Center, Department of Neurology, Boston Children's Hospital, Harvard Medical School, Boston, MA, United States [4]Berenson-Allen Center for Noninvasive Brain Stimulation and Division for Cognitive Neurology, Beth Israel Deaconess Medical Center, Harvard Medical School, Boston, MA, United States

OUTLINE

Neurotechnological Pediatric Neuropsych
DOI: https://doi.org/10.1016/B978-0-12-812777-3.00002-7

A WINDOW OF OPPORTUNITY

The brain continues to grow after birth, reaching approximately 80% of its adult size by the second year of life Dekaban, 1978; Knickmeyer et al., 2008). While cellular and molecular neurodevelopment processes start in utero, they continue in early life as neuronal circuitry expands and is fine-tuned, ultimately forming a functionally organized neural network (Tau and Peterson, 2010).

Neuroblasts and early neurons are seen in the subventricular zone of newborns, and neuronal migration is still observed around the anterior body of the lateral ventricle and in cerebral white matter up to several months after birth (Weickert et al., 2000; Kostovic and Rakic, 1980; Paredes et al., 2016). These and other data indicate that neurogenesis and neuronal migration continue throughout the postnatal period (Bhardwaj et al., 2006). These neurodevelopmental processes are accompanied by the continuing formation of cortical synaptic connections. Arborization of both pyramidal cells and inhibitory interneurons accelerates through the first 2 years of life and is accompanied by the expansion of cortical layers II and III relative to other layers (Mrzljak et al., 1990; Huttenlocher and Dabholkar, 1997; Landing et al., 2002). The time course of synaptogenesis is region-dependent. For example, synaptic density in the primary visual cortex reaches its peak much before the prefrontal cortex, despite starting at about the same time. Concurrent synaptic pruning processes also become increasingly active after birth. As with synaptogenesis, the timeline for pruning is also region-specific, starting with sensory and motor cortices after birth before progressing to the association cortices, corpus callosum, and beyond. This organization of neuronal circuitry through synaptogenesis, pruning, and subsequent activity-dependent synaptic remodeling is critical for normal brain function (Tau and Peterson, 2010; Levitt, 2003; Luhmann et al., 2016; Heck et al., 2008; Reiprich et al., 2005; Tagawa and Hirano, 2012; Mizuno et al., 2007; Heck et al., 2007; Inada et al., 2011; Jansen, 2017). Notably, whether these processes can be disrupted by neurostimulation is unknown.

Neuroglia also increases in both size and number after birth. Myelination occurs most rapidly during the first year of life and thereafter at a slower but steady pace. As with synaptogenesis and synaptic pruning, myelination too follows a region- and circuit-dependent time course. Myelination proceeds in a caudocranial direction, reaching the frontal lobes late in the first year. Sensory circuits myelinate first followed by motor and then association areas, and subcortical structures myelinate before cortical ones within given circuits (Tau and Peterson, 2010).

Development of sensory, motor, and cognitive function directly follows the completion of these region- and circuit-specific developmental processes. For instance, cortical remodeling and myelination of association areas may correspond to the development of inhibitory control over reflexive behavior as well as goal-directed behavior like reach-to-grasp by the third month after birth, and the appearance of hand-to-hand transfer around the sixth month of life may similarly reflect increasing coordination among sensory, motor, and association circuits (Tau and Peterson, 2010).

Relevant to brain stimulation in the young, critical neurodevelopmental processes of neurogenesis, apoptosis, migration, synaptogenesis, pruning, synaptic remodeling, and myelination are all at least partly activity-dependent (Tau and Peterson, 2010; Luhmann et al., 2016; Heck et al., 2008; Reiprich et al., 2005; Tagawa and Hirano, 2012; Mizuno et al., 2007; Mitew et al., 2016), with GABAergic signaling playing a critical role (Luhmann et al., 2016; Heck et al., 2007; Inada et al., 2011; Jansen, 2017). Thus, the developing brain may be particularly susceptible to external influences particularly from stimulation protocols that rely on the modulation of GABAergic signaling (Amadi et al., 2015; Huang et al., 2005).

THE YOUNG AND THE AMBIDEXTROUS

Maturation of human cortical excitability and corticospinal tract lateralization has been extensively studied using single pulse transcranial magnetic stimulation (spTMS) in both normal children as well as those with neuropsychiatric disease. spTMS is a noninvasive brain stimulation technique in which a single stimulus is delivered to the cortex to elicit an evoked response. When delivered to the motor cortex, the TMS (transcranial magnetic stimulation) pulse elicits a motor-evoked potential (MEP) in a contralateral limb muscle which is recorded via surface electromyography (EMG). MEP parameters can then be studied to obtain neurophysiologic insights (Hameed et al., 2017).

The stimulus intensity required for spTMS to generate MEPs (motor threshold; MT) progressively increases for the first 90 days after birth in humans, reaching a plateau phase that lasts until approximately

12 months of age and then decreases throughout childhood, reaching adult human levels by approximately age 15 years (Hameed et al., 2017; Saisanen et al., 2017; Kaye et al., 2017).

spTMS studies also indicate the presence of direct ipsilateral projections in neonates. Motor spTMS generates bilateral MEPs in this age group, and ipsilateral-evoked potentials exhibit shorter latencies than contralateral responses. These uncrossed corticospinal tracts are suppressed or lost with age, with a progressive decrease in ipsilateral MEP amplitude accompanied by increases in MT and MEP latency compared to contralateral (Rajapakse and Kirton, 2013; Hameed et al., 2017; Eyre, 2003; Eyre et al., 2001). While contralateral MEP latency progressively decreases during childhood and early adolescence (Saisanen et al., 2017; Kaye et al., 2017), short latency ipsilateral-evoked responses do not occur in normal subjects past infancy, even though ipsilateral projections persist in some adults (Muller et al., 1991; Muller and Homberg, 1992; Fietzek et al., 2000; Heinen et al., 1998; Muller et al., 1997; Nezu et al., 1999).

Unilateral brain injury early in life derails the normal corticospinal tract maturation patterns described above, with distinct results depending on the age at time of injury. Perinatal unilateral motor cortex or white matter injury, for example, results in persistence of ipsilateral and contralateral tracts arising from the undamaged hemisphere and eventually bilateral corticospinal connectivity in the contralesional hemisphere (Kirton et al., 2008). However, both healthy contralateral motor tracts in the lesional hemisphere as well as compensatory preservation of ipsilateral motor connections in the undamaged hemisphere appear to be absent in patients with acquired hemiplegia brought about by injury after the second year (Carr et al., 1993).

Proper interpretation of functional, behavioral, and clinical data obtained from younger patients using diagnostic or therapeutic neurostimulation protocols (e.g., functional motor mapping using navigated TMS) is therefore contingent upon a sound understanding of the maturational patterns of the underlying neurobiological and neurophysiological principles.

INSIGHTS FROM BASIC RESEARCH

Excitability Is a Hallmark of Immaturity

The basic mechanisms of neuronal excitability and plasticity upon which the diagnostic and therapeutic utility of neurotechnological approaches are predicated are primarily derived from preclinical animal model experiments (Hameed et al., 2017; An et al., 2012; Cao and Harris, 2012; Guerriero et al., 2015; Rakhade and Jensen, 2009; Sanchez and Jensen, 2001; Mix et al., 2015; Silverstein and Jensen, 2007). These insights into the cellular and molecular underpinning of neuronal function may well inform the adaptation of adult neurostimulation protocols to trials in children. For instance, given that the immature central nervous system is hyperexcitable and therefore vulnerable to seizures as indicated by experiments in rodent epilepsy models (Hameed et al., 2017; Rakhade and Jensen, 2009), and that seizures are a possible, if improbable, potential side effect of transcranial stimulation (Rosa et al., 2006), the maturational pattern of the excitation:inhibition (E:I) balance in the developing brain should be borne in mind when designing protocols for children (Table 2.1).

Neuronal activity is vital for normal central nervous system development in-utero and thus excitatory cortical and subcortical circuits outnumber inhibitory ones in the first few years of life (Sanchez and Jensen, 2001). However, with age, E:I balance shifts toward progressively lower cortical excitability, as the hyperexcitable state is not compatible with normal adult neuronal function. As discussed below, well-studied maturational trajectories of glutamate and GABA (γ-aminobutyric acid) receptor biology enable these shifts in excitability.

Glutamate clearance from the synaptic cleft into astrocytes via glutamate transporter 1 (GLT-1; called excitatory amino acid transporter 2, EAAT-2, in humans) is impaired in developing brains. Highly expressed in adult mammalian brains, GLT-1 provides much of the total glutamate clearance capacity and constitutes the primary glutamate removal mechanism from the synapse (Danbolt et al., 1992; Lehre and Danbolt, 1998). However, both GLT-1expression and rate of glutamate uptake by astrocytes are extremely low in rat hippocampus and neocortex for the first postnatal week (23–36 weeks gestation in humans) and do not reach normal adult levels till the 30th day after birth [human age: 4–11 years (Semple et al., 2013)] (Hanson et al., 2015; Ullensvang et al., 1997). Thus, synaptic glutamate concentrations may be higher in the immature brain.

Ionotropic glutamate receptors (*N*-methyl-D-aspartate receptors, NMDARs; alpha-amino-3-hydroxy-5-methylisoxazole-4-propionic acid receptors, AMPARs) consist of distinct subunits, and the relative abundance of individual subunits within the receptor, which varies with age, has important functional consequences (Hameed et al., 2017; Guerriero et al., 2015). NMDARs are heterotetramers consisting of an NR1 subunit and varying combinations of NR2 and NR3 subunit isoforms and, when activated by glutamate, are a major source of calcium influx into the intracellular space at synapses. The calcium-dependent intracellular signaling cascades triggered by NMDAR activation are critical for synaptic remodeling and likely for any long-term potentiation (LTP)–like and long-term depression (LTD)–like effects desired of therapeutic

TABLE 2.1 Maturational Trajectory of E:I Balance

Signaling system		Developmental stage		
		Immature	Mature	Dysmature
		↑ **Activity-dependent neural development**	↑ **Inhibitory tone**	**Deranged neural networks and E:I ratio**
Glutamatergic excitation	Glutamate transport	↓ GLT-1	↑ GLT-1	↓ GLT-1 (TBI, Isch)
		↑ Synaptic glutamate concentration	↓ Synaptic glutamate concentration	
	NMDAR	↑ $NR2_B$:$NR2_A$	↑ $NR2_A$:$NR2_B$	↑ Hippocampal $NR2_B$:$NR2_A$ (TBI, Isch)
		↑ Ca^{2+} influx/EPSP	↓ Ca^{2+} influx/EPSP	
		↓ Mg^{2+} block	↑ Mg^{2+} block	↓ Cortical NR1, $NR2_A$, $NR2_B$ (TBI)
	AMPAR	↓ GluR2 subunit	↑ GluR2 subunit	↑ Hippocampal GluR1:GluR2 (Isch)
		↑ AMPAR Ca^{2+} permeability	↓ AMPAR Ca^{2+} permeability	
GABAergic inhibition	Cl^- gradient	↑ NKCC1:KCC2	↓NKCC1:KCC2	↑ NKCC1:KCC2
		Depolarizing $GABA_A$	*Hyperpolarizing* $GABA_A$	*Depolarizing* $GABA_A$ (TBI, Isch, TSC, DS, FXS)
	$GABA_AR$	↓ α_1	↑ α_1	↓ Hippocampal α_1, α_4, δ, γ_2 (TBI)
		↓ Zolpidem, BZD response	↑ Zolpidem, BZD response	↓ Cortical α_1, γ_2 (TBI, Isch.)
		Prolonged PSPs	Short, adult IPSPs	
		↓ γ_2	↑ γ_2	↓ α_{1-4}, β_{1-2}, δ, γ_{1-2}
		↑ Zn^{2+} sensitivity	↓ Zn^{2+} sensitivity	↓ Tonic & Phasic $GABA_A$ signaling

E:I, Excitatory:inhibitory; *GLT-1*, Glutamate transporter 1; *TBI*, Traumatic brain injury; *FXS*, Fragile X syndrome; *TBI*, Traumatic brain injury; *NMDAR*, *N*-Methyl-D-aspartate receptors; *AMPAR*, Alpha-amino-3-hydroxy-5-methylisoxazole-4-propionic acid receptor; *BZD*, Benzodiazepine; *PSP*, postsynaptic potential; *IPSP*, inhibitory postsynaptic potential.

neurological interventions (Sui et al., 2014; Tawfik et al., 2010; Chaieb et al., 2015). However, excessive NMDAR signaling can also contribute to glutamate excitotoxicity and subsequent neuropathology that results from runaway glutamate-mediated neuronal activation, such as epileptogenesis (Rakhade and Jensen, 2009).

NMDARs in the developing mammalian brain predominantly contain $NR2_B$ subunits and have a longer current decay once activated, leading to greater calcium influx per action potential, as compared to the $NR2_A$-predominant NMDARs found in mature brains (Guerriero et al., 2015; Silverstein and Jensen, 2007; Monyer et al., 1994). Similarly, $NR2_C$, $NR2_D$, and $NR3_A$ subunit isoforms are also highly expressed in NMDARs in the first 2 weeks after birth in rodents (human age: birth to 2 years), and their abundance reduces the efficacy of the voltage-dependent NMDAR magnesium block normally present in fully developed brains (Blanke and VanDongen, 2009).

AMPARs, the other ionotropic receptor type, are also heterotetramers of differing combinations of GluR1, GluR2, GluR3, and GluR4 subunits. GluR2-containing AMPARs are impermeable to calcium and constitute the majority of the AMPAR population in adult brains. GluR2 is only well expressed in the developing rodent brain after the 21st day of life (human age: 2–3 years), and therefore activation of immature AMPARs in rodents less than 3 weeks of age leads to calcium influx into postsynaptic terminals (Hameed et al., 2017; Silverstein and Jensen, 2007).

Increased NMDAR activation which lasts longer per action potential, along with the abundance of calcium-permeable AMPARs, in immature mammalian brains thus results in greater calcium influx into neurons as compared to adults. Concurrent immaturity of normal adult mechanisms for neocortical synaptic glutamate clearance until the fourth week of life (human age 4–11 years) impairs removal of glutamate from the synaptic cleft and permits extrasynaptic receptor activation and synaptic crosstalk. While the resultant hyperexcitable state is critical for developmental plasticity and synaptogenesis, it is accompanied by lower effective thresholds for seizures and excitotoxic neuronal injury and should therefore be a consideration during the design of diagnostic and therapeutic modalities (Hameed et al., 2017; Guerriero et al., 2015; Rakhade and Jensen, 2009; Sanchez and Jensen, 2001; Silverstein and Jensen, 2007; Hanson et al., 2015; Sanchez and Jensen, 2006).

Less well elucidated is the maturational trajectory of metabotropic glutamate receptors (mGluRs), G-protein-coupled receptors that employ intracellular second messenger systems to modulate neuronal excitability and synaptic transmission. Presynaptic mGluR1α subtype expression peaks by the 9th day of life, particularly on CA1 hippocampal interneurons, followed by maturation of mGluR2, mGluR3, and mGluR5 expression by the 15th day of life in rodents (human age: 0–24 months)

(Avallone et al., 2006). The resistance of the developing brain to seizure-induced damage despite heightened seizure susceptibility may at least in part be due to an abundance of presynaptic mGluR1α receptors, which upon activation may facilitate presynaptic release of the inhibitory neurotransmitter GABA (Avallone et al., 2006)

mGluRs may have additional consequences for neurotechnological applications. The abundance of presynaptic mGluR1α early in life may plausibly contribute to the relatively high threshold reported for noninvasive brain stimulation-mediated activation in young brains (Hameed et al., 2017), although this may be better explained by incomplete myelination in early life. Recent evidence also indicates that the LTD-like effect induced by cathodal transcranial direct current stimulation is primarily mediated by mGluR5 receptors (Sun et al., 2016), and thus direct current stimulation protocols may be less if at all effective during the first 2 years after birth in humans.

Restraint Comes With Age

The powerful inhibitory mechanisms normally found in the adult mammalian central nervous system may also be compromised in childhood as it is not until the end of the first month after birth in rats (human age: 4–11 years) that glutamic acid decarboxylase (GAD—the enzyme responsible for GABA synthesis) and GABA receptor (GABAR) expression reaches adult levels (Rakhade and Jensen, 2009; Silverstein and Jensen, 2007; Swann et al., 1989).

GABA signaling is mediated via GABARs. $GABA_A$ receptors ($GABA_AR$; one of two major GABAR subtypes, most well studied) are pentameric transmembrane proteins with 19 subunit isoforms which determine the functional properties of the receptor. The expression and subunit composition of these receptors are tightly regulated both spatially and temporally, with $\alpha1$ and $\gamma2$-containing $GABA_ARs$ localized synaptically, and $\alpha4$, $\alpha5$, $\alpha6$, and δ subunits predominantly located peri- or extrasynaptically (Lee and Maguire, 2014). $GABA_AR$ activation mediates two distinct forms of inhibition: phasic and tonic.

$GABA_AR$ activation triggers the opening of a chloride ion–selective pore within the receptor and allows chloride ions to move down their concentration gradient and into neurons. The increased chloride conductance hyperpolarizes the membrane, inhibiting the firing of new action potentials. KCC2, a neuron-specific transporter that utilizes the transmembrane potassium ion gradient generated by the Na^+/K^+ ATP pump to export chloride ions in exchange for potassium, establishes and maintains this gradient, keeping intracellular chloride concentrations low (Sanchez and Jensen, 2001; Silverstein and Jensen, 2007). Phasic GABA inhibition is mediated by the activation of synaptic $GABA_A$ receptors, which regulate direct neuronal communication, and this transient inhibitory action shapes the temporal dynamics of neuronal networks. In addition, high-affinity extrasynaptic $GABA_ARs$ respond to extremely small concentrations of extrasynaptic or "ambient" GABA, resulting from spillover from the synapse or leaking from astrocytes, and provide a tonic GABA inhibition. This regulates the excitability of neuronal circuits and neuronal firing and contributes to a "shunting inhibition" where increases in membrane conductance increase the decrement of voltage with distance (Lee and Maguire, 2014; Cellot and Cherubini, 2013).

Although $GABA_ARs$ are expressed in immature brains, they have different subunit concentrations compared to adult receptors, and thus different functional characteristics. Age-related and region-specific changes in $GABA_AR$ subunit composition continue through neurodevelopment (Galanopoulou, 2008). At birth, $\alpha 1$ expression is low throughout the rodent brain and starts to increase rapidly after P6, reaching adult levels by P20. In contrast, $\alpha 2$ expression follows an opposite developmental trajectory, being prominently expressed in several brain regions at birth and decreasing thereafter. Increased $\alpha 1$ expression to adult levels is associated with increased sensitivity to pharmaceutical interventions such as zolpidem and benzodiazepines, and short duration adult-type inhibitory postsynaptic potentials (Hameed et al., 2017; Galanopoulou, 2008; Kapur and Macdonald, 1999; Bosman et al., 2005). The developmental switch in α subunit expression in rat hippocampal dentate and increased expression of $\gamma 2$ subunits also correlates with decreasing sensitivity to zinc (Brooks-Kayal et al., 2001). Zinc is a modulator of synaptic activity and neuronal plasticity in both development and adulthood and plays a particularly important role in the control of GABAergic transmission in immature neurons. Stimulation-dependent release of zinc from presynaptic vesicles in hyperexcitable immature networks may thus help control excessive GABAergic depolarization, in a region-specific pattern in the developing brain (Galanopoulou, 2008). The impact of the loss of sensitivity to zinc in mature neurons is lessened as GABAergic inhibition is more efficient.

GABAergic signaling precedes glutamatergic transmission developmentally, and GABAergic interneurons may provide an early source of activity in otherwise silent networks. GABA signaling itself potentially undergoes major changes during nervous system development. According to one prevailing theory, increased expression of the chloride *importer* NKCC1 and decreased expression of KCC2 relative to the adult rat brain in the first week after birth in rodents (human age: 36–40 week gestation) reverses chloride gradients (Plotkin et al., 1997; Rivera et al., 1999; Ganguly et al., 2001). This difference in intracellular concentration results in an inverse of the adult $GABA_AR$ function, with activation causing a net *efflux* of chloride from the neuron through the receptor, leading to membrane depolarization and paradoxically *enhancing* excitability during this stage in development (Sanchez and Jensen, 2001; Silverstein and Jensen, 2007; LoTurco et al., 1995). Increasing expression of neuronal KCC2 over the first 21 postnatal days in rodents (human age: 0–4; 4 years) gradually establishes adult chloride concentration gradients and correlates with the gradual increase in GABAergic inhibitory tone. Most of the data supporting this excitatory GABA theory has been obtained from in vitro brain slice experiments, however, and has been questioned as potentially being an artifact of experimental conditions, slices being susceptible to deficient energy metabolism and neuronal damage (Zilberter, 2016). One recent in vivo study has shown that GABA is a predominantly depolarizing neurotransmitter at the neuronal level while concurrently being inhibitory at the network level during early cortical development (Kirmse et al., 2015), whereas another has confirmed the general inhibitory effect of GABAergic action in intact neonatal brain (Valeeva et al., 2016). The cause of the observed differences in GABA signaling in vivo and in vitro has not yet been identified.

Developmental trajectories of tonic GABA signaling also differ by region. In the dentate gyrus, tonic signaling is provided as early as the third day of life in mice by $GABA_A$Rs rich in α_5 subunits and peaks in the second week after birth with an increase in the expression of $GABA_A$Rs with δ subunits, before declining (Holter et al., 2010). Tonic conductance mediated by α_5 and δ subunit-containing $GABA_A$Rs is also present in layer V pyramidal cells of the somatosensory cortex of newborn nice, is largest immediately after birth, and then decreases during the second postnatal week (Sebe et al., 2010). This decline may possibly be due to the increased activity of GABA transporters enhancing clearance of GABA from the extracellular space, and to the reduced regional expression of δ subunits with age (Cellot and Cherubini, 2013). On the other hand, tonic $GABA_A$-mediated currents increase with age in the visual cortex, cerebellum, and ventrobasal relay neurons in the thalamus (Cellot and Cherubini, 2013).

As mentioned earlier, the effects of noninvasive brain stimulation in adults depend at least in part on modulation of GABAergic inhibitory signaling (Amadi et al., 2015; Huang et al., 2005). As such, the maturational trajectory of GABA signaling may make the effect of neuromodulation on the immature brain difficult to predict and must be an important consideration when stimulating the developing nervous system. For example, theta burst stimulation (TBS), a specific protocol of repetitive TMS, relies on altered cortical GABAergic inhibition and may not work as predicted in the very immature brain (Huang et al., 2005; Trippe et al., 2009; Labedi et al., 2014).

In summary, immature mammalian brains are more susceptible to seizures given reduced glutamate clearance, decreased voltage-dependent magnesium block of the NMDAR, altered glutamate receptor subunit composition, and incomplete GABA-mediated inhibition. However, not all preclinical research indicates that brain stimulation poses heightened risks in the developing brain. Young rodent brains have been shown to be less vulnerable than adult ones to neuronal damage after seizures. For instance, while exposure to kainic acid causes neuronal death in the hippocampus and limbic system and results in the formation of recurrent excitatory neuronal circuits in adult rodents (Ben-Ari, 1985; Epsztein et al., 2005), developing brains do not exhibit any cell death following kainate-induced seizures regardless of dose (Nitecka et al., 1984; Tremblay et al., 1984). One possible explanation for this neuroprotection in immature brains may be the absence of well-developed hippocampal mossy fiber terminals which help sustain epileptic seizures by amplifying excitatory signaling (Ben-Ari, 2013). In addition, while GABA depolarizes neuronal membranes in the immature brain, unlike glutamate it clamps the voltage close to the reversal potential for chloride, creating a "shunting" effect that may prevent excessive calcium influx through voltage-dependent calcium channels (Safiulina et al., 2010). Recurrent seizures during neurodevelopment, however, may still disrupt critical milestones and result in aberrant networks, long-term cognitive and behavioral changes, and lower seizure thresholds in adulthood (Ben-Ari, 2013).

Thus, the maturational trajectory of the E:I equilibrium and the effect of stimulation on the developing brain should be considered and accounted for in the design and implementation of diagnostic and therapeutic pediatric neurotechnologies. This is particularly true for pediatric noninvasive neurostimulation, as seizures, a plausible—if unlikely—side effect, may have distinct effects on the young brain.

The Impressionable Years

Noninvasive brain stimulation protocols are designed with the assumption that the basic molecular and cellular mechanisms that affect the desired neuromodulatory changes are the same, or similar to, the neurobiological processes underpinning the brain's innate capacity to register, process, and store external stimuli as memories or alterations in nervous function. Such plasticity depends on LTP or LTD of synaptic strength, and thus knowledge of the neurodevelopmental regulation of synaptic plasticity gained from animals may inform the design of pediatric neurostimulation and neuromodulation protocols.

Clinicians and investigators are now able to reliably elicit LTP-like potentiation and LTD-like depression of human corticospinal responses using transcranial stimulation (Fitzgerald et al., 2006; Nitsche and Paulus, 2011); however, study of the developmental timeline of these phenomena is only just beginning. Translational research, where isolated rodent brain slices can be used to reliably reproduce and study use-dependent enhancement or depression of synaptic strength (LTP/LTD) (Bear and Malenka, 1994; Malenka and Bear, 2004; Lomo, 2003), may be used to gain valuable insights here as well. However, it is worth remembering that these isolated slices are susceptible to non-physiologic milieu such as deficient energy metabolism and neuronal damage and extrapolating clinical relevance from animal LTP/LTD studies can be challenging.

The rodent brain exhibits multiple, distinct, region specific maturational trajectories of synaptic plasticity. Adult mice exhibit a greater capacity for LTP in the visual cortex and compared to 4–5-week-old juvenile mice (human age: 4–11 years (Kirkwood et al., 1997). In contrast, LTP is present at birth in the rat barrel cortex, increases thereafter to reach maximal levels at P3–5 (humans: 32–36 weeks' gestation), and then falls away to undetectable levels by the second week of life (human age: 0–24 months) (An et al., 2012). TBS first induces LTP in isolated rat hippocampal slices at P12 (0–24 months in humans), with the response improving up to the 35th day after birth (humans: 11 years) (Cao and Harris, 2012). On the other hand, developing rats have both enhanced LTD and a lower threshold for LTD induction. Hippocampal LTD progressively declines from maximal levels at P14 (human age <2 years) and reaches adult levels by postnatal day 35 (human age: 4–11 years) (Dudek and Bear, 1993; Lante et al., 2006).

Differences in neuroplasticity between immature and mature brains have not been studied well in vivo. One study, however, reports that intermittent TBS (iTBS) functionally modifies cortical GABAergic parvalbumin positive (PV +) interneurons in rats only after the 32nd postnatal day (human age: 4–11 years), with a peak effect by the 40th day after birth (human age: 12–18 years). This age-dependence may be due to immature PV + cells lacking the degree of synaptic connectivity necessary to receive the level of excitatory input required to be modulated by iTBS. The development and modulation of such input to the PV + interneurons is mediated by the perineuronal nets surrounding them, a specialized extracellular matrix that preferentially enwraps these cells and facilitates their inhibitory capacity, and a robust PNN is not evident until well into development (Mix et al., 2015).

It should be borne in mind, however, that these data only reflect the absolute change in magnitude of the evoked postsynaptic response, not the induction threshold, which may well be lower in the immature brain.

Dysmaturity in Disease

Immature neuronal or network biology and physiology may be for one of two reasons: either the brain is chronologically immature or, more clinically relevant, appropriate maturation did not occur, and the brain remains pathologically *dysmature* into adulthood due to genetic or environmental insults. Deranged neurobiology has been identified in a range of neurodevelopmental and neurological syndromes

Animal models and limited data from human subjects indicate that inhibitory dysfunction due to preexisting or acquired dysmaturity of the GABAergic inhibitory system is of major pathophysiological importance in autism spectrum disorders, genetic and acquired epilepsies, trauma and post-traumatic sequelae, and associated conditions (Braat and Kooy, 2015a). Therefore, as neurotechnologies are developed that target a variety of neurological disease states, including those that primarily impact young patients, their use in adults with region-specific neurophysiological derangements may also be considered.

Acquired Disorders

Several rodent traumatic brain injury models show that the adult brain regresses to an immature state following injury, altering expression of several transporters. Focal cerebral ischemic injury and subsequent reperfusion in rats causes an increase in cortical NKCC1 and a concomitant decrease in KCC2 (Wang et al., 2014; Yan et al., 2001), and a similar NKCC1/KCC2 switch accompanied by deranged neuronal chloride homeostasis is seen as early as 1 day after closed head injury (CHI) in adult mice (Wang et al., 2017). Bumetanide administration is seen to be protective in both models, presumably by virtue of blocking NKCC1 (Wang et al., 2014; Yan et al., 2001; Wang et al., 2017). Injury also affects GLT-1 expression, with neocortical protein levels decreasing 24–72 hours after controlled cortical impact in rats and 1 week after fluid percussion injury (FPI) (van Landeghem et al., 2001; Goodrich et al., 2013).

Injury also affects receptor subunit expression in the adult brain. Increased hippocampal NMDAR $NR2_B/NR2_A$ and AMPAR GluR1/GluR2 ratios are observed after transient global ischemia in adult rats (Han et al., 2016). CHI in adult mice results in upregulated expression of $NR2_B$ subunits in hippocampal NMDARs due to increased tyrosine phosphorylation of the subunit. Administration of tyrosine kinase inhibitors following injury returns $NR2_B$ levels to normal and is associated with improved functional outcomes (Schumann et al., 2008). In contrast, a decrease in neocortical NMDAR NR1, $NR2_A$, $NR2_B$, and AMPAR GluR1 subunits is seen following CHI in mice (Schumann et al., 2008), and lower $NR2_B$ expression as well as phosphorylation is observed at the injury site after FPI in rats (Park et al., 2013). Brain injury in adult rats also causes time-dependent changes in $GABA_AR$ subunit expression, with hippocampal $\alpha1$, $\alpha4$, δ, and $\gamma2$ expression decreased 7 days after FPI (Raible et al., 2012; Gibson et al., 2010), and lower $\alpha1$ and $\gamma2$

levels seen in both the cortex and hippocampus 48 hours after global ischemic injury (Montori et al., 2012).

Genetic Dysmaturity

Decreased KCC2 levels and a higher NKCC1:KCC2 ratio are seen in the CSF of human patients with Rett syndrome (Duarte et al., 2013), mimicking the expression pattern seen in the immature brain. Specific mutations in $GABA_AR$ subunits have also been associated with epileptic and behavioral disorders—some older patients with Dravet syndrome who lack the typical voltage-gated sodium channel mutation instead exhibit $GABA_ARs$ deficient in $\alpha1$ subunit, reminiscent of chronologically immature $\alpha1$-deficient $GABA_ARs$ with slow kinetics and reduced zolpidem and benzodiazepine sensitivity. Similar mutations are also seen in early infantile epileptic encephalopathy (Braat and Kooy, 2015a; Carvill et al., 2014). Cerebellar $\alpha1$ subunit expression is also reduced in schizophrenia and major depression, as is expression in superior frontal and parietal cortices and cerebellum in autism (Fatemi and Folsom, 2015).

Tuberous sclerosis complex (TSC) is another neurological disorder characterized by E:I imbalance and GABAergic dysfunction. TSC is a genetic neurodevelopmental disorder (mutation of TSC1 or TSC2, which code for inhibitors of mechanistic target of rapamycin, a central cell-growth control pathway), and patients often develop benign growths called "tubers," early-life refractory epilepsy, and fall on the autism spectrum. Tubers removed from human TSC patients exhibit decreased $GABA_AR$ $\alpha1$ subunit expression which explains the benzodiazepine insensitivity seen in these patients. Also seen is increased NKCC1 and decreased KCC2 expression resulting in excitatory $GABA_AR$, again reminiscent of chronologically immature brains, and bumetanide (an NKCC1 antagonist) administration to tubers significantly decreases excitatory GABA action as seen on patch-clamp recordings (Talos et al., 2012).

Similarly, fragile X syndrome (FXS) is an inherited disease characterized by decreased or absent expression of the fragile X mental-retardation protein (FMRP) required for normal neurological development due to mutations in the fragile X mental retardation (Fmr1) gene on the X chromosome. Patients are at a higher risk of developing epilepsy and suffer from intellectual disabilities and neuropsychiatric disorders including autism. Animal models of FXS, juvenile Fmr1 knockout mice, exhibit a delayed excitatory-to-inhibitory switch in GABA signaling in the somatosensory cortex. Specifically, GABAergic signaling results in the depolarization rather than hyperpolarization of neuronal membrane potential in Fmr1 knockout mice but not wild-type controls 10 days after birth (human: term infant). At the same timepoint, NKCC1 expression is increased while KCC2 expression is not affected in the somatosensory cortex of mutants compared to controls. Hippocampal KCC2 expression, however, is decreased in mutant mice on up to postnatal day 30 (human age 4–11 years). The normal developmental excitatory-to-inhibitory GABA switch is restored in neonatal knockout mice when pregnant female mice are pretreated with bumetanide, confirming that the immature NKCC1:KCC2 ratio is a major driving force behind the GABAergic dysfunction, at least in this animal model (He et al., 2014; Braat and Kooy, 2015b). Significantly, decreased expression of α1-4, β1-2, δ, and γ1-2 subunits of $GABA_ARs$ in these animals compounds this dysfunction, with the receptors responsible for both phasic and tonic GABA signaling being dysmature (Braat et al., 2015; D'Hulst et al., 2006). $GABA_B$ receptor ($GABA_BR$) activation appears to upregulate FMRP expression in neurons (Zhang et al., 2015), and this and other GABA-centric therapies to ameliorate the dysfunction may have therapeutic benefits for patients with FXS (Zeidler et al., 2017; Lozano et al., 2014).

CONCLUSION: A MOVING TARGET

The development of novel neurotechnologies targeting a variety of suboptimally treated neurological diseases of the developing central nervous system is likely to be accompanied by a unique set of challenges related to a changing molecular biology that does not stabilize until late adolescence, or perhaps later. As discussed above, critical neurodevelopmental processes are all at least partly activity-dependent, with GABAergic signaling playing a critical role. Thus, the developing brain may be particularly susceptible to external influences from stimulation protocols such as TMS that rely on the modulation of GABAergic signaling. Moreover, seizures are a realistic, if improbable, side-effect of noninvasive brain stimulation. Immature mammalian brains are more susceptible to seizures, and while they are less vulnerable to seizure-induced damage, recurrent seizures may still disrupt critical milestones and result in aberrant networks, long-term cognitive and behavioral changes, and lower seizure thresholds in adulthood.

Proper interpretation of functional, behavioral, and clinical data obtained from younger patients using diagnostic or therapeutic neurostimulation protocols is also contingent upon a sound understanding of the maturational patterns of underlying neurobiological and neurophysiological principles. For example, direct current stimulation protocols may be less effective during the first 2 years after birth in humans due to mGluR5 receptor immaturity, and the relatively high threshold reported for noninvasive brain stimulation-mediated activation in young brains may be explained by incomplete myelination in early life. In addition, TBS relies on altered cortical GABAergic inhibition and may not work as predicted in the very immature brain.

The authors hope that the data discussed above provide insights that merit consideration during the design and implementation of these technologies. For instance, further research is required to investigate the effects of age-related differences in basic neurological mechanisms on the safety and efficacy of brain stimulation in the pediatric brain.

Acknowledgment

This work was supported by grants from the Boston Children's Hospital Translational Research Program (AR).

Conflicts of Interest

AR is a cofounder and consults for Neuro'motion, has consulted or consults for Cavion, Cyberonics, NeuroRex, Roche, Sage, and is a coinventor of a patent for real-time integration of TMS and EEG. AR receives or has received research funding from Brainsway, CREmedical, Fisher-Wallace, Eisai, Neuropace, Roche, Sage, Takeda. MQH, UD, and PM have nothing to declare.

References

Amadi, U., Allman, C., Johansen-Berg, H., Stagg, C.J., 2015. The homeostatic interaction between anodal transcranial direct current stimulation and motor learning in humans is related to GABAA activity. Brain Stimul. 8 (5), 898–905.

An, S., Yang, J.W., Sun, H., Kilb, W., Luhmann, H.J., 2012. Long-term potentiation in the neonatal rat barrel cortex in vivo. J. Neurosci. 32 (28), 9511–9516.

Avallone, J., Gashi, E., Magrys, B., Friedman, L.K., 2006. Distinct regulation of metabotropic glutamate receptor (mGluR1 alpha) in the developing limbic system following multiple early-life seizures. Exp. Neurol. 202 (1), 100–111.

Bear, M.F., Malenka, R.C., 1994. Synaptic plasticity: LTP and LTD. Curr. Opin. Neurobiol. 4 (3), 389–399.

Ben-Ari, Y., 1985. Limbic seizure and brain damage produced by kainic acid: mechanisms and relevance to human temporal lobe epilepsy. Neuroscience 14 (2), 375–403.

Ben-Ari, Y., 2013. The developing cortex. Handb. Clin. Neurol. 111, 417–426.

Bhardwaj, R.D., Curtis, M.A., Spalding, K.L., et al., 2006. Neocortical neurogenesis in humans is restricted to development. Proc. Natl. Acad. Sci. U.S.A. 103 (33), 12564–12568.

Blanke, M.L., VanDongen, A.M.J., 2009. Activation mechanisms of the NMDA receptor. In: Van Dongen, A.M. (Ed.), Biology of the NMDA Receptor. CRC Press/Taylor & Francis, Boca Raton, FL.

Bosman, L.W., Heinen, K., Spijker, S., Brussaard, A.B., 2005. Mice lacking the major adult GABAA receptor subtype have normal number of synapses, but retain juvenile IPSC kinetics until adulthood. J. Neurophysiol. 94 (1), 338–346.

Braat, S., Kooy, R.F., 2015a. The GABAA receptor as a therapeutic target for neurodevelopmental disorders. Neuron 86 (5), 1119–1130.
Braat, S., Kooy, R.F., 2015b. Insights into GABAAergic system deficits in fragile X syndrome lead to clinical trials. Neuropharmacology 88, 48–54.
Braat, S., D'Hulst, C., Heulens, I., et al., 2015. The GABAA receptor is an FMRP target with therapeutic potential in fragile X syndrome. Cell Cycle 14 (18), 2985–2995.
Brooks-Kayal, A.R., Shumate, M.D., Jin, H., Rikhter, T.Y., Kelly, M.E., Coulter, D.A., 2001. Gamma-aminobutyric acid(A) receptor subunit expression predicts functional changes in hippocampal dentate granule cells during postnatal development. J. Neurochem. 77 (5), 1266–1278.
Cao, G., Harris, K.M., 2012. Developmental regulation of the late phase of long-term potentiation (L-LTP) and metaplasticity in hippocampal area CA1 of the rat. J. Neurophysiol. 107 (3), 902–912.
Carr, L.J., Harrison, L.M., Evans, A.L., Stephens, J.A., 1993. Patterns of central motor reorganization in hemiplegic cerebral palsy. Brain 116 (Pt 5), 1223–1247.
Carvill, G.L., Weckhuysen, S., McMahon, J.M., et al., 2014. GABRA1 and STXBP1: novel genetic causes of Dravet syndrome. Neurology 82 (14), 1245–1253.
Cellot, G., Cherubini, E., 2013. Functional role of ambient GABA in refining neuronal circuits early in postnatal development. Front. Neural Circuits 7, 136.
Chaieb, L., Antal, A., Paulus, W., 2015. Transcranial random noise stimulation-induced plasticity is NMDA-receptor independent but sodium-channel blocker and benzodiazepines sensitive. Front. Neurosci. 9, 125.
Danbolt, N.C., Storm-Mathisen, J., Kanner, B.I., 1992. An [Na + + K +] coupled L-glutamate transporter purified from rat brain is located in glial cell processes. Neuroscience 51 (2), 295–310.
Dekaban, A.S., 1978. Changes in brain weights during the span of human life: relation of brain weights to body heights and body weights. Ann. Neurol. 4 (4), 345–356.
D'Hulst, C., De Geest, N., Reeve, S.P., et al., 2006. Decreased expression of the GABAA receptor in fragile X syndrome. Brain Res. 1121 (1), 238–245.
Duarte, S.T., Armstrong, J., Roche, A., et al., 2013. Abnormal expression of cerebrospinal fluid cation chloride cotransporters in patients with Rett syndrome. PLoS ONE. 8 (7), e68851.
Dudek, S.M., Bear, M.F., 1993. Bidirectional long-term modification of synaptic effectiveness in the adult and immature hippocampus. J. Neurosci. 13 (7), 2910–2918.
Epsztein, J., Represa, A., Jorquera, I., Ben-Ari, Y., Crepel, V., 2005. Recurrent mossy fibers establish aberrant kainate receptor-operated synapses on granule cells from epileptic rats. J. Neurosci. 25 (36), 8229–8239.
Eyre, J.A., 2003. Development and plasticity of the corticospinal system in man. Neural. Plast. 10 (1–2), 93–106.
Eyre, J.A., Taylor, J.P., Villagra, F., Smith, M., Miller, S., 2001. Evidence of activity-dependent withdrawal of corticospinal projections during human development. Neurology 57 (9), 1543–1554.
Fatemi, S.H., Folsom, T.D., 2015. GABA receptor subunit distribution and FMRP-mGluR5 signaling abnormalities in the cerebellum of subjects with schizophrenia, mood disorders, and autism. Schizophr Res. 167 (1–3), 42–56.
Fietzek, U.M., Heinen, F., Berweck, S., et al., 2000. Development of the corticospinal system and hand motor function: central conduction times and motor performance tests. Dev. Med. Child Neurol. 42 (4), 220–227.
Fitzgerald, P.B., Fountain, S., Daskalakis, Z.J., 2006. A comprehensive review of the effects of rTMS on motor cortical excitability and inhibition. Clin. Neurophysiol. 117 (12), 2584–2596.

Galanopoulou, A.S., 2008. GABA(A) receptors in normal development and seizures: friends or foes? Curr. Neuropharmacol. 6 (1), 1–20.

Ganguly, K., Schinder, A.F., Wong, S.T., Poo, M., 2001. GABA itself promotes the developmental switch of neuronal GABAergic responses from excitation to inhibition. Cell 105 (4), 521–532.

Gibson, C.J., Meyer, R.C., Hamm, R.J., 2010. Traumatic brain injury and the effects of diazepam, diltiazem, and MK-801 on GABA-A receptor subunit expression in rat hippocampus. J. Biomed. Sci. 17, 38.

Goodrich, G.S., Kabakov, A.Y., Hameed, M.Q., Dhamne, S.C., Rosenberg, P.A., Rotenberg, A., 2013. Ceftriaxone treatment after traumatic brain injury restores expression of the glutamate transporter, GLT-1, reduces regional gliosis, and reduces post-traumatic seizures in the rat. J. Neurotrauma 30 (16), 1434–1441.

Guerriero, R.M., Giza, C.C., Rotenberg, A., 2015. Glutamate and GABA imbalance following traumatic brain injury. Curr. Neurol. Neurosci. Rep. 15 (5), 27.

Hameed, M.Q., Dhamne, S.C., Gersner, R., et al., 2017. Transcranial magnetic and direct current stimulation in children. Curr. Neurol. Neurosci. Rep. 17 (2), 11.

Han, X.J., Shi, Z.S., Xia, L.X., et al., 2016. Changes in synaptic plasticity and expression of glutamate receptor subunits in the CA1 and CA3 areas of the hippocampus after transient global ischemia. Neuroscience 327, 64–78.

Hanson, E., Armbruster, M., Cantu, D., et al., 2015. Astrocytic glutamate uptake is slow and does not limit neuronal NMDA receptor activation in the neonatal neocortex. Glia 63 (10), 1784–1796.

He, Q., Nomura, T., Xu, J., Contractor, A., 2014. The developmental switch in GABA polarity is delayed in fragile X mice. J. Neurosci. 34 (2), 446–450.

Heck, N., Kilb, W., Reiprich, P., et al., 2007. GABA-A receptors regulate neocortical neuronal migration in vitro and in vivo. Cereb. Cortex 17 (1), 138–148.

Heck, N., Golbs, A., Riedemann, T., Sun, J.J., Lessmann, V., Luhmann, H.J., 2008. Activity-dependent regulation of neuronal apoptosis in neonatal mouse cerebral cortex. Cereb. Cortex 18 (6), 1335–1349.

Heinen, F., Fietzek, U.M., Berweck, S., Hufschmidt, A., Deuschl, G., Korinthenberg, R., 1998. Fast corticospinal system and motor performance in children: conduction proceeds skill. Pediatr. Neurol. 19 (3), 217–221.

Holter, N.I., Zylla, M.M., Zuber, N., Bruehl, C., Draguhn, A., 2010. Tonic GABAergic control of mouse dentate granule cells during postnatal development. Eur. J. Neurosci. 32 (8), 1300–1309.

Huang, Y.Z., Edwards, M.J., Rounis, E., Bhatia, K.P., Rothwell, J.C., 2005. Theta burst stimulation of the human motor cortex. Neuron 45 (2), 201–206.

Huttenlocher, P.R., Dabholkar, A.S., 1997. Regional differences in synaptogenesis in human cerebral cortex. J. Comp. Neurol. 387 (2), 167–178.

Inada, H., Watanabe, M., Uchida, T., et al., 2011. GABA regulates the multidirectional tangential migration of GABAergic interneurons in living neonatal mice. PLoS ONE 6 (12), e27048.

Jansen, L.A., 2017. Making connections with GABA. Epilepsy. Curr. 17 (6), 377–378.

Kapur, J., Macdonald, R.L., 1999. Postnatal development of hippocampal dentate granule cell gamma-aminobutyric $acid_A$ receptor pharmacological properties. Mol. Pharmacol. 55 (3), 444–452.

Kaye, H.L., Block, G., Jannati, A., et al., 2017. Maturation of Motor Cortex Excitability in Children With Focal Epilepsy as Measured by Navigated Transcranial Magnetic Stimulation. Society for Neuroscience Annual Meeting, Washington, DC.

Kirkwood, A., Silva, A., Bear, M.F., 1997. Age-dependent decrease of synaptic plasticity in the neocortex of alpha CaMKII mutant mice. Proc. Natl. Acad. Sci. U.S.A. 94 (7), 3380–3383.

Kirmse, K., Kummer, M., Kovalchuk, Y., Witte, O.W., Garaschuk, O., Holthoff, K., 2015. GABA depolarizes immature neurons and inhibits network activity in the neonatal neocortex in vivo. Nat. Commun. 6, 7750.

Kirton, A., Chen, R., Friefeld, S., Gunraj, C., Pontigon, A.M., Deveber, G., 2008. Contralesional repetitive transcranial magnetic stimulation for chronic hemiparesis in subcortical paediatric stroke: a randomised trial. Lancet Neurol. 7 (6), 507–513.

Knickmeyer, R.C., Gouttard, S., Kang, C., et al., 2008. A structural MRI study of human brain development from birth to 2 years. J. Neurosci. 28 (47), 12176–12182.

Kostovic, I., Rakic, P., 1980. Cytology and time of origin of interstitial neurons in the white matter in infant and adult human and monkey telencephalon. J. Neurocytol. 9 (2), 219–242.

Labedi, A., Benali, A., Mix, A., Neubacher, U., Funke, K., 2014. Modulation of inhibitory activity markers by intermittent theta-burst stimulation in rat cortex is NMDA-receptor dependent. Brain Stimul. 7 (3), 394–400.

Landing, B.H., Shankle, W.R., Hara, J., Brannock, J., Fallon, J.H., 2002. The development of structure and function in the postnatal human cerebral cortex from birth to 72 months: changes in thickness of layers II and III co-relate to the onset of new age-specific behaviors. Pediatr. Pathol. Mol. Med. 21 (3), 321–342.

Lante, F., Cavalier, M., Cohen-Solal, C., Guiramand, J., Vignes, M., 2006. Developmental switch from LTD to LTP in low frequency-induced plasticity. Hippocampus 16 (11), 981–989.

Lee, V., Maguire, J., 2014. The impact of tonic GABAA receptor-mediated inhibition on neuronal excitability varies across brain region and cell type. Front. Neural Circuits 8, 3.

Lehre, K.P., Danbolt, N.C., 1998. The number of glutamate transporter subtype molecules at glutamatergic synapses: chemical and stereological quantification in young adult rat brain. J. Neurosci. 18 (21), 8751–8757.

Levitt, P., 2003. Structural and functional maturation of the developing primate brain. J. Pediatr. 143 (4 Suppl), S35–S45.

Lomo, T., 2003. The discovery of long-term potentiation. Philos. Trans. R. Soc. Lond. B: Biol. Sci. 358 (1432), 617–620.

LoTurco, J.J., Owens, D.F., Heath, M.J., Davis, M.B., Kriegstein, A.R., 1995. GABA and glutamate depolarize cortical progenitor cells and inhibit DNA synthesis. Neuron 15 (6), 1287–1298.

Lozano, R., Hare, E.B., Hagerman, R.J., 2014. Modulation of the GABAergic pathway for the treatment of fragile X syndrome. Neuropsychiatr. Dis. Treat. 10, 1769–1779.

Luhmann, H.J., Sinning, A., Yang, J.W., et al., 2016. Spontaneous neuronal activity in developing neocortical networks: from single cells to large-scale interactions. Front. Neural Circuits 10, 40.

Malenka, R.C., Bear, M.F., 2004. LTP and LTD: an embarrassment of riches. Neuron 44 (1), 5–21.

Mitew, S., Xing, Y.L., Merson, T.D., 2016. Axonal activity-dependent myelination in development: Insights for myelin repair. J. Chem. Neuroanat. 76 (Pt A), 2–8.

Mix, A., Hoppenrath, K., Funke, K., 2015. Reduction in cortical parvalbumin expression due to intermittent theta-burst stimulation correlates with maturation of the perineuronal nets in young rats. Dev. Neurobiol. 75 (1), 1–11.

Mizuno, H., Hirano, T., Tagawa, Y., 2007. Evidence for activity-dependent cortical wiring: formation of interhemispheric connections in neonatal mouse visual cortex requires projection neuron activity. J. Neurosci. 27 (25), 6760–6770.

Montori, S., Dos Anjos, S., Poole, A., et al., 2012. Differential effect of transient global ischaemia on the levels of gamma-aminobutyric acid type A (GABA(A)) receptor subunit mRNAs in young and older rats. Neuropathol. Appl. Neurobiol. 38 (7), 710–722.

Monyer, H., Burnashev, N., Laurie, D.J., Sakmann, B., Seeburg, P.H., 1994. Developmental and regional expression in the rat brain and functional properties of four NMDA receptors. Neuron 12 (3), 529–540.

Mrzljak, L., Uylings, H.B., Van Eden, C.G., Judas, M., 1990. Neuronal development in human prefrontal cortex in prenatal and postnatal stages. Prog. Brain. Res. 85, 185–222.

Muller, K., Homberg, V., 1992. Development of speed of repetitive movements in children is determined by structural changes in corticospinal efferents. Neurosci. Lett. 144 (1–2), 57–60.

Muller, K., Homberg, V., Lenard, H.G., 1991. Magnetic stimulation of motor cortex and nerve roots in children. Maturation of cortico-motoneuronal projections. Electroencephalogr. Clin. Neurophysiol. 81 (1), 63–70.

Muller, K., Kass-Iliyya, F., Reitz, M., 1997. Ontogeny of ipsilateral corticospinal projections: a developmental study with transcranial magnetic stimulation. Ann. Neurol. 42 (5), 705–711.

Nezu, A., Kimura, S., Takeshita, S., 1999. Topographical differences in the developmental profile of central motor conduction time. Clin. Neurophysiol. 110 (9), 1646–1649.

Nitecka, L., Tremblay, E., Charton, G., Bouillot, J.P., Berger, M.L., Ben-Ari, Y., 1984. Maturation of kainic acid seizure-brain damage syndrome in the rat. II. Histopathological sequelae. Neuroscience 13 (4), 1073–1094.

Nitsche, M.A., Paulus, W., 2011. Transcranial direct current stimulation--update 2011. Restor. Neurol. Neurosci. 29 (6), 463–492.

Paredes, M.F., James, D., Gil-Perotin, S., et al., 2016. Extensive migration of young neurons into the infant human frontal lobe. Science 354 (6308), pii: aaf7073.

Park, Y., Luo, T., Zhang, F., et al., 2013. Downregulation of Src-kinase and glutamate-receptor phosphorylation after traumatic brain injury. J. Cereb. Blood. Flow. Metab. 33 (10), 1642–1649.

Plotkin, M.D., Snyder, E.Y., Hebert, S.C., Delpire, E., 1997. Expression of the Na-K-2Cl cotransporter is developmentally regulated in postnatal rat brains: a possible mechanism underlying GABA's excitatory role in immature brain. J. Neurobiol. 33 (6), 781–795.

Raible, D.J., Frey, L.C., Cruz Del Angel, Y., Russek, S.J., Brooks-Kayal, A.R., 2012. GABA (A) receptor regulation after experimental traumatic brain injury. J. Neurotrauma 29 (16), 2548–2554.

Rajapakse, T., Kirton, A., 2013. Non-invasive brain stimulation in children: applications and future directions. Transl. Neurosci. 4 (2), 217–233.

Rakhade, S.N., Jensen, F.E., 2009. Epileptogenesis in the immature brain: emerging mechanisms. Nat. Rev. Neurol. 5 (7), 380–391.

Reiprich, P., Kilb, W., Luhmann, H.J., 2005. Neonatal NMDA receptor blockade disturbs neuronal migration in rat somatosensory cortex in vivo. Cereb. Cortex 15 (3), 349–358.

Rivera, C., Voipio, J., Payne, J.A., et al., 1999. The K + /Cl − co-transporter KCC2 renders GABA hyperpolarizing during neuronal maturation. Nature 397 (6716), 251–255.

Rosa, M.A., Picarelli, H., Teixeira, M.J., Rosa, M.O., Marcolin, M.A., 2006. Accidental seizure with repetitive transcranial magnetic stimulation. J. ECT. 22 (4), 265–266.

Safiulina, V.F., Caiati, M.D., Sivakumaran, S., Bisson, G., Migliore, M., Cherubini, E., 2010. Control of GABA release at mossy fiber-CA3 connections in the developing hippocampus. Front. Synaptic Neurosci. 2, 1.

Saisanen, L., Julkunen, P., Lakka, T., Lindi, V., Kononen, M., Maatta, S., 2018. Development of corticospinal motor excitability and cortical silent period from mid-childhood to adulthood—a navigated TMS study. Neurophysiol Clin. 48 (2), 65–75.

Sanchez, R.M., Jensen, F.E., 2001. Maturational aspects of epilepsy mechanisms and consequences for the immature brain. Epilepsia 42 (5), 577–585.

Sanchez, R.M., Jensen, F.E., 2006. Modeling hypoxia-induced seizures and hypoxic encephalopathy in the neonatal period. In: Pitkanen, A., Moshe, S.L., Schwartzkroin, P.A. (Eds.), Models of Seizures and Epilepsy. Elsevier, San Diego, CA.

Schumann, J., Alexandrovich, G.A., Biegon, A., Yaka, R., 2008. Inhibition of NR2B phosphorylation restores alterations in NMDA receptor expression and improves functional recovery following traumatic brain injury in mice. J. Neurotrauma 25 (8), 945–957.
Sebe, J.Y., Looke-Stewart, E.C., Estrada, R.C., Baraban, S.C., 2010. Robust tonic GABA currents can inhibit cell firing in mouse newborn neocortical pyramidal cells. Eur. J. Neurosci. 32 (8), 1310–1318.
Semple, B.D., Blomgren, K., Gimlin, K., Ferriero, D.M., Noble-Haeusslein, L.J., 2013. Brain development in rodents and humans: identifying benchmarks of maturation and vulnerability to injury across species. Prog. Neurobiol. 106–107, 1–16.
Silverstein, F.S., Jensen, F.E., 2007. Neonatal seizures. Ann. Neurol. 62 (2), 112–120.
Sui, L., Huang, S., Peng, B., Ren, J., Tian, F., Wang, Y., 2014. Deep brain stimulation of the amygdala alleviates fear conditioning-induced alterations in synaptic plasticity in the cortical-amygdala pathway and fear memory. J. Neural Transm. 121 (7), 773–782.
Sun, Y., Lipton, J.O., Boyle, L.M., et al., 2016. Direct current stimulation induces mGluR5-dependent neocortical plasticity. Ann. Neurol. 80 (2), 233–246.
Swann, J.W., Brady, R.J., Martin, D.L., 1989. Postnatal development of GABA-mediated synaptic inhibition in rat hippocampus. Neuroscience 28 (3), 551–561.
Tagawa, Y., Hirano, T., 2012. Activity-dependent callosal axon projections in neonatal mouse cerebral cortex. Neural. Plast. Article ID 797295, 10 pp.
Talos, D.M., Sun, H., Kosaras, B., et al., 2012. Altered inhibition in tuberous sclerosis and type IIb cortical dysplasia. Ann. Neurol. 71 (4), 539–551.
Tau, G.Z., Peterson, B.S., 2010. Normal development of brain circuits. Neuropsychopharmacology 35 (1), 147–168.
Tawfik, V.L., Chang, S.Y., Hitti, F.L., et al., 2010. Deep brain stimulation results in local glutamate and adenosine release: investigation into the role of astrocytes. Neurosurgery 67 (2), 367–375.
Tremblay, E., Nitecka, L., Berger, M.L., Ben-Ari, Y., 1984. Maturation of kainic acid seizure-brain damage syndrome in the rat. I. Clinical, electrographic and metabolic observations. Neuroscience 13 (4), 1051–1072.
Trippe, J., Mix, A., Aydin-Abidin, S., Funke, K., Benali, A., 2009. Theta burst and conventional low-frequency rTMS differentially affect GABAergic neurotransmission in the rat cortex. Exp. Brain Res. 199 (3–4), 411–421.
Ullensvang, K., Lehre, K.P., Storm-Mathisen, J., Danbolt, N.C., 1997. Differential developmental expression of the two rat brain glutamate transporter proteins GLAST and GLT. Eur. J. Neurosci. 9 (8), 1646–1655.
Valeeva, G., Tressard, T., Mukhtarov, M., Baude, A., Khazipov, R., 2016. An optogenetic approach for investigation of excitatory and inhibitory network GABA actions in mice expressing channelrhodopsin-2 in GABAergic neurons. J. Neurosci. 36 (22), 5961–5973.
van Landeghem, F.K., Stover, J.F., Bechmann, I., et al., 2001. Early expression of glutamate transporter proteins in ramified microglia after controlled cortical impact injury in the rat. Glia 35 (3), 167–179.
Wang, G., Huang, H., He, Y., Ruan, L., Huang, J., 2014. Bumetanide protects focal cerebral ischemia-reperfusion injury in rat. Int. J. Clin. Exp. Pathol. 7 (4), 1487–1494.
Wang, F., Wang, X., Shapiro, L.A., et al., 2017. NKCC1 up-regulation contributes to early post-traumatic seizures and increased post-traumatic seizure susceptibility. Brain. Struct. Funct. 222 (3), 1543–1556.
Weickert, C.S., Webster, M.J., Colvin, S.M., et al., 2000. Localization of epidermal growth factor receptors and putative neuroblasts in human subependymal zone. J. Comp. Neurol. 423 (3), 359–372.
Yan, Y., Dempsey, R.J., Sun, D., 2001. Na + –K + –Cl – cotransporter in rat focal cerebral ischemia. J. Cereb. Blood. Flow. Metab. 21 (6), 711–721.

Zeidler, S., de Boer, H., Hukema, R.K., Willemsen, R., 2017. Combination therapy in fragile X syndrome; possibilities and pitfalls illustrated by targeting the mGluR5 and GABA pathway simultaneously. Front. Mol. Neurosci. 10, 368.

Zhang, W., Xu, C., Tu, H., et al., 2015. GABAB receptor upregulates fragile X mental retardation protein expression in neurons. Sci. Rep. 5, 10468.

Zilberter, M., 2016. Reality of inhibitory GABA in neonatal brain: time to rewrite the textbooks? J. Neurosci. 36 (40), 10242–10244.

CHAPTER

3

Environmental Stimulation Modulating the Pathophysiology of Neurodevelopmental Disorders

Mari A. Kondo[1,2] *and Anthony J. Hannan*[3,4]

[1]School of Psychiatry, University of New South Wales, Sydney, NSW, Australia [2]John Curtin School of Medical Research, Australian National University, Canberra, ACT, Australia [3]Florey Institute of Neuroscience and Mental Health, Melbourne Brain Centre, University of Melbourne, Parkville, VIC, Australia [4]Department of Anatomy and Neuroscience, University of Melbourne, Parkville, VIC, Australia

OUTLINE

Neurotechnological Pediatric Neuropsych
DOI: https://doi.org/10.1016/B978-0-12-812777-3.00003-9

INTRODUCTION: THE IMPACT OF EARLY LIFE DEPRIVATION ON BRAIN DEVELOPMENT

The human brain is an extraordinary organ. In addition to governing basic bodily functions, it is responsible for all of the higher order cognitive and social behaviors that we consider to be key characteristics of our species. Although the brain is assembled in accordance with a genetic blueprint, an individual's environment has a profound effect on the brain's development and function. The human brain is particularly immature at birth, compared to most other mammals, and postnatal sensory inputs are essential for the brain to develop correctly. Postnatal brain development is then increasingly influenced by experience-dependent neural plasticity.

The visual system has been studied extensively as a model of experience-dependent neural plasticity. The electrophysiological work by Hubel and Wiesel showed that there are critical periods during which neural inputs need to be experienced for normal brain function to develop (Hubel and Wiesel, 1963; Wiesel and Hubel, 1963a,b). They showed in the cat that the effect of depriving visual stimuli from one eye for approximately 3 months differed depending on whether it was from birth, 1–2 months postnatal, or in adulthood. In cats monocularly deprived from birth, despite the structure of the eye and optical nerve being normal, the visual cortex was unable to respond to input from the previously deprived eye, preventing the brain from synthesizing the input from the two eyes as binocular vision. No such impairment was observed in cats monocularly deprived as adults, while animals that had experienced a few weeks of visual stimulus prior to monocular deprivation showed a lesser degree of behavioral and cellular impairment. In humans, congenital cataracts and strabismus (misalignment of the eyes preventing them from simultaneously focusing on the same target) should be surgically corrected in early childhood to minimize lasting impairment to binocular vision, depth perception, and sensory and motor development (Berardi et al., 2015; Hensch, 2004; Sarwar and Waqar, 2013). Although the brain has the capability to process sensory information from birth, this ability only develops to its full potential if an individual is exposed to appropriate stimuli during critical periods when the brain is particularly plastic in response to incoming stimuli.

Similar to the critical periods for the maturation of sensory processing, the brain also has sensitive periods for psychosocial development. Early work examining the effect of different types and durations of social deprivation in rhesus monkeys demonstrated that adverse psychosocial environments can impact development leading to lasting social behavioral problems (Harlow and Harlow, 1962). The importance of our postnatal everyday environment for neurodevelopment is evidenced by data from children raised in deprived institutional settings or experiencing other severe childhood adversities. The Bucharest Early Intervention Project (BEIP), which started in 2001, is particularly informative since it is a randomized control trial comparing the effects of foster care to continual institutional care (Sheridan et al., 2010). Children were enrolled in the study from an average age of 22 months (baseline: ranging from 6 to 30 months) and assessed on measures of cognitive, social, emotional, and neural development (Zeanah et al., 2003). When brain electrical activity was measured by electroencephalogram (EEG) at baseline, prior to separating the institutionalized children into groups, those who had ever been institutionalized had relatively decreased alpha and beta band activity and increased theta band activity compared to age-matched children who had been living with their biological families since birth. A follow-up at 42 months found no effect of foster care but a follow-up at 8 years of age (when the foster care group had been with their foster families for 5.5–7.5 years) showed that the EEG profile of children fostered before 24 months of age had higher alpha band activity (associated with brain maturation and increases in attention) and were comparable to the never institutionalized children (Vanderwert et al., 2010). An analysis of neuronal networks at 8 years of age found aberrant connectivity in children who remained institutionalized. Although those who had been placed in foster care also displayed aberrant connectivity, it was less pronounced, suggesting a positive effect of intervention (Stamoulis et al., 2017).

Early-life adverse psychosocial environments are strongly associated with neurodevelopmental and psychiatric disorders (McLaughlin et al., 2014; Sheridan et al., 2010). Deleterious prenatal environments, including factors affecting the mother, such as infections, severe psychological stress, nutritional state, medications, alcohol and drugs, as well as birth complications, are known to affect fetal brain development and are also implicated in psychiatric and neurodevelopmental disorders (Sale et al., 2014). Attention-deficit/hyperactivity disorder (ADHD) is particularly prevalent in children who have ever been institutionalized, compared to community controls (McLaughlin et al., 2010). In the BEIP, the prevalence of ADHD at 54 months of age was over 7 times higher in children who had ever been institutionalized (27%) and fostering did not have a significant ameliorative effect (Zeanah et al., 2009). The rate of internalizing disorders, including depression and anxiety disorders, was significantly lower in children in foster care (22.0%) compared to those in institutional care (44.2%) (Zeanah et al., 2009). At baseline, cognitive ability was also significantly deficient in children who had ever been institutionalized. Follow ups at 42 and 54 months of age showed that the children placed into foster care at an earlier age had better cognitive outcomes (Nelson et al., 2007). However, the follow-up at 12 years of age showed that foster care had minimal impact on memory and executive function, with children who had ever been institutionalized continuing to perform more poorly than the never institutionalized children (Bick et al., 2017).

USING ENVIRONMENTAL STIMULATION TO TREAT BRAIN DISORDERS

The BEIP is a real-life experiment examining whether placing a young child in a foster care home environment can mitigate the effects of institutional early-life psychosocial deprivation on brain development. Due to ethical limitations, questions remain about the sensitive period for cognitive, social, and emotional development in humans, and there are other limitations to the study. Nonetheless, results from the BEIP suggest that environmental interventions in humans can rescue some of the functional deficits caused by early deprivation of essential stimuli. It is not yet clear if the effects are lasting or whether they are due to compensatory mechanisms, such as learning, or a genuine rescue at the network, cellular, and molecular levels.

The developing brain is particularly sensitive to its environment. This not only gives it a massive learning advantage over the adult brain but also makes it more vulnerable to abuse and deprivation of essential stimuli. The modulation of synapse number and strength, and more broadly neural networks, through activity-dependent means is crucial for brain development, with a large body of evidence suggesting that neurotrophins, including brain-derived neurotrophic factor (BDNF), play an important role in this process (Lu, 2003; McOmish and Hannan, 2007). After the closure of critical/sensitive periods, the strength and number of synaptic connections, and the growth and organization of neurons, become stabilized to prevent high levels of plasticity, protecting the structure and function of the mature brain (Hensch, 2005). However, exciting research is demonstrating that neural plasticity can be boosted later in life using genetic, pharmacological, and environmental interventions, leading the way for improved treatments targeting the causes underlying neurodevelopmental and pediatric psychiatric disorders (Bavelier et al., 2010; Berardi et al., 2015).

An experimental paradigm called environmental enrichment has shown that interacting with a stimulating and varied environment as a part of everyday life has positive effects on the brains of healthy animals as well as disease models. Environmental enrichment involves controlled modification of an animal's home environment, increasing novelty and complexity relative to standard housing conditions (objects of varying textures, sizes, and shapes are added and regularly changed; animals are group-housed to allow naturalistic social interactions and access to running-wheels is given for increased physical activity) (reviewed in Kondo and Hannan, 2016). The enriched housing conditions facilitate enhanced sensory, cognitive, and motor stimulation and have been shown to slow down disease onset and/or progression in many disease models including those previously thought to be archetypal, genetically mediated diseases such as Huntington's disease (HD) and Down syndrome (Laviola et al., 2008; Martinez-Cue et al., 2002; Nithianantharajah and Hannan, 2006; Spires et al., 2004b; van Dellen et al., 2000). Work in mouse models of HD, a neurodegenerative disease, showed that environmental enrichment can rescue physical symptoms, such as motor coordination and depressive-like behavior, as well as BDNF deficits in the hippocampus and striatum (Hockly et al., 2002; Pang et al, 2009; Spires et al., 2004b; van Dellen et al., 2000).

Although genetic and pharmacological interventions may have side effects or other practical considerations for human use, environmental interventions similar to the environmental enrichment paradigm have been used safely and successfully, as documented in the BEIP and other studies. There are positive reports of recovery from visual impairment using environmental manipulations such as fast-paced computer games. Interestingly, in addition to the improvements in amblyopic vision, positive effects were observed in a range of skills such as perception, visuomotor coordination, spatial cognition, attention and decision-making, with beneficial effects being reported up to 2 years after the end of the intervention (Bavelier et al., 2010). A big challenge for therapeutic brain remodeling is to induce neural plasticity in the target regions without negatively impacting other regions. With environmental interventions, the animal's experience filters the stimuli leading to physiologically coordinated plastic changes with less side effects than methods that may induce plasticity more effectively, but do so indiscriminately. Environmental modulation is an under-researched and under-utilized tool that has the potential to improve overall health and well-being, particularly if we use a broad definition and consider the impact of lifestyle changes.

Repetitive transcranial magnetic stimulation (rTMS), an exogenous, direct form of brain stimulation, is gaining traction as a safe and effective form of treatment for antidepressant-resistant major depressive disorder (Pridmore and Pridmore, 2018). TMS (transcranial magnetic stimulation) is also being trialed for treatment in a range of brain disorders including frontotemporal dementia, schizophrenia, and fibromyalgia (Antczak et al., 2018; Dougall et al., 2015; Fitzgibbon et al., 2018). Only superficial cortical areas can be stimulated directly by TMS; however, functional changes are seen in interconnected cortical and subcortical regions. It seems to reset the synchrony of thalamocortical oscillations allowing the reemergence of intrinsic rhythms in various brain regions which lead to changes in neurotransmission, neural plasticity, and cerebral blood flow (Leuchter et al., 2013). Although there is some debate surrounding the mechanism of action for TMS, it appears to induce stochastic resonance in the human brain; the introduction of "noise" allows better detection of signals that would otherwise be below the threshold for detection/signal transduction (Schwarzkopf et al., 2011). Environmental enrichment can in fact be considered a natural form of brain stimulation. It increases neurogenesis and neurotrophin expression and has been shown to induce plasticity by acting on the same factors targeted by pharmacological interventions (Berardi et al., 2015; Kempermann et al., 1997; Laviola et al., 2008; Nithianantharajah and Hannan, 2006; Rampon et al., 2000; Sale et al., 2014; Spires et al., 2004a; van Praag et al., 2000). Although further research is required, it may be that the additional stimuli experienced in enriched environments function like "noise" in TMS. It is hoped that direct brain stimulation-based treatments will become effective stand-alone treatments for neurodevelopmental and psychiatric disorders, or beneficial treatments that synergistically combine with environmental modulation/enrichment. It is exciting that combination treatments using rTMS and behavioral therapies are showing promise in major depressive disorders (Donse et al., 2018; Russo et al., 2018). Next, we will discuss the pathophysiology and examine the impact of environmental enrichment in models of an archetypal monogenic neurodevelopmental disorder, Rett syndrome (RTT), and schizophrenia, one of the most common and devastating multigenic neurodevelopmental disorders.

RETT SYNDROME: PATHOGENIC MEDIATORS AND ENVIRONMENTAL MODULATORS

The Pathophysiology of Rett Syndrome

RTT is an early childhood onset neurodevelopmental disorder that causes severe physical, cognitive, psychological, and autonomic disability. RTT was first described in 1966 by Andreas Rett and mainly affects females. The incidence of RTT is 1 in 10,000–15,000 live births (Amir et al., 1999). There is currently no cure for RTT, and treatment involves the management of symptoms to ease discomfort and maximize functionality. The vast majority of classic RTT cases are caused by mutations in the methyl-CpG-binding protein 2 (*MECP2*) gene on the X chromosome (Amir et al., 1999; Motil et al., 2008; Vacca et al., 2001). The typical disease trajectory for RTT is an apparently normal birth followed by developmental stagnation then loss of acquired skills such as gross and fine motor control, speech, communication skills, and cognitive function (Engerstrom, 1992; Hagberg et al., 1983). At later stages of the disease, autonomic function is affected, and respiratory problems, sleep disturbances, scoliosis, dyspraxia, cardiac arrhythmia, and seizures may develop (Byard, 2006). The brains of RTT children weigh about 30% less than normal at any given age (Belichenko et al., 1994); however, there is no evidence of neuronal or glial atrophy, demyelination or gliosis (Jellinger et al., 1988). The dendritic arbors of cortical pyramidal cells are reduced as are the density of dendritic spines in the frontal cortex and hippocampus (Armstrong et al., 1998; Belichenko et al., 1994; Chapleau et al., 2009). Seizures and disturbed EEGs are common (Glaze et al., 1998).

There are several mouse models of RTT, which recapitulate the genetic mutations, symptoms, and neurophysiology of patients (Katz et al., 2012). Impairments have been observed in the brains of these mice at molecular, cellular, synaptic, and neural circuitry levels. Cortical hypoconnectivity and hypoexcitability have been found to coexist with hyperexcitability in the brainstem and hippocampus (Dani et al., 2005; Kron et al., 2012; Moretti et al., 2006; Taneja et al., 2009). The brains of RTT patients and mutant mice are described as having immature characteristics, such as smaller or less complex dendritic arbors and synaptic connections (Kishi and Macklis, 2004, 2005). This is unsurprising, as increasing levels of MeCP2 expression in the brain are associated with neuronal maturation, particularly in the hippocampus, cortex, cerebellum, and brainstem, the brain regions primarily affected in RTT (Kishi and Macklis, 2004). MeCP2 may also be involved in formation, maturation, and/or pruning of synapses (Chao et al., 2007; Kishi and Macklis, 2005; Moretti et al., 2006; Zoghbi, 2003). Indeed, impairments in long-term potentiation (LTP) in the hippocampus as well as primary motor and primary somatosensory cortex of mutant mice suggest that MeCP2 plays an important role in induction of synaptic plasticity (Asaka et al., 2006; Moretti et al., 2006). Since RTT-like symptoms can be induced in adult mice by loss of *Mecp2* and reversed when *Mecp2* is reactivated in mutant mice this suggests that in the mouse models at least, neurons are not irrevocably damaged by the absence of MeCP2 during development (Gadalla et al., 2011; Giacometti et al., 2007; Guy et al., 2007; Jugloff et al., 2008).

Behavioral Abnormalities in Rett Syndrome

Although classed by some as an autistic spectrum disorder, RTT patients lack many of the impairments in social interaction associated with autism. In a rare controlled study which took physical disability into account and utilized a standardized measure of behavioral and emotional problems (the Developmental Behavior Checklist), RTT girls were significantly less likely to engage in antisocial behaviors, compared to girls with severe mental retardation, and showed significantly less core autistic behaviors such as aloofness, avoiding eye contact, and not responding to other's feelings, when compared with autistic children (Mount et al., 2003). In a separate study matching RTT adults with severe intellectual disability (ID), autistic adults with ID and ID-only controls, Matson et al. (2008) concluded that problematic behaviors, although higher than control, were much less common in RTT compared to the autism group. In addition, the RTT patients were significantly less impaired on measures of social skills than those with autism.

The behavioral difficulties faced by RTT patients do not seem to stem from a lack of desire to socialize, but from other impairments that make social participation difficult. Substantial physical disability is one of these hurdles with the regression phase of the disorder causing loss of the ability to coordinate movements. Most patients lost functional hand use within a few months of stereotypy onset (Vignoli et al., 2009). A study of women with RTT (aged 22–44 years) found that 20% had never walked and 80% were immobile at the time of the study (Witt-Engerstrom and Hagberg, 1990). In addition, RTT patients face difficulties in cognition, anxiety, and mood. In a study that took into consideration the difficulties of evaluating intellectual function in people with severe physical and communication impairments, RTT patients with some preserved speech were found to have an intelligence quotient (IQ) of 50 or under (Zappella et al., 1998). It has also been reported that up to 70% of children with RTT display low mood and over 75% have frequent episodes of anxiety (Sansom et al., 1993).

Despite the multitude of difficulties including limited verbal communication and physical disability, RTT patients enjoy social interaction through methods of self-expression such as eye- and finger-pointing and gestures (Lavas et al., 2006). If therapies could be developed that reverse or prevent some of the regression characteristic of RTT, patients would experience a substantial improvement in quality of life.

MeCP2 and Epigenetics

The vast majority (85%–95%) of classic RTT cases are attributed to mutations in *MECP2* (Moretti and Zoghbi, 2006; Motil et al., 2008; Vacca et al., 2001). Many of the mutations cause complete or partial loss of function of the MeCP2 protein (Yusufzai and Wolffe, 2000). *MECP2* is made up of 4 exons that produce two isoforms (Amir et al., 2005; Mnatzakanian et al., 2004). Large deletions involving exons 3 and 4, contained in both isoforms, are sometimes detected in classic RTT patients (Erlandson et al., 2003). These are similar to the mutations in *Mecp2*-null mutant mouse models. Dysregulation of MeCP2 levels has also been reported in several disorders other than RTT, including Angelman syndrome, Prader–Willi syndrome, Down syndrome, autism, and ADHD (Nagarajan et al., 2006; Samaco et al., 2008; Swanberg et al., 2009). The MeCP2 protein is relatively well conserved across vertebrates indicating its importance in biological functions (Lewis et al., 1992; Weaving et al., 2005; Yusufzai and Wolffe, 2000). It is involved in methylation-dependent transcriptional regulation of target genes through chromatin remodeling by recruiting histone deacetylases and corepressors, RNA splicing, repression of retrotransposons and binds unmethylated DNA to remodel chromatin independently of histone-modifying proteins (Buschdorf and Stratling, 2004; Chahrour et al., 2008; Georgel et al., 2003; Jones et al., 1998; Nikitina et al., 2007a,b; Yasui et al., 2007; Young et al., 2005; Yu et al., 2001). The primary role of MeCP2 in neurodevelopmental disorders is thought to be the transcriptional repression of neuronal activity-dependent target genes with methylated promoters (Chen et al., 2003; De Filippis et al., 2013; Martinowich et al., 2003; McGill et al., 2006; Tao et al., 2009; Zhou et al., 2006).

In multicellular eukaryotes, gene silencing through methylation of 5′CpG sequences in DNA is a common mechanism of epigenetic regulation (Jaenisch and Bird, 2003; Klose and Bird, 2003; Klose and Bird, 2006). Environmental experiences can become hardwired and influence gene expression by affecting epigenetic markings, such as DNA methylation, resulting in long-term behavioral and cognitive effects (Bredy et al., 2003a,b; Meaney and Szyf, 2005). MeCP2's function as a DNA methylation-dependent transcriptional repressor allows it to act as a bridge between environmentally mediated methylation patterns and gene expression.

Regulation of *BDNF* by MeCP2

BDNF is essential for neuronal survival, differentiation, and synaptic plasticity during development and in adulthood (Lu, 2003). Normal MeCP2 regulates expression of the *bdnf* gene in a neuronal activity-dependent manner. In resting neurons, MeCP2 is bound to methylated CpGs near *bdnf* promoter IV in rodents (exon nomenclature from Aid et al., 2007). Depolarization of cultured cortical neurons causes MeCP2 to dissociate from the *bdnf* gene promoter. The displacement of MeCP2 from the promoter appears to be caused by a combination of cytosine demethylation and MeCP2 phosphorylation (Chen et al., 2003; Martinowich et al., 2003). These findings contributed to the now prevailing view that DNA methylation-related chromatin remodeling is dynamic and important for activity-dependent gene regulation, and therefore critical for synapse development and neural plasticity. An examination of *bdnf* conditional mutant mice revealed that loss of BDNF causes RTT-relevant pathologies including smaller brain size, smaller CA2 neurons, smaller glomerulus size, and a characteristic hindlimb-clasping phenotype (Chang et al., 2006). Furthermore, symptomatic *Mecp2* mutant mice only had 70% of the wild-type level of BDNF in whole brain extract. Deletion of *bdnf* in *Mecp2*-null mice caused an earlier onset of RTT-like symptoms, while BDNF overexpression in the *Mecp2* mutant mouse extended the lifespan, rescued a locomotor defect, and reversed an electrophysiological deficit (Chang et al., 2006). These results provided in vivo evidence for an interaction between *Mecp2* and *bdnf*, demonstrating the physiological significance of altered BDNF expression/signaling in RTT disease progression.

The Therapeutic Effects of Environmental Stimulation in Preclinical Models of Rett Syndrome

Mouse models of human disorders are valuable tools for understanding disease mechanisms as well as trialing potential treatments. *Mecp2*-mutant mouse models have enabled us to study the consequences of deficient MeCP2 protein and how this results in the RTT phenotype. The first genome-wide *Mecp2*-null mice were created by deleting exon 3 (Chen et al., 2001), and exons 3 and 4 of *Mecp2* (Guy et al., 2001). The hemizygous null males developed a stiff, uncoordinated gait, hindlimb clasping, and irregular breathing, followed by rapid weight loss and death at approximately 10 weeks of age (Chen et al., 2001; Guy et al., 2001). Brain-specific knockout of *Mecp2*, using the Nestin-Cre transgene, showed that the mutant phenotype is caused by MeCP2 deficiency in the central nervous system (Chen et al., 2001; Guy et al., 2001). Deletion of *Mecp2*, by CamK-Cre93-mediated recombination (activated in postnatal forebrain, hippocampus, brainstem, and only marginally in cerebellum), demonstrated that the phenotype is caused by MeCP2 deficiency in neurons, not glia, and that a deficit in postmitotic neurons is sufficient to produce disease (Chen et al., 2001).

Using an independently created null model (*Mecp2*^{tm1Tam}), Pelka et al. (2006) found that mutant males have a motor coordination deficit and cerebellar motor learning deficit. The mixed genetic background extended the lifespan of the male mice enabling behavioral testing which was not possible using the Chen and Guy models. The female heterozygous *Mecp2*^{tm1Tam} mutant mice have construct validity for RTT and approximately 50% normal MeCP2. They display behavioral symptoms as well as corticosterone and BDNF dysregulation that are similar to those displayed by RTT patients. We investigated whether environmental enrichment can delay symptoms in hemizygous and heterozygous mutant mice, using age-matched wild-type and mutant littermates. Based on the proposed role of BDNF in the pathogenesis of RTT and the importance of neurotrophins in experience-dependent neuronal plasticity, BDNF protein levels in specific brain regions were also assessed. Exposing the heterozygous female mutant mice to postweaning environmental enrichment caused a substantial rescue of disease symptoms: motor coordination, anxiety, depressive-like behaviors, basal corticosterone and hippocampal BDNF levels (Kondo et al., 2008, 2016). By comparing different housing conditions, we also found that the positive effect of enrichment on heterozygous mutants was due to the combined effects of voluntary physical exercise and sensory/cognitive stimulation (Kondo et al., 2016). Compared to the female heterozygous mutants, behavioral deficits were more severe in the *Mecp2* hemizygous-null male mutants. They also showed no symptom improvement with postweaning enrichment, highlighting the importance of even partial MeCP2 presence in mediating the central nervous system's response to positive external stimuli after weaning.

Mecp2 mutant mice have abnormal stress–response regulation (Kondo et al., 2016; McGill et al., 2006; Nuber et al., 2005). The observation that *Mecp2* conditional knockout mice were unable to readily adjust to new social situations supports a role for MeCP2 in responding to altered physiological states (Fyffe et al., 2008). Enrichment appeared to increase resilience to stress in the heterozygous mice (Kondo et al., 2016). The concept of enrichment increasing the brain's resilience to stress or injury is referred to as brain and cognitive reserve (Nithianantharajah and Hannan, 2009).

Trials Using Environmental Modulation in Rett Syndrome Patients

The positive results from the heterozygous female mice suggest that a specially tailored, enhanced environment approach may benefit patients helping them to restore functionality and increase independence leading to a better quality of life. There have been a small number of patient intervention studies that have attempted to treat RTT using mental and/or physical stimulation. Improvement in purposeful hand use and language comprehension was reported following music therapy (Yasuhara and Sugiyama, 2001) and physical therapies, including physiotherapy, is considered beneficial for improving movement and quality of life (Downs et al., 2009; Jacobsen et al., 2001; Piazza et al., 1993). A study in four RTT girls with independent mobility found that daily exercise for 2 months on a modified treadmill improved resting heart rate and general functional abilities (Lotan et al., 2004). Most recently, a 6-month environmental enrichment intervention in 12 young girls with RTT was found to improve gross motor skills and increase blood levels of BDNF (Downs et al., 2018).

The mouse and human studies show that appropriate forms of brain and physical stimulation can reduce symptoms. However, current environmental methods on their own are insufficient, particularly for severe neurodevelopmental disorders like RTT. Further clinical studies are needed to identify the optimal forms of environmental stimulation and the best ways to maintain the effects of enhanced sensory, cognitive and physical stimulation in the face of continued MeCP2 deficit. One way forward may be to combine environmental therapy, including direct brain stimulation, with drug treatments such as fluoxetine (a selective serotonin-reuptake inhibitor used to treat depression) which induces transcription of *Mecp2* in mice (Cassel et al., 2006).

SCHIZOPHRENIA: PATHOGENIC MEDIATORS AND ENVIRONMENTAL MODULATORS

The Pathophysiology of Schizophrenia

Schizophrenia is a devastating neurodevelopmental and psychiatric disorder which affects approximately 1% of the population. It consists of a complex combination of psychotic "positive symptoms" (including hallucinations and delusions), "negative symptoms" (including affective dysfunction), and cognitive deficits (which are most resistant to current antipsychotic drugs). This brain disorder generally strikes individuals in the primes of their lives, in adolescence and young adulthood.

Schizophrenia is currently inadequately treated, as antipsychotic medications generally are ineffective for cognitive deficits (which have a major negative impact on quality of life), show variable efficacy for psychotic symptoms, and have major negative side-effect profiles. The burden of disease in schizophrenia, encompassing both sufferers and their families, is one of the highest of all brain disorders, and there is a large unmet need for new preventative and therapeutic approaches.

Schizophrenia is a complex polygenetic disorder, whose genetics has only recently begun to be elucidated via the power of genome-wide association studies utilizing large international cohorts (Schizophrenia Working Group of the Psychiatric Genomics Consortium, 2014). However, there remains substantial "missing heritability," which requires further genetic, environmental, and epigenetic research, in order to fully comprehend such complex polygenic disorders (Eichler et al., 2010; Hannan, 2010, 2018; Manolio et al., 2009). Evidence from twin studies and other epidemiology indicates that genetic factors contribute approximately half of the risk for schizophrenia, and environmental factors contribute the other half. The environmental factors are not well defined but can include stress, maternal infection, cannabis use, and possibly various other experience-dependent modifiers such as cognitive stimulation and physical activity. One of the key goals in schizophrenia research is to understand how genetic and environmental factors combine in pathogenic pathways, and to define associated gene–environment interactions (Burrows and Hannan, 2013; McGrath et al., 2011). This in turn will lead to new therapeutic approaches.

This genetic and environmental complexity makes schizophrenia very challenging to model preclinically. However, despite the difficulties of capturing genetic construct validity of a polygenic disorder in an animal model, face and predictive validity can be robustly established. The most commonly utilized preclinical models of schizophrenia have involved genetically targeted mouse models. These can be combined with environmental factors to model gene–environment interactions, as described in the next section.

Environmental Stimuli Modulating Schizophrenia Endophenotypes in Preclinical Models

The first evidence that environmental stimulation could be beneficial in a preclinical model of schizophrenia was provided using environmental enrichment of gene-edited mice (McOmish et al., 2008). This study was performed in phospholipase C-beta 1 (PLC-b1) knockout mice which show abnormalities of cortical development that are highly relevant to schizophrenia (Hannan et al., 2001). These mice were also shown to exhibit both face and predictive validity for schizophrenia (McOmish et al., 2008). Furthermore, these schizophrenia-like endophenotypes were rescued by environmental enrichment (McOmish et al., 2008). This may have direct clinical relevance to cognitive stimulation and physical activity interventions in schizophrenia.

This initial study was later followed up in a second mouse model of schizophrenia affecting an upstream component of this signaling pathway, metabotropic glutamate receptor 5 (mGlu5) knockout mice (Burrows et al., 2015). It was striking that the cognitive deficits in the mGlu5 knockout mice were rescued by environmental enrichment. This is clinically relevant, as the cognitive deficits of schizophrenia generally remain untreated by antipsychotic drugs.

BENEFICIAL EFFECTS OF ENVIRONMENTAL STIMULATION IN OTHER PRECLINICAL MODELS OF NEURODEVELOPMENTAL DISORDERS

Whilst we have focused on RTT and schizophrenia, preclinical models of other neurodevelopmental disorders have been shown to respond positively to environmental enrichment. A mouse model of fragile X syndrome, which has a null mutation in the *Fmr1* gene, was shown to be beneficially impacted by environmental enrichment (Meredith et al., 2007; Restivo et al., 2005). Furthermore, therapeutic impacts have also been observed in a mouse model of Down syndrome (Begenisic et al., 2015; Dierssen et al., 2003; Martinez-Cue et al., 2002). Whilst current preclinical models of autism spectrum disorder (ASD) have variable levels of construct and face validity, some preclinical and clinical findings support the use of environmental enrichment as a potentially beneficial intervention for ASD (Hill-Yardin and Hannan, 2013).

In these preclinical models of neurodevelopmental disorders, as well as RTT and schizophrenia described above, many of the effects of environmental enrichment may reflect those which occur in the healthy (wild-type) brain. However, in some cases, the neurodevelopmental disorder responds differentially (e.g., partially) to environmental enrichment, which is where appropriate experimental design is crucial. These experiments have identified gene–environment interactions which not only reveal the therapeutic potential of environmental enrichment but also its beneficial impacts on the typically developing mammalian brain.

It should also be noted that we have focused on environmental stimulation interventions applied during the postnatal (and in particular postweaning) period in rodent models. However, there is evidence that such interventions can have effects not only in utero, but also preconceptually, via transgenerational epigenetic mechanisms. The transgenerational epigenetic inheritance of acquired traits is an exciting new field which has major implications for the understanding, prevention, and treatment of brain disorders, particularly those which are mediated by neurodevelopmental dysfunction (Pang et al., 2017; Yeshurun and Hannan, 2018).

CONCLUSIONS

It is clear that environmental stimulation (sensory, cognitive, and physical activity interventions) can have strong therapeutic effects in preclinical models of neurodevelopmental disorders, including RTT and schizophrenia. These findings have implications for the way in which cognitive stimulation and physical activity interventions might be used as part of a comprehensive treatment package to treat neurodevelopmental and psychiatric disorders. However, the experiments discussed above, in which specific molecular and cellular mechanisms have been elucidated, also have implications for the development of other interventions. In particular, understanding the therapeutic effects of environmental enrichment will guide the development of enviromimetics (novel therapeutics which mimic or enhance the beneficial effects of cognitive stimulation and physical activity), including direct brain stimulation and drug treatments (Hannan, 2004; McOmish and Hannan, 2007). Although there has been limited exploration of brain stimulation-based treatments for RTT, the successful use of rTMS in combination with behavioral treatments in major depressive disorders suggests that this promising approach could be trialed in other disorders. These, and other, novel therapeutic approaches are urgently needed for these devastating disorders of brain development and function.

References

Aid, T., Kazantseva, A., Piirsoo, M., Palm, K., Timmusk, T., 2007. Mouse and rat BDNF gene structure and expression revisited. J. Neurosci. Res. 85 (3), 525–535. Available from: https://doi.org/10.1002/jnr.21139.

Amir, R.E., Van den Veyver, I.B., Wan, M., Tran, C.Q., Francke, U., Zoghbi, H.Y., 1999. Rett syndrome is caused by mutations in X-linked MECP2, encoding methyl-CpG-binding protein 2. Nat. Genet. 23 (2), 185–188. Available from: https://doi.org/10.1038/13810.

Amir, R.E., Fang, P., Yu, Z., Glaze, D.G., Percy, A.K., Zoghbi, H.Y., et al., 2005. Mutations in exon 1 of MECP2 are a rare cause of Rett syndrome. J. Med. Genet. 42 (2), e15. Available from: https://doi.org/10.1136/jmg.2004.026161. 42/2/e15 [pii].

Antczak, J., Kowalska, K., Klimkowicz-Mrowiec, A., Wach, B., Kasprzyk, K., Banach, M., et al., 2018. Repetitive transcranial magnetic stimulation for the treatment of cognitive impairment in frontotemporal dementia: an open-label pilot study. Neuropsychiatr. Dis. Treat. 14, 749–755. Available from: https://doi.org/10.2147/NDT.S153213.

Armstrong, D.D., Dunn, K., Antalffy, B., 1998. Decreased dendritic branching in frontal, motor and limbic cortex in Rett syndrome compared with trisomy 21. J. Neuropathol. Exp. Neurol. 57 (11), 1013–1017.

Asaka, Y., Jugloff, D.G., Zhang, L., Eubanks, J.H., Fitzsimonds, R.M., 2006. Hippocampal synaptic plasticity is impaired in the Mecp2-null mouse model of Rett syndrome. Neurobiol. Dis. 21 (1), 217–227. Available from: https://doi.org/10.1016/j.nbd.2005.07.005 [pii] S0969-9961(05)00204-4.

Bavelier, D., Levi, D.M., Li, R.W., Dan, Y., Hensch, T.K., 2010. Removing brakes on adult brain plasticity: from molecular to behavioral interventions. J. Neurosci. 30 (45), 14964–14971. Available from: https://doi.org/10.1523/JNEUROSCI.4812-10.2010.

Begenisic, T., Sansevero, G., Baroncelli, L., Cioni, G., Sale, A., 2015. Early environmental therapy rescues brain development in a mouse model of Down syndrome. Neurobiol. Dis. 82, 409–419. Available from: https://doi.org/10.1016/j.nbd.2015.07.014.

Belichenko, P.V., Oldfors, A., Hagberg, B., Dahlstrom, A., 1994. Rett syndrome: 3-D confocal microscopy of cortical pyramidal dendrites and afferents. Neuroreport 5 (12), 1509–1513.

Berardi, N., Sale, A., Maffei, L., 2015. Brain structural and functional development: genetics and experience. Dev. Med. Child Neurol. 57 (Suppl. 2), 4–9. Available from: https://doi.org/10.1111/dmcn.12691.

Bick, J., Zeanah, C.H., Fox, N.A., Nelson, C.A., 2017. Memory and executive functioning in 12-year-old children with a history of institutional rearing. Child Dev. Available from: https://doi.org/10.1111/cdev.12952.

Bredy, T.W., Grant, R.J., Champagne, D.L., Meaney, M.J., 2003a. Maternal care influences neuronal survival in the hippocampus of the rat. Eur. J. Neurosci. 18 (10), 2903–2909. Available from: http://doi.org/10.1046/j.1460-9568.2003.02965.x.

Bredy, T.W., Humpartzoomian, R.A., Cain, D.P., Meaney, M.J., 2003b. Partial reversal of the effect of maternal care on cognitive function through environmental enrichment. Neuroscience 118 (2), 571–576 [pii]S0306452202009181.

Burrows, E.L., Hannan, A.J., 2013. Decanalization mediating gene–environment interactions in schizophrenia and other psychiatric disorders with neurodevelopmental etiology. Front. Behav. Neurosci. 7, 157. Available from: https://doi.org/10.3389/fnbeh.2013.00157.

Burrows, E.L., McOmish, C.E., Buret, L.S., Van den Buuse, M., Hannan, A.J., 2015. Environmental enrichment ameliorates behavioral impairments modeling schizophrenia in mice lacking metabotropic glutamate receptor 5. Neuropsychopharmacology 40 (8), 1947–1956. Available from: https://doi.org/10.1038/npp.2015.44.

Buschdorf, J.P., Stratling, W.H., 2004. A WW domain binding region in methyl-CpG-binding protein MeCP2: impact on Rett syndrome. J. Mol. Med. 82 (2), 135–143. Available from: https://doi.org/10.1007/s00109-003-0497-9.

Byard, R.W., 2006. Forensic issues and possible mechanisms of sudden death in Rett syndrome. J. Clin. Forensic Med. 13 (2), 96–99. Available from: https://doi.org/10.1016/j.jcfm.2005.08.013 [pii] S1353-1131(05)00152-5.

Cassel, S., Carouge, D., Gensburger, C., Anglard, P., Burgun, C., Dietrich, J.B., et al., 2006. Fluoxetine and cocaine induce the epigenetic factors MeCP2 and MBD1 in adult rat brain. Mol. Pharmacol. 70 (2), 487–492. Available from: https://doi.org/10.1124/mol.106.022301.

Chahrour, M., Jung, S.Y., Shaw, C., Zhou, X., Wong, S.T., Qin, J., et al., 2008. MeCP2, a key contributor to neurological disease, activates and represses transcription. Science 320 (5880), 1224–1229. Available from: https://doi.org/10.1126/science.1153252.

Chang, Q., Khare, G., Dani, V., Nelson, S., Jaenisch, R., 2006. The disease progression of Mecp2 mutant mice is affected by the level of BDNF expression. Neuron 49 (3), 341–348. Available from: https://doi.org/10.1016/j.neuron.2005.12.027.

Chao, H.T., Zoghbi, H.Y., Rosenmund, C., 2007. MeCP2 controls excitatory synaptic strength by regulating glutamatergic synapse number. Neuron 56 (1), 58–65. Available from: https://doi.org/10.1016/j.neuron.2007.08.018.

Chapleau, C.A., Calfa, G.D., Lane, M.C., Albertson, A.J., Larimore, J.L., Kudo, S., et al., 2009. Dendritic spine pathologies in hippocampal pyramidal neurons from Rett syndrome brain and after expression of Rett-associated MECP2 mutations. Neurobiol. Dis. 35 (2), 219–233. Available from: https://doi.org/10.1016/j.nbd.2009.05.001.

Chen, R.Z., Akbarian, S., Tudor, M., Jaenisch, R., 2001. Deficiency of methyl-CpG binding protein-2 in CNS neurons results in a Rett-like phenotype in mice. Nature Genet. 27 (3), 327–331. Available from: https://doi.org/10.1038/85906.

Chen, W.G., Chang, Q., Lin, Y., Meissner, A., West, A.E., Griffith, E.C., et al., 2003. Derepression of BDNF transcription involves calcium-dependent phosphorylation of MeCP2. Science 302 (5646), 885–889. Available from: https://doi.org/10.1126/science.1086446.

Dani, V.S., Chang, Q., Maffei, A., Turrigiano, G.G., Jaenisch, R., Nelson, S.B., 2005. Reduced cortical activity due to a shift in the balance between excitation and inhibition in a mouse model of Rett syndrome. Proc. Natl. Acad. Sci. U.S.A 102 (35), 12560–12565. Available from: https://doi.org/10.1073/pnas.0506071102.

De Filippis, B., Ricceri, L., Fuso, A., Laviola, G., 2013. Neonatal exposure to low dose corticosterone persistently modulates hippocampal mineralocorticoid receptor expression and improves locomotor/exploratory behaviour in a mouse model of Rett syndrome. Neuropharmacology 68, 174–183. Available from: https://doi.org/10.1016/j.neuropharm.2012.05.048.

Dierssen, M., Benavides-Piccione, R., Martinez-Cue, C., Estivill, X., Florez, J., Elston, G.N., et al., 2003. Alterations of neocortical pyramidal cell phenotype in the Ts65Dn mouse model of Down syndrome: effects of environmental enrichment. Cereb. Cortex 13 (7), 758–764.

Donse, L., Padberg, F., Sack, A.T., Rush, A.J., Arns, M., 2018. Simultaneous rTMS and psychotherapy in major depressive disorder: clinical outcomes and predictors from a large naturalistic study. Brain Stimul. 11 (2), 337–345. Available from: https://doi.org/10.1016/j.brs.2017.11.004.

Dougall, N., Maayan, N., Soares-Weiser, K., McDermott, L.M., McIntosh, A., 2015. Transcranial magnetic stimulation (TMS) for schizophrenia. Cochrane Database Syst. Rev. (8), CD006081. Available from: https://doi.org/10.1002/14651858.CD006081.pub2.

Downs, J., Bergman, A., Carter, P., Anderson, A., Palmer, G.M., Roye, D., et al., 2009. Guidelines for management of scoliosis in Rett syndrome patients based on expert consensus and clinical evidence. Spine (Phila Pa 1976) 34 (17), E607–E617. Available from: https://doi.org/10.1097/BRS.0b013e3181a95ca4.

Downs, J., Rodger, J., Li, C., Tan, X., Hu, N., Wong, K., et al., 2018. Environmental enrichment intervention for Rett syndrome: an individually randomised stepped wedge trial. Orphanet J. Rare Dis. 13 (1), 3. Available from: https://doi.org/10.1186/s13023-017-0752-8.

Eichler, E.E., Flint, J., Gibson, G., Kong, A., Leal, S.M., Moore, J.H., et al., 2010. Missing heritability and strategies for finding the underlying causes of complex disease. Nat. Rev. Genet. 11 (6), 446–450. Available from: https://doi.org/10.1038/nrg2809.

Engerstrom, I.W., 1992. Rett syndrome: the late infantile regression period—a retrospective analysis of 91 cases. Acta Paediatr. 81 (2), 167–172.

Erlandson, A., Samuelsson, L., Hagberg, B., Kyllerman, M., Vujic, M., Wahlstrom, J., 2003. Multiplex ligation-dependent probe amplification (MLPA) detects large deletions in the

MECP2 gene of Swedish Rett syndrome patients. Genet. Test. 7 (4), 329–332. Available from: https://doi.org/10.1089/109065703322783707.

Fitzgibbon, B.M., Hoy, K.E., Knox, L.A., Guymer, E.K., Littlejohn, G., Elliot, D., et al., 2018. Evidence for the improvement of fatigue in fibromyalgia: a 4-week left dorsolateral prefrontal cortex repetitive transcranial magnetic stimulation randomized-controlled trial. Eur. J. Pain . Available from: https://doi.org/10.1002/ejp.1213.

Fyffe, S.L., Neul, J.L., Samaco, R.C., Chao, H.T., Ben-Shachar, S., Moretti, P., et al., 2008. Deletion of Mecp2 in Sim1-expressing neurons reveals a critical role for MeCP2 in feeding behavior, aggression, and the response to stress. Neuron 59 (6), 947–958. Available from: https://doi.org/10.1016/j.neuron.2008.07.030 [pii]S0896-6273(08)00629-6.

Gadalla, K.K., Bailey, M.E., Cobb, S.R., 2011. MeCP2 and Rett syndrome: reversibility and potential avenues for therapy. Biochem. J. 439 (1), 1–14. Available from: https://doi.org/10.1042/BJ20110648.

Georgel, P.T., Horowitz-Scherer, R.A., Adkins, N., Woodcock, C.L., Wade, P.A., Hansen, J.C., 2003. Chromatin compaction by human MeCP2. Assembly of novel secondary chromatin structures in the absence of DNA methylation. J. Biol. Chem. 278 (34), 32181–32188. Available from: https://doi.org/10.1074/jbc.M305308200.

Giacometti, E., Luikenhuis, S., Beard, C., Jaenisch, R., 2007. Partial rescue of MeCP2 deficiency by postnatal activation of MeCP2. Proc. Natl. Acad. Sci. U.S.A. 104 (6), 1931–1936. Available from: https://doi.org/10.1073/pnas.0610593104.

Glaze, D.G., Schultz, R.J., Frost, J.D., 1998. Rett syndrome: characterization of seizures versus non-seizures. Electroencephalogr. Clin. Neurophysiol. 106 (1), 79–83.

Guy, J., Hendrich, B., Holmes, M., Martin, J.E., Bird, A., 2001. A mouse Mecp2-null mutation causes neurological symptoms that mimic Rett syndrome. Nat. Genet. 27 (3), 322–326. Available from: https://doi.org/10.1038/85899.

Guy, J., Gan, J., Selfridge, J., Cobb, S., Bird, A., 2007. Reversal of neurological defects in a mouse model of Rett syndrome. Science 315 (5815), 1143–1147. Available from: https://doi.org/10.1126/science.1138389.

Hagberg, B., Aicardi, J., Dias, K., Ramos, O., 1983. A progressive syndrome of autism, dementia, ataxia, and loss of purposeful hand use in girls: Rett's syndrome: report of 35 cases. Ann. Neurol. 14 (4), 471–479. Available from: https://doi.org/10.1002/ana.410140412.

Hannan, A.J., 2004. Molecular mediators, environmental modulators and experience-dependent synaptic dysfunction in Huntington's disease. Acta Biochim. Pol. 51 (2), 415–430. Available from: http://doi.org/035001415.

Hannan, A.J., 2010. Tandem repeat polymorphisms: modulators of disease susceptibility and candidates for 'missing heritability'. Trends Genet. 26 (2), 59–65. Available from: https://doi.org/10.1016/j.tig.2009.11.008.

Hannan, A.J., 2018. Tandem repeats mediating genetic plasticity in health and disease. Nat. Rev. Genet. 19 (5), 286–298. Available from: https://doi.org/10.1038/nrg.2017.115.

Hannan, A.J., Blakemore, C., Katsnelson, A., Vitalis, T., Huber, K.M., Bear, M., et al., 2001. PLC-beta1, activated via mGluRs, mediates activity-dependent differentiation in cerebral cortex. Nat. Neurosci. 4 (3), 282–288. Available from: https://doi.org/10.1038/85132.

Harlow, H.F., Harlow, M., 1962. Social deprivation in monkeys. Sci. Am. 207, 136–146.

Hensch, T.K., 2004. Critical period regulation. Annu. Rev. Neurosci. 27, 549–579. Available from: https://doi.org/10.1146/annurev.neuro.27.070203.144327.

Hensch, T.K., 2005. Critical period plasticity in local cortical circuits. Nat. Rev. Neurosci. 6 (11), 877–888. Available from: https://doi.org/10.1038/nrn1787.

Hill-Yardin, E.L., Hannan, A.J., 2013. Translating preclinical environmental enrichment studies for the treatment of autism and other brain disorders: comment on Woo and Leon (2013). Behav. Neurosci. 127 (4), 606–609. Available from: https://doi.org/10.1037/a0033319.

Hockly, E., Cordery, P.M., Woodman, B., Mahal, A., van Dellen, A., Blakemore, C., et al., 2002. Environmental enrichment slows disease progression in R6/2 Huntington's disease mice. Ann. Neurol. 51 (2), 235–242. Available from: https://doi.org/10.1002/ana.10094.

Hubel, D.H., Wiesel, T.N., 1963. Receptive fields of cells in striate cortex of very young, visually inexperienced kittens. J. Neurophysiol. 26, 994–1002. Available from: https://doi.org/10.1152/jn.1963.26.6.994.

Jacobsen, K., Viken, A., von Tetzchner, S., 2001. Rett syndrome and ageing: a case study. Disabil. Rehabil. 23 (3-4), 160–166.

Jaenisch, R., Bird, A., 2003. Epigenetic regulation of gene expression: how the genome integrates intrinsic and environmental signals. Nat. Genet. 33 (Suppl), 245–254. Available from: https://doi.org/10.1038/ng1089.

Jellinger, K., Armstrong, D., Zoghbi, H.Y., Percy, A.K., 1988. Neuropathology of Rett syndrome. Acta Neuropathol. 76 (2), 142–158.

Jones, P.L., Veenstra, G.J., Wade, P.A., Vermaak, D., Kass, S.U., Landsberger, N., et al., 1998. Methylated DNA and MeCP2 recruit histone deacetylase to repress transcription. Nat. Genet. 19 (2), 187–191. Available from: https://doi.org/10.1038/561.

Jugloff, D.G., Vandamme, K., Logan, R., Visanji, N.P., Brotchie, J.M., Eubanks, J.H., 2008. Targeted delivery of an Mecp2 transgene to forebrain neurons improves the behavior of female Mecp2-deficient mice. Hum. Mol. Genet. 17 (10), 1386–1396. Available from: https://doi.org/10.1093/hmg/ddn026.

Katz, D.M., Berger-Sweeney, J.E., Eubanks, J.H., Justice, M.J., Neul, J.L., Pozzo-Miller, L., et al., 2012. Preclinical research in Rett syndrome: setting the foundation for translational success. Dis. Model Mech. 5 (6), 733–745. Available from: https://doi.org/10.1242/dmm.011007.

Kempermann, G., Kuhn, H.G., Gage, F.H., 1997. More hippocampal neurons in adult mice living in an enriched environment. Nature 386 (6624), 493–495. Available from: https://doi.org/10.1038/386493a0.

Kishi, N., Macklis, J.D., 2004. MECP2 is progressively expressed in post-migratory neurons and is involved in neuronal maturation rather than cell fate decisions. Mol. Cell. Neurosci. 27 (3), 306–321. Available from: https://doi.org/10.1016/j.mcn.2004.07.006 [pii]S1044-7431(04)00170-8.

Kishi, N., Macklis, J.D., 2005. Dissecting MECP2 function in the central nervous system. J. Child Neurol. 20 (9), 753–759.

Klose, R., Bird, A., 2003. Molecular biology. MeCP2 repression goes nonglobal. Science 302 (5646), 793–795. Available from: https://doi.org/10.1126/science.1091762. 302/5646/793 [pii].

Klose, R.J., Bird, A.P., 2006. Genomic DNA methylation: the mark and its mediators. Trends Biochem. Sci. 31 (2), 89–97. Available from: https://doi.org/10.1016/j.tibs.2005.12.008 [pii]S0968-0004(05)00352-X.

Kondo, M.A., Hannan, A.J., 2016. Gene–environment interactions in the etiology of psychiatric and neurodevelopmental disorders. In: Sale, A. (Ed.), Environmental Experience and Plasticity of the Developing Brain. John Wiley & Sons, Inc, Hoboken, New Jersey, pp. 47–72. Available from: https://doi.org/10.1002/9781118931684.ch3.

Kondo, M., Gray, L.J., Pelka, G.J., Christodoulou, J., Tam, P.P., Hannan, A.J., 2008. Environmental enrichment ameliorates a motor coordination deficit in a mouse model of Rett syndrome—Mecp2 gene dosage effects and BDNF expression. Eur. J. Neurosci. 27 (12), 3342–3350. Available from: https://doi.org/10.1111/j.1460-9568.2008.06305.x.

Kondo, M.A., Gray, L.J., Pelka, G.J., Leang, S.K., Christodoulou, J., Tam, P.P., et al., 2016. Affective dysfunction in a mouse model of Rett syndrome: Therapeutic effects of environmental stimulation and physical activity. Dev. Neurobiol. 76 (2), 209–224. Available from: https://doi.org/10.1002/dneu.22308.

Kron, M., Howell, C.J., Adams, I.T., Ransbottom, M., Christian, D., Ogier, M., et al., 2012. Brain activity mapping in Mecp2 mutant mice reveals functional deficits in forebrain circuits, including key nodes in the default mode network, that are reversed with ketamine treatment. J. Neurosci. 32 (40), 13860–13872. Available from: https://doi.org/10.1523/JNEUROSCI.2159-12.2012.

Lavas, J., Slotte, A., Jochym-Nygren, M., van Doorn, J., Engerstrom, I.W., 2006. Communication and eating proficiency in 125 females with Rett syndrome: the Swedish Rett Center Survey. Disabil. Rehabil. 28 (20), 1267–1279.

Laviola, G., Hannan, A.J., Macri, S., Solinas, M., Jaber, M., 2008. Effects of enriched environment on animal models of neurodegenerative diseases and psychiatric disorders. Neurobiol. Dis. 31 (2), 159–168. Available from: https://doi.org/10.1016/j.nbd.2008.05.001 [pii]S0969-9961(08)00091-0.

Leuchter, A.F., Cook, I.A., Jin, Y., Phillips, B., 2013. The relationship between brain oscillatory activity and therapeutic effectiveness of transcranial magnetic stimulation in the treatment of major depressive disorder. Front. Hum. Neurosci. 7, 37. Available from: https://doi.org/10.3389/fnhum.2013.00037.

Lewis, J.D., Meehan, R.R., Henzel, W.J., Maurer-Fogy, I., Jeppesen, P., Klein, F., et al., 1992. Purification, sequence, and cellular localization of a novel chromosomal protein that binds to methylated DNA. Cell 69 (6), 905–914 [pii]0092-8674(92)90610-O.

Lotan, M., Isakov, E., Merrick, J., 2004. Improving functional skills and physical fitness in children with Rett syndrome. J. Intellect. Disabil. Res. 48 (Pt 8), 730–735. Available from: https://doi.org/10.1111/j.1365-2788.2003.00589.x [pii]JIR589.

Lu, B., 2003. BDNF and activity-dependent synaptic modulation. Learn. Mem. 10 (2), 86–98. Available from: https://doi.org/10.1101/lm.54603.

Manolio, T.A., Collins, F.S., Cox, N.J., Goldstein, D.B., Hindorff, L.A., Hunter, D.J., et al., 2009. Finding the missing heritability of complex diseases. Nature 461 (7265), 747–753. Available from: https://doi.org/10.1038/nature08494.

Martinez-Cue, C., Baamonde, C., Lumbreras, M., Paz, J., Davisson, M.T., Schmidt, C., et al., 2002. Differential effects of environmental enrichment on behavior and learning of male and female Ts65Dn mice, a model for Down syndrome. Behav. Brain Res. 134 (1-2), 185–200 [pii]S0166432802000268.

Martinowich, K., Hattori, D., Wu, H., Fouse, S., He, F., Hu, Y., et al., 2003. DNA methylation-related chromatin remodeling in activity-dependent BDNF gene regulation. Science 302 (5646), 890–893. Available from: https://doi.org/10.1126/science.1090842.

Matson, J.L., Dempsey, T., Wilkins, J., 2008. Rett syndrome in adults with severe intellectual disability: exploration of behavioral characteristics. Eur. Psychiatry 23 (6), 460–465. Available from: https://doi.org/10.1016/j.eurpsy.2007.11.008.

McGill, B.E., Bundle, S.F., Yaylaoglu, M.B., Carson, J.P., Thaller, C., Zoghbi, H.Y., 2006. Enhanced anxiety and stress-induced corticosterone release are associated with increased Crh expression in a mouse model of Rett syndrome. Proc. Natl. Acad. Sci. U.S.A. 103 (48), 18267–18272. Available from: https://doi.org/10.1073/pnas.0608702103.

McGrath, J.J., Hannan, A.J., Gibson, G., 2011. Decanalization, brain development and risk of schizophrenia. Transl. Psychiatry 1, e14. Available from: https://doi.org/10.1038/tp.2011.16.

McLaughlin, K.A., Fox, N.A., Zeanah, C.H., Sheridan, M.A., Marshall, P., Nelson, C.A., 2010. Delayed maturation in brain electrical activity partially explains the association between early environmental deprivation and symptoms of attention-deficit/hyperactivity disorder. Biol. Psychiatry 68 (4), 329–336. Available from: https://doi.org/10.1016/j.biopsych.2010.04.005.

McLaughlin, K.A., Sheridan, M.A., Lambert, H.K., 2014. Childhood adversity and neural development: deprivation and threat as distinct dimensions of early experience. Neurosci. Biobehav. Rev. 47, 578–591. Available from: https://doi.org/10.1016/j.neubiorev.2014.10.012.

McOmish, C.E., Hannan, A.J., 2007. Enviromimetics: exploring gene environment interactions to identify therapeutic targets for brain disorders. Expert. Opin. Ther. Targets 11 (7), 899–913. Available from: https://doi.org/10.1517/14728222.11.7.899.

McOmish, C.E., Burrows, E., Howard, M., Scarr, E., Kim, D., Shin, H.S., et al., 2008. Phospholipase C-beta1 knockout mice exhibit endophenotypes modeling schizophrenia which are rescued by environmental enrichment and clozapine administration. Mol. Psychiatry 13 (7), 661–672. Available from: https://doi.org/10.1038/sj.mp.4002046.

Meaney, M.J., Szyf, M., 2005. Maternal care as a model for experience-dependent chromatin plasticity?. Trends Neurosci. 28 (9), 456–463. Available from: https://doi.org/10.1016/j.tins.2005.07.006.

Meredith, R.M., Holmgren, C.D., Weidum, M., Burnashev, N., Mansvelder, H.D., 2007. Increased threshold for spike-timing-dependent plasticity is caused by unreliable calcium signaling in mice lacking fragile X gene FMR1. Neuron 54 (4), 627–638. Available from: https://doi.org/10.1016/j.neuron.2007.04.028.

Mnatzakanian, G.N., Lohi, H., Munteanu, I., Alfred, S.E., Yamada, T., MacLeod, P.J., et al., 2004. A previously unidentified MECP2 open reading frame defines a new protein isoform relevant to Rett syndrome. Nat. Genet. 36 (4), 339–341. Available from: https://doi.org/10.1038/ng1327.

Moretti, P., Zoghbi, H.Y., 2006. MeCP2 dysfunction in Rett syndrome and related disorders. Curr. Opin. Genet. Dev. 16 (3), 276–281. Available from: https://doi.org/10.1016/j.gde.2006.04.009.

Moretti, P., Levenson, J.M., Battaglia, F., Atkinson, R., Teague, R., Antalffy, B., et al., 2006. Learning and memory and synaptic plasticity are impaired in a mouse model of Rett syndrome. J. Neurosci. 26 (1), 319–327. Available from: https://doi.org/10.1523/JNEUROSCI.2623-05.2006.

Motil, K.J., Ellis, K.J., Barrish, J.O., Caeg, E., Glaze, D.G., 2008. Bone mineral content and bone mineral density are lower in older than in younger females with Rett syndrome. Pediatr. Res. 64 (4), 435–439. Available from: https://doi.org/10.1203/PDR.0b013e318180ebcd.

Mount, R.H., Hastings, R.P., Reilly, S., Cass, H., Charman, T., 2003. Towards a behavioral phenotype for Rett syndrome. Am. J. Ment. Retard. 108 (1), 1–12.

Nagarajan, R.P., Hogart, A.R., Gwye, Y., Martin, M.R., LaSalle, J.M., 2006. Reduced MeCP2 expression is frequent in autism frontal cortex and correlates with aberrant MECP2 promoter methylation. Epigenetics 1 (4), e1–e11.

Nelson 3rd, C.A., Zeanah, C.H., Fox, N.A., Marshall, P.J., Smyke, A.T., et al., 2007. Cognitive recovery in socially deprived young children: the Bucharest Early Intervention Project. Science 318 (5858), 1937–1940. Available from: https://doi.org/10.1126/science.1143921.

Nikitina, T., Ghosh, R.P., Horowitz-Scherer, R.A., Hansen, J.C., Grigoryev, S.A., Woodcock, C.L., 2007a. MeCP2-chromatin interactions include the formation of chromatosome-like structures and are altered in mutations causing Rett syndrome. J. Biol. Chem. 282 (38), 28237–28245. Available from: https://doi.org/10.1074/jbc.M704304200.

Nikitina, T., Shi, X., Ghosh, R.P., Horowitz-Scherer, R.A., Hansen, J.C., Woodcock, C.L., 2007b. Multiple modes of interaction between the methylated DNA binding protein MeCP2 and chromatin. Mol. Cell. Biol. 27 (3), 864–877. Available from: https://doi.org/10.1128/MCB.01593-06.

Nithianantharajah, J., Hannan, A.J., 2006. Enriched environments, experience-dependent plasticity and disorders of the nervous system. Nat. Rev. Neurosci. 7 (9), 697–709. Available from: https://doi.org/10.1038/nrn1970.

Nithianantharajah, J., Hannan, A.J., 2009. The neurobiology of brain and cognitive reserve: mental and physical activity as modulators of brain disorders. Prog. Neurobiol. 89 (4), 369–382. Available from: https://doi.org/10.1016/j.pneurobio.2009.10.001.

Nuber, U.A., Kriaucionis, S., Roloff, T.C., Guy, J., Selfridge, J., Steinhoff, C., et al., 2005. Up-regulation of glucocorticoid-regulated genes in a mouse model of Rett syndrome. Hum. Mol. Genet. 14 (15), 2247–2256. Available from: https://doi.org/10.1093/hmg/ddi229.

Pang, T.Y., Du, X., Zajac, M.S., Howard, M.L., Hannan, A.J., 2009. Altered serotonin receptor expression is associated with depression-related behavior in the R6/1 transgenic mouse model of Huntington's disease. Hum. Mol. Genet. 18 (4), 753–766. Available from: https://doi.org/10.1093/hmg/ddn385.

Pang, T.Y.C., Short, A.K., Bredy, T.W., Hannan, A.J., 2017. Transgenerational paternal transmission of acquired traits: Stress-induced modification of the sperm regulatory transcriptome and offspring phenotypes. Curr. Opin. Behav. Sci. 14, 140–147. Available from: https://doi.org/10.1016/j.cobeha.2017.02.007.

Pelka, G.J., Watson, C.M., Radziewic, T., Hayward, M., Lahooti, H., Christodoulou, J., et al., 2006. Mecp2 deficiency is associated with learning and cognitive deficits and altered gene activity in the hippocampal region of mice. Brain 129 (Pt 4), 887–898. Available from: https://doi.org/10.1093/brain/awl022.

Piazza, C.C., Anderson, C., Fisher, W., 1993. Teaching self-feeding skills to patients with Rett syndrome. Dev. Med. Child Neurol. 35 (11), 991–996.

Pridmore, S., Pridmore, W., 2018. Repetitive transcranial magnetic stimulation in the treatment of depression. Aust. J. Gen. Pract. 47 (3), 122–125.

Rampon, C., Jiang, C.H., Dong, H., Tang, Y.P., Lockhart, D.J., Schultz, P.G., et al., 2000. Effects of environmental enrichment on gene expression in the brain. Proc. Natl. Acad. Sci. U.S.A. 97 (23), 12880–12884. Available from: https://doi.org/10.1073/pnas.97.23.12880.

Restivo, L., Ferrari, F., Passino, E., Sgobio, C., Bock, J., Oostra, B.A., et al., 2005. Enriched environment promotes behavioral and morphological recovery in a mouse model for the fragile X syndrome. Proc. Natl. Acad. Sci. U.S.A. 102 (32), 11557–11562. Available from: https://doi.org/10.1073/pnas.0504984102.

Russo, G.B., Tirrell, E., Busch, A., Carpenter, L.L., 2018. Behavioral activation therapy during transcranial magnetic stimulation for major depressive disorder. J. Affect. Disord. 236, 101–104. Available from: https://doi.org/10.1016/j.jad.2018.04.108.

Sale, A., Berardi, N., Maffei, L., 2014. Environment and brain plasticity: towards an endogenous pharmacotherapy. Physiol. Rev. 94 (1), 189–234. Available from: https://doi.org/10.1152/physrev.00036.2012.

Samaco, R.C., Fryer, J.D., Ren, J., Fyffe, S., Chao, H.T., Sun, Y., et al., 2008. A partial loss of function allele of methyl-CpG-binding protein 2 predicts a human neurodevelopmental syndrome. Hum. Mol. Genet. 17 (12), 1718–1727. Available from: https://doi.org/10.1093/hmg/ddn062.

Sansom, D., Krishnan, V.H., Corbett, J., Kerr, A., 1993. Emotional and behavioural aspects of Rett syndrome. Dev. Med. Child Neurol. 35 (4), 340–345.

Sarwar, H., Waqar, S., 2013. Surgery for infantile esotropia: is timing everything? J. Perioper. Pract. 23 (5), 107–109.

Schizophrenia Working Group of the Psychiatric Genomics Consortium, 2014. Biological insights from 108 schizophrenia-associated genetic loci. Nature 511 (7510), 421–427. Available from: https://doi.org/10.1038/nature13595.

Schwarzkopf, D.S., Silvanto, J., Rees, G., 2011. Stochastic resonance effects reveal the neural mechanisms of transcranial magnetic stimulation. J. Neurosci. 31 (9), 3143–3147. Available from: https://doi.org/10.1523/JNEUROSCI.4863-10.2011.

Sheridan, M., Drury, S., McLaughlin, K., Almas, A., 2010. Early institutionalization: neurobiological consequences and genetic modifiers. Neuropsychol. Rev. 20 (4), 414–429. Available from: https://doi.org/10.1007/s11065-010-9152-8.

Spires, T.L., Grote, H.E., Garry, S., Cordery, P.M., Van Dellen, A., Blakemore, C., et al., 2004a. Dendritic spine pathology and deficits in experience-dependent dendritic plasticity in R6/1 Huntington's disease transgenic mice. Eur. J. Neurosci. 19 (10), 2799–2807. Available from: https://doi.org/10.1111/j.0953-816X.2004.03374.x.

Spires, T.L., Grote, H.E., Varshney, N.K., Cordery, P.M., van Dellen, A., Blakemore, C., et al., 2004b. Environmental enrichment rescues protein deficits in a mouse model of Huntington's disease, indicating a possible disease mechanism. J. Neurosci. 24 (9), 2270–2276. Available from: https://doi.org/10.1523/JNEUROSCI.1658-03.2004. 24/9/2270.

Stamoulis, C., Vanderwert, R.E., Zeanah, C.H., Fox, N.A., Nelson, C.A., 2017. Neuronal networks in the developing brain are adversely modulated by early psychosocial neglect. J. Neurophysiol. 118 (4), 2275–2288. Available from: https://doi.org/10.1152/jn.00014.2017.

Swanberg, S.E., Nagarajan, R.P., Peddada, S., Yasui, D.H., LaSalle, J.M., 2009. Reciprocal co-regulation of EGR2 and MECP2 is disrupted in Rett syndrome and autism. Hum. Mol. Genet. 18 (3), 525–534. Available from: https://doi.org/10.1093/hmg/ddn380.

Taneja, P., Ogier, M., Brooks-Harris, G., Schmid, D.A., Katz, D.M., Nelson, S.B., 2009. Pathophysiology of locus ceruleus neurons in a mouse model of Rett syndrome. J. Neurosci. 29 (39), 12187–12195. Available from: https://doi.org/10.1523/JNEUROSCI.3156-09.2009.

Tao, J., Hu, K., Chang, Q., Wu, H., Sherman, N.E., Martinowich, K., et al., 2009. Phosphorylation of MeCP2 at Serine 80 regulates its chromatin association and neurological function. Proc. Natl. Acad. Sci. U.S.A. 106 (12), 4882–4887. Available from: https://doi.org/10.1073/pnas.0811648106.

Vacca, M., Filippini, F., Budillon, A., Rossi, V., Mercadante, G., Manzati, E., et al., 2001. Mutation analysis of the MECP2 gene in British and Italian Rett syndrome females. J. Mol. Med. 78 (11), 648–655.

van Dellen, A., Blakemore, C., Deacon, R., York, D., Hannan, A.J., 2000. Delaying the onset of Huntington's in mice. Nature 404 (6779), 721–722. Available from: https://doi.org/10.1038/35008142.

van Praag, H., Kempermann, G., Gage, F.H., 2000. Neural consequences of environmental enrichment. Nat. Rev. Neurosci. 1 (3), 191–198. Available from: https://doi.org/10.1038/35044558.

Vanderwert, R.E., Marshall, P.J., Nelson 3rd, C.A., Zeanah, C.H., Fox, N.A., 2010. Timing of intervention affects brain electrical activity in children exposed to severe psychosocial neglect. PLoS ONE 5 (7), e11415. Available from: https://doi.org/10.1371/journal.pone.0011415.

Vignoli, A., La Briola, F., Canevini, M.P., 2009. Evolution of stereotypies in adolescents and women with Rett syndrome. Mov. Disord. 24 (9), 1379–1383. Available from: https://doi.org/10.1002/mds.22595.

Weaving, L.S., Ellaway, C.J., Gecz, J., Christodoulou, J., 2005. Rett syndrome: clinical review and genetic update. J. Med. Genet. 42 (1), 1–7. Available from: https://doi.org/10.1136/jmg.2004.027730.

Wiesel, T.N., Hubel, D.H., 1963a. Effects of visual deprivation on morphology and physiology of cells in the cats lateral geniculate body. J. Neurophysiol. 26, 978–993. Available from: https://doi.org/10.1152/jn.1963.26.6.978.

Wiesel, T.N., Hubel, D.H., 1963b. Single-cell responses in striate cortex of kittens deprived of vision in one eye. J. Neurophysiol. 26, 1003–1017. Available from: https://doi.org/10.1152/jn.1963.26.6.1003.

Witt-Engerstrom, I., Hagberg, B., 1990. The Rett syndrome: gross motor disability and neural impairment in adults. Brain Dev. 12 (1), 23–26.

Yasuhara, A., Sugiyama, Y., 2001. Music therapy for children with Rett syndrome. Brain Dev. 23 (Suppl. 1), S82–S84 [pii]S0387760401003369.

Yasui, D.H., Peddada, S., Bieda, M.C., Vallero, R.O., Hogart, A., Nagarajan, R.P., et al., 2007. Integrated epigenomic analyses of neuronal MeCP2 reveal a role for long-range

interaction with active genes. Proc. Natl. Acad. Sci. U.S.A. 104 (49), 19416–19421. Available from: https://doi.org/10.1073/pnas.0707442104.
Yeshurun, S., Hannan, A.J., 2018. Transgenerational epigenetic influences of paternal environmental exposures on brain function and predisposition to psychiatric disorders. Mol. Psychiatry. Available from: https://doi.org/10.1038/s41380-018-0039-z.
Young, J.I., Hong, E.P., Castle, J.C., Crespo-Barreto, J., Bowman, A.B., Rose, M.F., et al., 2005. Regulation of RNA splicing by the methylation-dependent transcriptional repressor methyl-CpG binding protein 2. Proc. Natl. Acad. Sci. U.S.A. 102 (49), 17551–17558. Available from: https://doi.org/10.1073/pnas.0507856102.
Yu, F., Zingler, N., Schumann, G., Stratling, W.H., 2001. Methyl-CpG-binding protein 2 represses LINE-1 expression and retrotransposition but not Alu transcription. Nucleic Acids Res. 29 (21), 4493–4501.
Yusufzai, T.M., Wolffe, A.P., 2000. Functional consequences of Rett syndrome mutations on human MeCP2. Nucleic Acids Res. 28 (21), 4172–4179.
Zappella, M., Gillberg, C., Ehlers, S., 1998. The preserved speech variant: a subgroup of the Rett complex: a clinical report of 30 cases. J. Autism. Dev. Disord. 28 (6), 519–526.
Zeanah, C.H., Nelson, C.A., Fox, N.A., Smyke, A.T., Marshall, P., Parker, S.W., et al., 2003. Designing research to study the effects of institutionalization on brain and behavioral development: the Bucharest Early Intervention Project. Dev. Psychopathol. 15 (4), 885–907.
Zeanah, C.H., Egger, H.L., Smyke, A.T., Nelson, C.A., Fox, N.A., Marshall, P.J., et al., 2009. Institutional rearing and psychiatric disorders in Romanian preschool children. Am. J. Psychiatry 166 (7), 777–785. Available from: https://doi.org/10.1176/appi.ajp.2009.08091438.
Zhou, Z., Hong, E.J., Cohen, S., Zhao, W.N., Ho, H.Y., Schmidt, L., et al., 2006. Brain-specific phosphorylation of MeCP2 regulates activity-dependent Bdnf transcription, dendritic growth, and spine maturation. Neuron 52 (2), 255–269. Available from: https://doi.org/10.1016/j.neuron.2006.09.037 [pii] S0896-6273(06)00775-6.
Zoghbi, H.Y., 2003. Postnatal neurodevelopmental disorders: meeting at the synapse? Science 302 (5646), 826–830. Available from: https://doi.org/10.1126/science.1089071. 302/5646/826 [pii].

CHAPTER

4

Ethics of Device-Based Treatments in Pediatric Neuropsychiatric Disorders

Nick J. Davis

Department of Psychology, Manchester Metropolitan University, Manchester, United Kingdom

OUTLINE

Neurotechnological Pediatric Neuropsych

DOI: https://doi.org/10.1016/B978-0-12-812777-3.00004-0

INTRODUCTION

Rise of Child Neuropsychology

The incidence and the prevalence of neuropsychiatric disorders is rising around the world. This is true for both adults and children and adolescents. A recent survey of incidence trends in Denmark found increasing incidence of several disorders amongst children born in the 1990s (Atladóttir et al., 2007), and a single-shot survey in the United Kingdom found that 9.5% of the young people had one or more psychological disorders (Ford et al., 2003). It is estimated that a third of young people will experience a period poor mental health during their life (Merikangas et al., 2009). One recent study observed that young people born in the 1990s were very much more likely to visit a clinician for psychiatric reasons than were people born in the 1960s; however, the prevalence of severe symptoms such as suicidal ideation was lower in the 1990s cohort; this suggests an increasing comfort with discussing mental health issues and a greater readiness on the clinicians' part to offer a diagnosis based on less-severe presentations (Twenge, 2014).

The use of the term "disorder" is itself somewhat problematic. Many conditions were once labeled medical disorders that now we would not think aberrant. For example, homosexuality was treated as a mental illness for a considerable stretch of the last two centuries and was only removed from the Diagnostic and Statistical Manual of Mental Disorders (DSM) in 1973. Similarly, in the United Kingdom "old age" (or "senility") was regarded as a cause of death in itself until the 1980s, when guidance was issued that a more specific cause should be noted. New (so-called) disorders emerge regularly in the literature. For example, Diagnostic and Statistical Manual Version 5 of the American Psychiatric Association (DSM-5) (American Psychiatric Association, 2013) includes "Internet Gaming Disorder" as a topic for future research, despite low evidence for a stable etiology or for harmful effects on the person (Weinstein et al., 2017). The issue of "medicalization" owes as much to sociology as it does to medical science, with pressure coming from more engaged, and consumer-minded, users of health care (Ballard and Elston, 2005).

Traditional Treatments

Most health-care authorities recommend a stepped pathway for treating neuropsychiatric disorders, from least invasive to most radical. For example, for depressive disorders, patients will typically start with a course of talking therapy; if this is unsuccessful, the patient may be prescribed a mild antidepressant, or a stronger drug if the mild drug does not show good results. If the patient does not respond well to drug treatments, the patient may be considered for more radical treatments, including electroconvulsive therapy or deep-brain stimulation (DBS). The principle of the stepped pathway is to deliver the treatment with the least potential harm, while still delivering an effective level of benefit in most cases; this, in most cases, mirrors the relative burden of the treatments on the health-care service. Stepped care pathways allow for patients to "skip" steps either through their own choice or at a clinician's recommendation (Schrijvers et al., 2012). However, the determination about a person's progress through the pathway relies on some assumptions about the efficacy and about the cost effectiveness of the treatments (Bower and Gilbody, 2005). Specialized pediatric stepped care pathways must be mindful of the desire of parents to be involved in decisions related to a child's care (e.g., Campo et al., 2005).

Talking Therapy

The least invasive step in a care pathway is usually to offer some form of talking therapy, such as counseling or psychoanalysis. Do such therapies work in younger people? Pattison and Harris (2006) reviewed studies that used a range of therapeutic approaches for treating a variety of problems faced by people aged 3–18 years. They found that talking or creative therapies had some evidence of utility for some problems, and for children within some age ranges, but that the paucity of data made it difficult to draw firm conclusions. Therapeutic approaches such as cognitive behavioral therapy require a patient to understand her or his own cognitive biases, so they can be changed (Beck, 1979). However, this approach requires a degree of metacognition that gradually develops through a person's life (e.g., Weil et al., 2013).

Pharmacology

Given the patchy efficacy of talking therapies, the next option for treating neuropsychiatric disorders in children is pharmacological treatment. Drug treatments for children are generally derived from treatments that are developed for adults. There is a reluctance amongst pharmaceutical companies to test drugs on children; so many therapies delivered to children are given "off-label," meaning that they have never been specifically approved for pediatric use (Conroy et al., 2000). This lack of testing means that possible adverse effects may arise at the clinic that would otherwise have been detected at the licensing stage. Although difficult to measure, it is thought that toxicity for off-label drugs is higher than for drugs tested in pediatric trials (Turner et al., 1999).

The practical and ethical difficulties of drug testing in children mean that promising therapies may be missed due to caution or financial disincentive. Conversely, the use of adult-tested drugs in pediatric settings means that children are exposed to heightened risks. This uncertainty means that an option that sits somewhere in the scale of risk between talking therapies and drug treatments may be of some practical benefit. This is where device-based treatments sit: they promise a more direct intervention into neural processes, while at the same time being associated with lower overall risk.

DEVICE-BASED APPROACHES TO NEUROPSYCHIATRIC TREATMENT

History

Every advance in science and technology has encouraged people to attempt to use that advance to treat some kind of disorder. The discovery and harnessing of electrical energy in the 18th and 19th centuries led to numerous proposals for electrotherapy, with hospitals in London developing apparatus to deliver electrical energy from the 1760s (Colwell, 1922). It is not clear what these apparatus were used for; however, the famous experiments of Galvani in inducing muscle twitches in the leg of a dead frog are likely to have inspired the use of electricity for reanimating dead or dying limbs. A wide variety of dangerous-looking apparatus intended for domestic uses were developed and sold during the 19th century in particular, until devices that claim to have medical uses were brought under regulation in the 20th century (Basford, 2001).

Magnetic energy has also been used in treatments of varying efficacy, although the relative lack of understanding of magnetism, and the comparative difficulty in generating the force, has meant that magnetic therapy lagged behind electrical therapy for much of history. Early uses of magnetic energy included (supposed) treatments for impotence, baldness, paralysis, and infectious disease (Basford, 2001). Presumably the use of magnets to withdraw metallic foreign objects from the bodies of soldiers was more successful. Magnetic therapies were subjected to scrutiny at the end of the 19th century and were found wanting (Peterson and Kennelly, 1892). Static magnetic fields are unlikely to produce physiological effects, although it has recently been claimed that static magnets can alter cortical excitability (Oliviero et al., 2011).

Types of Device

There are different classes of a device that have a potentially psychotherapeutic application. Maslen et al. (2014a), in assessing the regulatory landscape that applies to cognitive enhancement devices, divide devices into the following four classes: (1) transcranial current stimulators, (2) transcranial magnetic stimulators, (3) cranial electrotherapy stimulators, and (4) neurofeedback devices. In addition to these four classes, we should also add the surgical technique of deep-brain stimulation. However, of the classes considered by Maslen et al., cranial electrotherapy is least well understood, and despite some evidence of effects on the brain from its use (Feusner et al., 2012), it is not clear that it possesses sufficient efficacy or controllability to be a useful therapy, and so it will not be considered further here.

Transcranial electric stimulation (tES) involves the application of weak electric currents between two or more electrodes placed on the head. The most common form of tES, tDCS uses direct current. Currents on the order of 1–2 mA pass between the electrodes via the head, with some of the current reaching the brain (Vöröslakos et al., 2018). Electrical charge accumulates on the surface between the highly conducting cerebrospinal fluid and the highly insulating gray matter of the brain, which modulates the activity of electrically sensitive cells in a polarity-dependent manner. Cells near the negative electrode (cathode) of the circuit become hyperpolarized and therefore less likely to generate action potentials, while those near the positive electrode (anode) become relatively depolarized and so more likely to generate action potentials. tDCS is not thought to cause neuronal firing in itself but biases the likelihood of a region of tissue to become active in response to an input. Allowing the current to flow for several minutes leads to longer lasting effects on the brain, with effects detectable some tens of minutes after the end of stimulation. The mechanism of this persistence is not clear but seems to rely on long-term potentiation-like mechanisms (Nitsche et al., 2003). Stimulation with alternating current or with randomly varying current (transcranial random noise stimulation) is also possible, although these modalities have not so far been used in pediatric neuropsychiatry. The different tissues of the head vary greatly in their electrical conductivity, and the skull in particular is a highly insulating medium. This means that the electric current flows diffusely through the cranial tissues, meaning that tES is by no means a spatially focal technique. In addition, the requirement for a long period of stimulation means that temporal precision is not possible. However, the greatest advantage of tDCS in particular is the seeming ease with which cortical excitability can be raised, leading to the possibility of functional enhancement (Coffman et al., 2014).

Transcranial magnetic stimulation (TMS) uses the principle of electromagnetic induction to generate brief pulses of magnetic energy. In the most commonly used devices, a figure-of-eight-shaped coil of wire receives an electric current. This current induces a magnetic field, which, due to the shape of the coil, is concentrated in a small volume. When placed against the head, the magnetic energy readily crosses the skull, and there induce electric currents in neural tissue. These currents cause cells to depolarize and to generate action potentials. A single pulse of TMS causes firing of a targeted patch of cells, then a brief refractory period; this effect can be used either to assess the excitability of a brain region (amplitude of the response) or to induce a brief "virtual lesion" (information received during the refractory period is not processed). Delivering multiple pulses to the same brain area, termed "repetitive TMS" or "rTMS," causes a buildup of effects, with the effect depending on the temporal pattern of the pulses. Common patterns include "low-frequency rTMS" such as 1 Hz protocols or continuous theta-burst stimulation, which induce long-lasting inhibition of cortical excitability and "high frequency rTMS" such as 10 Hz protocols or intermittent theta-burst stimulation, which induce long-lasting facilitation of cortical excitability (Maeda et al., 2000; Huang et al., 2005). Compared to tES, TMS is thought to be highly focal, with areas of the cortex around 1 cm^2 being targeted with most figure-of-eight-coil designs. In addition, single pulses can be targeted with sub-millisecond precision, making TMS a very important tool for research and clinical applications.

DBS is a surgical technique, whereby fine, needle-like electrodes are inserted into the brain and permanently implanted. The electrodes deliver electrical stimulation to the target brain area, which may enhance or disrupt signals in that region. DBS is remarkably successful in treating Parkinson's disease, which is primarily a disorder of the basal ganglia (Bronstein et al., 2011). This success has prompted researchers to apply DBS to other disorders, including obsessive–compulsive disorder (Abelson et al., 2005) and depression (Mayberg et al., 2005). To date the results of many of these trials have been less promising than hoped.

Finally, neurofeedback is a means of displaying a person's brain activity so they can adapt their behavior to optimize that activity. For example, a scale on a computer screen might display the spectral power of an EEG frequency band, such as the alpha band (8–12 Hz). Changing a mental state, such as by relaxing, may change the alpha power and therefore move the scale on the screen, leading to a reward for the participant (Enriquez-Geppert et al., 2017). EEG-neurofeedback has been used with some success in epilepsy and anxiety disorders (Egner and Sterman, 2006; Hammond, 2005). It is technically more challenging to derive neurofeedback signals from MRI; however, the amygdala offers a target for a substrate for real-time modulation of the emotion network (Johnston et al., 2010).

ETHICS IN TREATING THE DEVELOPING BRAIN

Ethical issues in treating neuropsychiatric disorders fall into four classes. In the first class there are problems in defining what counts as a "disorder" for the purpose of treatment. Second, there are concerns over the safety of particular techniques when applied to pediatric cases. Third, there are issues in establishing informed consent for a procedure, and finally we have issues to do with the potential for early-life interventions to have lifecourse-affecting outcomes.

Treatment Versus Enhancement

The techniques described in this chapter are focused on treatment of neuropsychiatric conditions. However, many conditions represent one end of a range of values, where clinicians may arbitrarily define a limit at which the measurement becomes pathological. To take a concrete example, intelligence quotient (IQ) scores are defined to be normally distributed, with a population mean of 100 and a standard deviation of 15. This distribution is illustrated in Fig. 4.1. This means that 68% of the population have an IQ score of within 15 points of the mean, and 95% of the people are within 30 points (i.e., from 70 to 130). People who

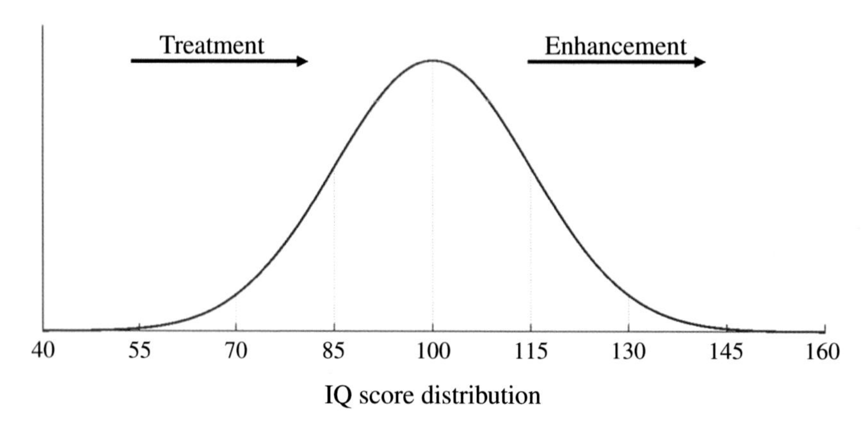

FIGURE 4.1 An idealized distribution of IQ scores in a population. IQ is defined to have a mean of 100 and a standard deviation of 15. Most people would express a desire to increase their IQ, which can be thought of as moving to the right on this figure. However, moving rightward from the "normal" range (85–115) is qualitatively different from moving rightward from the lower ranges (under 85). The former is often conceptualized as "cosmetic enhancement," whereas the latter could be seen as "treatment" since low intelligence may negatively affect a person's life. *IQ*, Intelligence quotient.

score 130 or above are thought to be exceptionally intelligent (indeed, this is the criterion for joining Mensa), while those with an IQ in the 1960s or below have exceptionally low intelligence. Cognitive intelligence predicts important lifecourse measures such as occupational attainment, mortality, and relationship stability (Roberts et al., 2007). Most people wish to increase their intelligence, or to put it crudely, to "move rightward" in the distribution. Moving rightward from the low end of the distribution may be seen as a treatment, especially if the person's low intelligence may represent a harm to herself, or to her future health. Conversely, moving rightward from the normal range (85 and above) is not so obviously necessary and may be thought of as an enhancement to a person's otherwise healthy capacity.

For a given procedure, the distinction between its use as a treatment and its use as an enhancement is rarely clear-cut. Although cognitive enhancement is not in itself harmful, there are potential harms to the person and to others from the overt or covert use of enhancement drugs and technology (Davis, 2017).

Safety Concerns

Brain stimulation with TMS or tDCS is often described as safe. However, there are some potential safety concerns when using NIBS in healthy people, and these concerns are elevated for younger people.

The greatest and most immediate risk from NIBS is that altering the excitability of a brain area may trigger a seizure. Seizure is a rare side effect of rTMS, with fewer than 30 reports across the literature (Dobek et al., 2015), three of which were adolescents who were receiving rTMS for the treatment of depression (Cullen et al., 2016; Hu et al., 2011; Chiramberro et al., 2013). To date only one seizure has been recorded in association with tDCS, and that was in an already highly complex pediatric case (Ekici, 2015). Neurofeedback has not so far been associated with a seizure event, possibly reflecting the endogenous nature of the brain state changes; however, neurofeedback can modulate seizure risk, as reflected in its possible using in reducing epileptic symptoms (Walker and Kozlowski, 2005). DBS is a surgical technique that is only applied to people with serious neurological conditions, so the safety concerns are less applicable for this technique; however, the neuroethics of DBS is a topic of active development (Maslen et al., 2018).

The smaller size of the young head means that NIBS protocols designed for adults may be unsafe when applied to a child. tDCS is thought to affect brain function by generating an electric field gradient at the surface of the brain. The amount of current that crosses the skull to reach that surface is only on the order of 25% in adults' heads (Huang et al., 2017; Vöröslakos et al., 2018). However, the thinner bone of the skull, which is an electrical insulator, means that relatively more current passes through the skull in children. Seemingly paradoxically, the intensity of a TMS pulse needed to generate a muscular response, the motor threshold, is higher in children than in adults (e.g., Mall et al., 2004). The relationship between dose and response for NIBS in children is complex, and protocols that base stimulation intensity purely on a percentage of motor threshold are likely to be missing a large number of factors that precisely determine the appropriate dose. It would therefore be potentially dangerous to take an adult NIBS protocol and to apply it directly to a child without a full understanding of the factors that diminish or enhance the transfer of energy to brain tissue, and from brain tissue to an overt response.

Of the different classes of risk that arise in neurostimulation, it is important to distinguish between immediate harms, such as a seizure or damage to the skin, and longer term harms. Davis (2017) discusses the different forms of harm that may result from the use of different forms of neuromodulation. The side effects discussed in this section fall into the class of "neurobiological harms;" however, within this class, we can see differences of consequence. A small burn to the skin arising from a poorly placed electrode will cause a small amount of discomfort and will rapidly improve. A stimulation-induced seizure will be immediately serious, in that the safety of the patient must be secured, but the long-term consequences may not be serious if the patient is monitored.

It is less clear how serious are the outcomes of procedures that aim to modulate the function of the brain over a longer time scale. For example, a neurofeedback intervention that has a long-term effect on endogenous brain oscillations may make the brain more susceptible to disorders of brain rhythms later in life. Novel neurointerventions will require monitoring to understand the consequences of the changes made in the young and developing brain.

Consent and Assent

A child under the age of majority cannot give consent for an intervention; consent must be given by a parent or guardian, or by another legally appropriate adult. However, in most cases the assent of the child must be sought before a procedure. This requirement makes for a more complex interaction between the physician and the patient, as the intervention must be explained in a way that allows both the child and the guardian to make a well-informed decision about benefits and risks.

Kuther (2003) has mapped the legal issues in determining the different levels of involvement in decision-making for children and adolescents. She finds that children of all ages are able to make rational choices, with varying levels of parental involvement, but with older adolescents showing more complex decision-making, and greater independence from the guardian's influence. There should therefore be a shift in how a physician involves the minor in decision-making, from assent-only in younger children, to a greater level of autonomy, and therefore consent-seeking, in older adolescents. At present, this balance is left to the judgment of the physician, which raises the possibility of tension between the physician, the guardian, and the child around the best course of treatment.

Lifelong Change

An intervention given to a child may affect the entire outcome of their life. This is of course the desired outcome for most treatments: neuropsychiatric conditions such as mood disorders, anxiety disorders, or eating disorders or neurodevelopmental conditions such as autism spectrum disorder or attention deficit–hyperactivity disorder can be severely life-limiting; so treating these disorders affords the child the chance to lead a flourishing life. However, treatments may also close off options for a person's future life. In an extreme case, an organ transplant performed in childhood is likely to commit the child to a lifetime of immunosuppressant treatment (e.g., Vincenti, 2003). Children exposed to neurological surgery have a greater risk of sensory or motor deficits or of cognitive problems and have an elevated seizure risk (Packer et al., 2003). The long duration of the sequelae of treatment in childhood means that particular care must be taken when weighing up the risks and benefits of an intervention. For most neuropsychiatric disorders, a long duration of a positive outcome is precisely what patients seek; however apart from a few isolated cases (e.g., Downar et al., 2012), rapid and complete remission from a condition is unknown. We do not know the long-term effects of neurostimulation, either in children or in adults, as most research protocols tend to avoid long-lasting effects, and few studies follow up their participants over more than a few months (Davis, 2014; Davis and van Koningsbruggen, 2013). Childhood interventions must therefore be approached with caution.

Summary

Children and adolescents are not "small adults." The developing anatomy of the child means that specific protocols must be developed to target brain regions of interest, in order to have a given level of outcome and to avoid potential harm. At the same time, we must recognize that children and adolescents must be involved in the decisions that affect their health care, which may affect the course of their life. Maslen et al. (2014b) have argued that the ethical issues that pertain in discussing neurological treatments and enhancements in adults have greater force in childhood cases, with some procedures best left until the child reaches the age of majority and can make decisions for him- or herself.

In the following sections, we will examine three conditions of neurological origin for which device-based treatments have been proposed. In these case studies, we will look at the current treatment options, and the possibilities afforded by neuromodulation. We will also look at the risks and benefits entailed by these treatments.

CASE STUDY: EATING DISORDERS

Eating disorders are a wide range of neuropsychiatric disorders, centered around an aberrant attitude to food consumption. The DSM-5 (American Psychiatric Association, 2013) defines three main groups of eating disorders. The most common disorder of this class is anorexia nervosa, characterized by a profound restriction of food intake, and a consequential drastic loss of body weight. Bulimia nervosa involves excessive food intake compensated for by "purges" (through vomiting or laxatives) and heightened concern for body image. Binge eating disorder is associated with large food intakes, but generally without purging. Other eating-related disorders may be classified as "other specified feeding or eating disorder" or OSFED (formerly "eating disorder—not otherwise specified," or EDNOS, in previous versions of the DSM).

Eating disorders are relatively uncommon in the general population, with an estimated incidence of around five cases of anorexia nervosa per 100,000 person-years (Smink et al., 2012). However, this increases sharply to around 110 per 100,000 person-years in girls aged 15–19 years (van Son et al., 2006). Bulimia nervosa appears to be decreasing in incidence, from around 12 to around 6.6 cases per 100,000 person-years between 1993 and the early 2000s (Smink et al., 2012), but again this figure is higher, at around 40 per 100,000 person-years, in girls aged 10–19 (Currin et al., 2005). The mortality rate of anorexia nervosa is very high, with one study estimating a crude mortality rate of 5.1 deaths per 1000 person-years, with a fifth of those deaths due to suicide (Arcelus et al., 2011). The high incidence, and the high risk of death, among younger people makes eating disorders a key target for improvements in health care.

The etiology of eating disorders is complex. Although criteria exist that allow clinicians to diagnose a specific disorder based on clusters of symptoms (e.g., DSM-5), the picture is complicated by the high rate of comorbid symptoms; one study estimated that 97% of the female inpatients receiving treatment for eating disorders also had some form of additional psychological symptom (Blinder et al., 2006). One popular view of the development of eating disorders is that they reflect a desire for control by young people who feel that they lack control elsewhere in their life. In one view, the developing young person attempts to establish an internal locus of control (Rotter, 1966) through one of the few means available to them, namely, restricting eating. The notion of "control" is not unitary and varies according to one's theoretical stance (Surgenor et al., 2002); however, it retains some explanatory power for understanding eating disorders and their persistence.

Neuromodulation in Eating Disorders

The recent changes in incidence and improvements in mortality for eating disorders appear to come from improvements in specialist care for people with specific conditions. Current methods of care include inpatient treatment of acute nutritional symptoms, and stepped levels of outpatient and community care to assist in adopting healthier lifestyles and in addressing associated psychological problems. Pharmacological treatments are not usually applied in eating disorders, although these may be appropriate if the psychological symptoms warrant such treatment; however, the efficacy of such treatments is uncertain, for example, in anorexia nervosa (Claudino et al., 2006; Kishi et al., 2012). More recently researchers and clinicians have begun to explore DBS and NIBS for eating disorders.

The interest in brain stimulation for eating disorders comes from fMRI results that show abnormal functional responses to emotional, reward-related or food-related stimuli. For example, in people with anorexia, the right caudate nucleus of basal ganglia and the medial frontal area are relatively over-responsive, and the left inferior parietal cortex relatively under-responsive, to food-related stimuli compared to those areas in healthy controls (Zhu et al., 2012). Of these targets, frontal and parietal areas are accessible, in principle, to NIBS, while the caudate nucleus is deep enough that DBS would be required to modulate activity in these regions. To date, few neuromodulation studies have shown convincing results, largely due to small-sized trials and experiments, and the position of the United Kingdom's standards body, the National Institute for Health and Care Excellence, is that the efficacy of neuromodulation in eating disorders is unproved (NICE, 2017).

Ethical Questions

Neuromodulation approaches to eating disorders raise several ethical problems. These principally relate to safety concerns around the intensity of NIBS and questions about the possibility of true informed consent for a procedure.

As discussed above, NIBS in children raises questions about the assumptions made when designing a protocol for NIBS. The different cranial anatomy of the child compared to the adult, in terms of absolute size, skull thickness, fat deposition, cortical folding, etc., means that the energy applied during TMS or tDCS may be distributed differently on the brain surface, possibly leading to higher doses than would be found in an adult (Davis, 2014; Kessler et al., 2013). Given the young age of many people with eating disorders, clinicians should bear in mind the smaller and thinner skull of the patient. However, restricted eating also leads to changes in bone density, in bone mineralization, and in the density of fatty tissue in the scalp and brain (Widdows and Davis, 2014). Bone density is known to be affected in people with eating disorders, with reduced density in anorexia (Misra et al., 2004) and in bulimia (Howat et al., 1989). People with eating disorders show lower bone mineralization, which changes the electrical permittivity, meaning that the skull offers less resistance to the applied current (Ivancich et al., 1992). Fatty tissue around the head is known to affect the delivery of tDCS to the brain. A study that modeled the electric field induced by tDCS in people with healthy levels of body fat versus people with obesity showed a complex relationship between fat and current density (Truong et al., 2013), with increased subcutaneous fat generally reducing brain-surface current density; however, the equivalent study in people with very low body fat does not exist.

The fundamentally psychological origin of eating disorders means that psychological comorbidities are frequently present (Herzog et al., 1992; Telch and Stice, 1998). The high co-occurrence of DSM Axis I and Axis II disorders (APA, 2013) means that clinicians must be careful to ascertain that the patient has sufficient capacity to understand the procedure and its consequences. People with eating disorders are frequently resistant to treatment, with clinicians facing difficulties in persuading patients to abandon ideals of body image, or to commit wholeheartedly to treatment (Vitousek et al., 1998). Given this, there is a serious risk that a person with an eating disorder may be more than usually likely to be unable to give informed consent or assent to an intervention.

Summary

Eating disorders represent a pernicious threat to the immediate and long-term health of young people. The high morbidity rate and the high probability of long-term health conditions following recovery that are associated with eating disorders, especially anorexia nervosa, means that innovative treatments are required to maximize the chance that an individual patient will respond well to treatment and will recover healthy eating behavior in the future. Noninvasive brain stimulation should therefore be explored more comprehensively as a treatment for eating disorders, with the caveat that there are considerable scientific and technical problems to solve before the technologies are suitable for widespread use.

CASE STUDY: ATTENTION DEFICIT–HYPERACTIVITY DISORDER

Attention deficit–hyperactivity disorder (ADHD) is a disorder of behavior that is associated with a range of symptoms including lack of focus on a task, inability to sit still, and deficits in social processing (American Psychiatric Association, 2013). The prevalence of ADHD is somewhat difficult to estimate, partly because of changing diagnostic criteria, and cultural differences in applying the criteria; however, a range of 5%–12% of the children emerges from worldwide studies (Faraone et al., 2003; Polanczyk et al., 2007). ADHD is commonly associated with young boys, and in fact the incidence of ADHD in boys is around three times higher than in girls, although this may be due in part to referral bias (Gaub and Carlson, 1997). It is increasingly recognized that adults show related symptoms, with a similar prevalence profile (Faraone and Biederman, 2005), but with seemingly worsening comorbid conditions (Anker et al., 2018). ADHD is associated with poor educational performance, even when the symptoms are treated (Loe and Feldman, 2007).

It is not known if there is an organic cause for ADHD. Genetic studies implicate a large number of genes that associate with aspects of the condition (Khan and Faraone, 2006), with particular interest in genes that differ between the sexes (Biederman et al., 2008). It seems that there is no one gene that predicts ADHD risk, but that a number of rare variants contribute to the etiology. There is limited evidence for anatomical differences in the ADHD and non-ADHD brain (e.g., Qiu et al., 2011), although it may be more productive to search for alterations in functional connectivity between regions, rather than for differences in architecture (Konrad and Eickhoff, 2010). Indeed, changes in brain-wide network structures may provide a better model for ADHD symptoms than any single-brain structure (Castellanos and Proal, 2012).

Neuromodulation in ADHD

The standard treatment for ADHD is a combination of therapeutic approaches and pharmacological treatments. Pharmacological treatment is recommended as a first treatment option for primary ADHD, with the most common prescription being methylphenidate (usually branded Ritalin), a stimulant that increases alertness. Methylphenidate is well tolerated for primary ADHD, but may worsen comorbid conditions such as tics, and interacts poorly with some antidepressants. As with many stimulants, there is a risk of dependence or addiction with methylphenidate, so patients should be monitored during their course of the drug (Kutcher et al., 2004). In addition, prolonged use of stimulants risks the development of respiratory or cardiovascular problems. It would therefore be clinically useful to find non-pharmacological treatment options for ADHD.

Neuromodulation in ADHD has not shown great promise so far. One early study in children used rTMS to the dorsolateral prefrontal cortex (DLPFC), with improvements in ADHD symptoms; however, this applied both to the active and to the sham sessions of the study, suggesting a placebo effect (Weaver et al., 2012). More promisingly, a study in adult ADHD suggested an improvement in measures of attention following a single session of rTMS to DLPFC, compared to baseline and to sham, although measures of mood and anxiety did not change (Bloch et al., 2010). tDCS has been used to modulate response accuracy in a go/no-go task, although no improvement was seen in a Stroop task, suggesting weak and non-generalized effects on response inhibition (Soltaninejad et al., 2015). A similar study showed a very small improvement in interference control in a flanker task (Breitling et al., 2016). Overall, the evidence for effective neuromodulation in improving ADHD symptoms is not available. It must also be taken into account that people with ADHD have a heightened incidence of seizure when off medication, with the medication somewhat mitigating this risk (Wiggs et al., 2018); magnetic or electric treatments that heighten seizure risk should therefore be used with caution.

A more promising approach for treating ADHD is neurofeedback. In the 1970s, it was noted that a child with hyperkinesia could be trained to modulate the sensorimotor EEG rhythm, around 12–14 Hz, leading to a modulation of hyperkinetic symptoms (Lubar and Shouse, 1976; Shouse and Lubar, 1979). Modern approaches to neurofeedback for ADHD focus on the ratio between the power of the theta rhythm (4–7 Hz), which is associated with relaxed, but non-sleepy, states, and the power of the beta rhythm (12–35 Hz), which is associated with maintenance of postural tone (Davis et al., 2012). For example, Gevensleben et al. (2009) found an improvement in ADHD ratings after a 4-week program of neurofeedback, using either theta/beta training or training of the slow cortical potential; this improvement was also present at a 6-month follow-up, compared to a control group (Gevensleben et al., 2010). Arns et al. (2009), in a meta-analysis, conclude that the theta/beta ratio may have some utility in determining treatment options for subgroups of children with ADHD, but that there is considerable heterogeneity in the existing studies, meaning that this ratio is not well enough characterized in the general population for diagnostic purposes.

Summary

ADHD is a disorder that adversely affects the early educational life of many children and in some people extends into adulthood. The increasing prevalence of the disorder means that an increasing number of people will require treatment. The treatment option of methylphenidate is effective in most cases of primary ADHD, but there are concerns about the side effects of pharmacological stimulants, especially when used over long periods. There is therefore a need to explore alternative treatments, with neurofeedback currently showing most promise, but with the possibility that neuromodulation may also be effective; in either case more research is needed into the best protocols. An important focus of future research should be on understanding the risks posed by novel neurotechnologies, and setting these against the possible harms of not treating ADHD or in persisting with the existing treatment options. Given the relatively benign nature of ADHD, compared to some other neuropsychiatric disorders, it may be felt that pursuing novel device-based treatments does not add sufficiently to the range of options.

CASE STUDY: AUTISM SPECTRUM DISORDER

Autism spectrum disorder (ASD) and ADHD share a number of core features, especially given the wide range of the autism spectrum. DSM-5 lists a cluster of symptoms associated with ASD, including deficits in social interactions, restricted interests, and repetitive behaviors, and where the deficits cannot be explained primarily by intellectual impairment (American Psychiatric Association, 2013).

The incidence of ASD is difficult to estimate but may be as high as around 30–60 cases per 10,000 (Rutter, 2005). A recent survey suggested that the incidence of autism diagnosis is increasing, however the symptom severity is dropping, suggesting that clinicians are offering an ASD diagnosis more readily now than in the past (Arvidsson et al., 2018). There is a marked difference in diagnoses for males versus females, with around 80% of the ASD cases being male (Fombonne, 1999). The reason for this difference is unclear, with possible explanations ranging from genetic factors, the protective effect of testosterone, and physicians' ascertainment bias (Werling and Geschwind, 2013).

Standard Therapies

There is no single all-inclusive therapy for ASD, largely due to the breadth of symptoms that may arise within the disorder. The lack of a mechanistic explanation for the etiology of ASD has meant that a vast number of speculative therapies have been attempted, with generally poor levels of success. These have included vitamin–mineral therapy (Nye and Brice, 2005), virtual reality interventions (Kandalaft et al., 2013), hormones (Roberts et al., 2001), and antipsychotic drugs (Posey et al., 2008). In most cases, the sample size, and therefore statistical power, has been low, and the results difficult to generalize.

The social aspects of ASD have been addressed using a variety of methods. People with ASD who attend mainstream school may benefit from peer-mediated intervention, where classmates are trained to help the child with ASD in social situations or to model socially appropriate behavior (Chan et al., 2009). The purported link between the hormone oxytocin and social behavior has also prompted research into its use in ASD (Hollander et al., 2007), although the mechanism that might drive the link has not been established (Quattrocki and Friston, 2014). Currently, the most successful therapeutic approach to autism is applied behavior analysis (ABA), where principles of learning and motivation are applied to daily activities (Virués-Ortega, 2010). The flexibility of the ABA approach means that therapists can adapt their treatment to the individual.

Neuromodulation in ASD

In ADHD, the upward modulation of theta-band power with neurofeedback has shown some success in promoting concentration and calmness. Neurofeedback approaches to ASD have taken the opposite strategy, of reducing theta activity and boosting beta power. For example, one study that took this approach found improvements in lab-based social functioning tasks, and in parental satisfaction (Kouijzer et al., 2009b); these benefits were also present some months after the intervention (Kouijzer et al., 2009a). However, despite a number of positive studies, there is considerable heterogeneity in studies into neurofeedback in ASD, and some distance to cover before neurofeedback could be considered a viable treatment option (Holtmann et al., 2011). More recently it has been suggested that adding nonneural physiological measures into an integrated biofeedback intervention may be more effective than pure neurofeedback (Friedrich et al., 2014).

Researchers have attempted to intervene more directly into the neural processes of autism through the use of TMS. Oberman et al. (2015) reviewed the studies that used TMS protocols as a therapeutic intervention in ASD. As with neurofeedback studies, the methodological approaches are very mixed in ASD research, and the sample sizes are rather small, and the results are therefore correspondingly difficult to rely upon. Many studies target the DLPFC, reporting improvements in certain cognitive or perceptual tasks, compared to "waitlist" control participants, or compared to pre-stimulation scores. Neither of these control methods is scientifically ideal and nor helps to compare post-intervention performance to that of people without ASD. A later review of TMS in ASD also calls for better standardization of methods and of control procedures (Oberman et al., 2016); however, the practicalities of TMS mean that it is difficult to create a perfect control condition (there is no "placebo" in TMS: Davis et al., 2013). More recently the researchers have attempted to use tDCS in ASD. The problems of small samples and of methodological heterogeneity are even more pressing here; two separate studies targeting the left DLPFC reported benefits from tDCS, whether the area was under the anode or under the cathode (Amatachaya et al., 2014; D'Urso et al., 2015).

Summary

The lack of a clear model of how autism affects the brain means that interventions that target the brain are necessarily exploratory. Neither neurofeedback nor TMS or tDCS have shown consistent promise in treating the core symptoms of ASD, although some of the constellatory symptoms may be susceptible to treatment through these modalities. As with ADHD there is an increasing rate of diagnosis of ASD, and parents of children with ASD are often perceived to be active advocates for the best care and treatment of their child, even where these demands conflict with the advice of experts or with service providers (Pierce and Tincani, 2007). The scientific basis of device-based treatment of ASD risks falling behind this demand, meaning that young people may be exposed to unnecessary risk from unproved treatments.

CONCLUSIONS

The incidence of many neuropsychiatric disorders is increasing in younger people. This rise is likely to be driven by an increased willingness of younger people and their families to discuss worrying symptoms and to seek help. However, this increase in diagnosis is also accompanied by an increasing demand for treatments. In adult cases, a treatment pathway would normally start with a course of talking therapy, before progressing to mild pharmacological treatment, to stronger treatments, and possibly to surgical intervention. In such a pathway the principle is to start with the least invasive treatment, or the treatment with the lowest risk of side effects, before attempting the next level of treatment that entails greater risks.

For many disorders, the treatment pathway breaks down. Talking therapies require a level of cognitive development that does not develop until later in childhood and adolescence. For some disorders, there is no good pharmacological option, or the risks of long-term treatment are undesirable. It is therefore desirable to find a new option for treating such disorders. Recently scientists and clinicians have turned to the brain stimulation options of transcranial magnetic or electric stimulation. These technologies offer a middle ground of minimal invasiveness and relative safety, with the possibility of beneficial effects in some disorders.

In the case of eating disorders, the existing treatment options are somewhat unsatisfying, and the risk to the person's acute and long-term health is great. This suggests that device-based treatments may have a role to play in treating these conditions, although there remain some technical questions about the best means of delivering stimulation to a cranial anatomy that looks different to that of a healthy person. Conversely, ADHD has a much more effective pharmacological treatment in the shape of methylphenidate. However, there are concerns over the long-term use of stimulants such as this, and such questions will become more pressing as the rate of diagnosis of ADHD goes up. It would therefore be sensible to pursue the promising effects shown by neurofeedback in addressing some of the symptoms of ADHD. The case of the autism spectrum is more difficult, as there is no reliably effective pharmacological or device-based treatment for the core symptoms nor is there a clear understanding of a possible neural basis for targeting an intervention; however, behavioral treatment (ABA) is effective in many cases and has no obvious side effects.

Overarching all these case studies is the need to develop a coherent framework for ethical analysis of neuropsychiatric treatments in children and young people. Doctors and patients understand that medical decision-making involves a weighing-up of risks and benefits, and this applies to decisions made on behalf of a child as much as it does to adult patients. However, in pediatric cases, the risks and benefits also apply to a longer future lifetime, and in some cases to effects on future opportunities. Pediatric cases also necessarily include questions about the patient's autonomy, since children lack the legal capacity to give their own consent; such cases therefore require careful consideration of the patient's ability to properly consider the options available and to offer clear assent to a procedure. Set against these concerns, we must also consider the possible benefits of early intervention into neuropsychiatric disorders. The three case studies considered here, autism, ADHD, and eating disorders, along with a host of other disorders that arise in childhood, all have detrimental effects on a young person's life, either through direct threat to the person's health, or through an effect on schooling and socialization. So the search for safe and effective treatments for these disorders must continue and should include device-based as well as behavioral and pharmacological interventions.

In conclusion, device-based treatments have some potential in treating children with neuropsychiatric disorders. There are some technical barriers to widespread adoption of such treatments, including the lack of good scientific basis for dosage-setting, clear understanding of disease mechanisms, and the usual difficulties of testing novel treatments on a young population. There are also a number of ethical considerations that apply in such cases, particularly the longer time course over which the outcomes of decisions are played out, and the difficulty of involving a child in decisions around their own treatment. Nevertheless there are some grounds for hope that treatments such as transcranial electric or magnetic stimulation, or neurofeedback or deep brain stimulation may offer new hope in cases where existing options are inadequate.

References

Abelson, J.L., Curtis, G.C., Sagher, O., Albucher, R.C., Harrigan, M., Taylor, S.F., et al., 2005. Deep brain stimulation for refractory obsessive–compulsive disorder. Biol. Psychiatry 57 (5), 510–516. Available from: https://doi.org/10.1016/j.biopsych.2004.11.042.

Amatachaya, A., Auvichayapat, N., Patjanasoontorn, N., Suphakunpinyo, C., Ngernyam, N., Aree-Uea, B., et al., 2014. Effect of anodal transcranial direct current stimulation on autism: a randomized double-blind crossover trial. Behav. Neurol. 2014, 173073. Available from: https://doi.org/10.1155/2014/173073.

American Psychiatric Association, 2013. Diagnostic and Statistical Manual of Mental Disorders:DSM-5, fifth ed. American Psychiatric Association, Washington, D.C.

Anker, E., Bendiksen, B., Heir, T., 2018. Comorbid psychiatric disorders in a clinical sample of adults with ADHD, and associations with education, work and social characteristics: a cross-sectional study. BMJ Open 8, e019700.

Arcelus, J., Mitchell, A., Wales, J., Nielson, S., 2011. Mortality rates in patients with anorexia nervosa and other eating disorders: a meta-analysis of 36 studies. Arch. Gen. Psychiatry 68, 724–731.

Arns, M., de Ridder, S., Strehl, U., Breteler, M., Coenen, A., 2009. Efficacy of neurofeedback treatment in ADHD: the effects on inattention, impulsivity and hyperactivity: a meta-analysis. Clin. EEG. Neurosci. 40 (3), 180–189. Available from: https://doi.org/10.1177/155005940904000311.

Arvidsson, O., Gillberg, C., Lichtenstein, P., Lundström, S., 2018. Secular changes in the symptom level of clinically diagnosed autism. J. Child Psychol. Psychiatry . Available from: https://doi.org/10.1111/jcpp.12864.

Atladóttir, H.O., Parner, E.T., Schendel, D., Dalsgaard, S., Thomsen, P.H., Thorsen, P., 2007. Time trends in reported diagnoses of childhood neuropsychiatric disorders: a Danish cohort study. Arch. Pediatr. Adolesc. Med. 161 (2), 193–198. Available from: https://doi.org/10.1001/archpedi.161.2.193.

Ballard, K., Elston, M., 2005. Medicalisation: a multi-dimensional concept. Soc. Theory Health 3, 228–241.

Basford, J.R., 2001. A historical perspective of the popular use of electric and magnetic therapy. Arch. Phys. Med. Rehabil. 82 (9), 1261–1269. Available from: https://doi.org/10.1053/apmr.2001.25905.

Beck, A., 1979. Cognitive Therapy and the Emotional Disorders. Penguin.

Biederman, J., Kim, J.W., Doyle, A.E., Mick, E., Fagerness, J., Smoller, J.W., et al., 2008. Sexually dimorphic effects of four genes (COMT, SLC6A2, MAOA, SLC6A4) in genetic associations of ADHD: a preliminary study. Am. J. Med. Genet. B: Neuropsychiatr. Genet. 147B (8), 1511–1518. Available from: https://doi.org/10.1002/ajmg.b.30874.

Blinder, B.J., Cumella, E.J., Sanathara, V.A., 2006. Psychiatric comorbidities of female inpatients with eating disorders. Psychosom. Med. 68 (3), 454–462. Available from: https://doi.org/10.1097/01.psy.0000221254.77675.f5.

Bloch, Y., Harel, E.V., Aviram, S., Govezensky, J., Ratzoni, G., Levkovitz, Y., 2010. Positive effects of repetitive transcranial magnetic stimulation on attention in ADHD subjects: a randomized controlled pilot study. World J. Biol. Psychiatry 11 (5), 755–758. Available from: https://doi.org/10.3109/15622975.2010.484466.

Bower, P., Gilbody, S., 2005. Stepped care in psychological therapies: access, effectiveness and efficiency. Narrative literature review. Br. J. Psychiatry 186, 11–17. Available from: https://doi.org/10.1192/bjp.186.1.11.

Breitling, C., Zaehle, T., Dannhauer, M., Bonath, B., Tegelbeckers, J., Flechtner, H.H., et al., 2016. Improving interference control in ADHD patients with transcranial direct current stimulation (tDCS). Front. Cell. Neurosci. 10, 72. Available from: https://doi.org/10.3389/fncel.2016.00072.

Bronstein, J.M., Tagliati, M., Alterman, R.L., Lozano, A.M., Volkmann, J., Stefani, A., et al., 2011. Deep brain stimulation for Parkinson disease: an expert consensus and review of key issues. Arch. Neurol. 68 (2), 165. Available from: https://doi.org/10.1001/archneurol.2010.260.

Campo, J., Shafer, S., Strohm, J., Lucas, A., Gelacek Cassesse, C., Shaeffer, D., et al., 2005. Pediatric behavioral health in primary care: a collaborative approach. J. Am. Psychiatr. Nurses. Assoc. 11, 276–282.

Castellanos, F.X., Proal, E., 2012. Large-scale brain systems in ADHD: beyond the prefrontal–striatal model. Trends Cogn. Sci. 16 (1), 17–26. Available from: https://doi.org/10.1016/j.tics.2011.11.007.

Chan, J., Lang, R., Rispoli, M., O'Reilly, M., Sigafoos, J., Cole, H., 2009. Use of peer-mediated interventions in the treatment of autism spectrum disorders: a systematic review. Res. Autism Spectr. Disord. 3, 876–889.

Chiramberro, M., Lindberg, N., Isometsä, E., Kähkönen, S., Appelberg, B., 2013. Repetitive transcranial magnetic stimulation induced seizures in an adolescent patient with major depression: a case report. Brain Stimulation: Basic Transl. Clin. Res. Neuromodulation 6 (5), 830–831.

Claudino, A.M., Hay, P., Lima, M.S., Bacaltchuk, J., Schmidt, U., Treasure, J., 2006. Antidepressants for anorexia nervosa. Cochrane Database Syst. Rev. (1), CD004365. Available from: https://doi.org/10.1002/14651858.CD004365.pub2.

Coffman, B.A., Clark, V.P., Parasuraman, R., 2014. Battery powered thought: enhancement of attention, learning, and memory in healthy adults using transcranial direct current stimulation. NeuroImage 85 (Pt 3), 895–908. Available from: https://doi.org/10.1016/j.neuroimage.2013.07.083.

Colwell, H., 1922. An Essay on the History of Electrotherapy and Diagnosis. William Heinemann, London, UK.

Conroy, S., Choonara, I., Impicciatore, P., Mohn, A., Arnell, H., Rane, A., et al., 2000. Survey of unlicensed and off label drug use in paediatric wards in European countries. European Network for Drug Investigation in Children. BMJ 320 (7227), 79–82.

Cullen, K.R., Jasberg, S., Nelson, B., Klimes-Dougan, B., Lim, K.O., Croarkin, P.E., 2016. Seizure induced by deep transcranial magnetic stimulation in an adolescent with depression. J. Child Adolesc. Psychopharmacol. 26 (7), 637–641.

Currin, L., Schmidt, U., Treasure, J., Jick, H., 2005. Time trends in eating disorder incidence. Br. J. Psychiatry 186, 132–135. Available from: https://doi.org/10.1192/bjp.186.2.132.

Davis, N., 2014. Transcranial stimulation of the developing brain: a plea for extreme caution. Front. Hum. Neurosci. 8, 600.

Davis, N., 2017. A taxonomy of harms inherent in cognitive enhancement. Front. Hum. Neurosci. 11, 63.

Davis, N., van Koningsbruggen, M., 2013. 'Non-invasive' brain stimulation is not non-invasive. Front. Syst. Neurosci. 7, 76. Available from: https://doi.org/10.3389/fnsys.2013.00076.

Davis, N., Tomlinson, S., Morgan, H., 2012. The role of beta-frequency neural oscillations in motor control. J. Neurosci. 32, 403–404.

Davis, N., Gold, E., Pascual-Leone, A., Bracewell, R., 2013. Challenges of proper placebo control for noninvasive brain stimulation in clinical and experimental applications. Eur. J. Neurosci. 38 (7), 2973–2977.

Dobek, C.E., Blumberger, D.M., Downar, J., Daskalakis, Z.J., Vila-Rodriguez, F., 2015. Risk of seizures in transcranial magnetic stimulation: a clinical review to inform consent process focused on bupropion. Neuropsychiatr. Dis. Treat. 11, 2975–2987. Available from: https://doi.org/10.2147/NDT.S91126.

Downar, J., Sankar, A., Giacobbe, P., Woodside, B., Colton, P., 2012. Unanticipated rapid remission of refractory bulimia nervosa, during high-dose repetitive transcranial magnetic stimulation of the dorsomedial prefrontal cortex: a case report. Front. Psychiatry 3, 30.

D'Urso, G., Bruzzese, D., Ferrucci, R., Priori, A., Pascotto, A., Galderisi, S., et al., 2015. Transcranial direct current stimulation for hyperactivity and noncompliance in autistic disorder. World J. Biol. Psychiatry 16 (5), 361–366. Available from: https://doi.org/10.3109/15622975.2015.1014411.

Egner, T., Sterman, M.B., 2006. Neurofeedback treatment of epilepsy: from basic rationale to practical application. Expert Rev. Neurotherap. 6 (2), 247–257. Available from: https://doi.org/10.1586/14737175.6.2.247.

Ekici, B., 2015. Transcranial direct current stimulation-induced seizure: analysis of a case. Clin. EEG Neurosci. 46 (2), 169. Available from: https://doi.org/10.1177/1550059414540647.

Enriquez-Geppert, S., Huster, R.J., Herrmann, C.S., 2017. EEG-neurofeedback as a tool to modulate cognition and behavior: a review tutorial. Front. Hum. Neurosci. 11, 51. Available from: https://doi.org/10.3389/fnhum.2017.00051.

Faraone, S.V., Biederman, J., 2005. What is the prevalence of adult ADHD? Results of a population screen of 966 adults. J. Atten. Disord. 9 (2), 384–391. Available from: https://doi.org/10.1177/1087054705281478.

Faraone, S.V., Sergeant, J., Gillberg, C., Biederman, J., 2003. The worldwide prevalence of ADHD: is it an American condition? World Psychiatry. 2 (2), 104–113.

Feusner, J.D., Madsen, S., Moody, T.D., Bohon, C., Hembacher, E., Bookheimer, S.Y., et al., 2012. Effects of cranial electrotherapy stimulation on resting state brain activity. Brain Behav. 2 (3), 211–220. Available from: https://doi.org/10.1002/brb3.45.

Fombonne, E., 1999. The epidemiology of autism: a review. Psychol. Med. 29 (4), 769–786.

Ford, T., Goodman, R., Meltzer, H., 2003. The British Child and Adolescent Mental Health Survey 1999: the prevalence of DSM-IV disorders. J. Am. Acad. Child Adolesc. Psychiatry 42 (10), 1203–1211. Available from: https://doi.org/10.1097/00004583-200310000-00011.

Friedrich, E.V., Suttie, N., Sivanathan, A., Lim, T., Louchart, S., Pineda, J.A., 2014. Brain-computer interface game applications for combined neurofeedback and biofeedback treatment for children on the autism spectrum. Front. Neuroeng. 7, 21. Available from: https://doi.org/10.3389/fneng.2014.00021.

Gaub, M., Carlson, C.L., 1997. Gender differences in ADHD: a meta-analysis and critical review. J. Am. Acad. Child Adolesc. Psychiatry 36 (8), 1036–1045. Available from: https://doi.org/10.1097/00004583-199708000-00011.

Gevensleben, H., Holl, B., Albrecht, B., Vogel, C., Schlamp, D., Kratz, O., et al., 2009. Is neurofeedback an efficacious treatment for ADHD? A randomised controlled clinical trial. J. Child Psychol. Psychiatry 50 (7), 780–789. Available from: https://doi.org/10.1111/j.1469-7610.2008.02033.x.

Gevensleben, H., Holl, B., Albrecht, B., Schlamp, D., Kratz, O., Studer, P., et al., 2010. Neurofeedback training in children with ADHD: 6-month follow-up of a randomised controlled trial. Eur. Child Adolesc. Psychiatry 19 (9), 715–724. Available from: https://doi.org/10.1007/s00787-010-0109-5.

Hammond, D.C., 2005. Neurofeedback with anxiety and affective disorders. Child Adolesc. Psychiatr. Clin. N. Am. 14 (1), 105–123. Available from: https://doi.org/10.1016/j.chc.2004.07.008. vii.

Herzog, D.B., Keller, M.B., Sacks, N.R., Yeh, C.J., Lavori, P.W., 1992. Psychiatric comorbidity in treatment-seeking anorexics and bulimics. J. Am. Acad. Child Adolesc. Psychiatry 31 (5), 810–818.

Hollander, E., Bartz, J., Chaplin, W., Phillips, A., Sumner, J., Soorya, L., et al., 2007. Oxytocin increases retention of social cognition in autism. Biol. Psychiatry 61 (4), 498–503. Available from: https://doi.org/10.1016/j.biopsych.2006.05.030.

Holtmann, M., Steiner, S., Hohmann, S., Poustka, L., Banaschewski, T., Bölte, S., 2011. Neurofeedback in autism spectrum disorders. Dev. Med. Child Neurol. 53 (11), 986–993. Available from: https://doi.org/10.1111/j.1469-8749.2011.04043.x.

Howat, P.M., Varner, L.M., Hegsted, M., Brewer, M.M., Mills, G.Q., 1989. The effect of bulimia upon diet, body fat, bone density, and blood components. J. Am. Diet. Assoc. 89 (7), 929–934.

Hu, S.H., Wang, S.S., Zhang, M.M., Wang, J.W., Hu, J.B., Huang, M.L., et al., 2011. Repetitive transcranial magnetic stimulation-induced seizure of a patient with adolescent-onset depression: a case report and literature review. J. Int. Med. Res. 39 (5), 2039–2044.

Huang, Y., Edwards, M., Rounis, E., Bhatia, K., Rothwell, J., 2005. Theta burst stimulation of the human motor cortex. Neuron 45, 201–206.

Huang, Y., Liu, A.A., Lafon, B., Friedman, D., Dayan, M., Wang, X., et al., 2017. Measurements and models of electric fields in the in vivo human brain during transcranial electric stimulation. eLife 6. Available from: https://doi.org/10.7554/eLife.18834.

Ivancich, A., Grigera, J., Muravchik, C., 1992. Electric properties of natural and demineralized bones. Dielectric properties up to 1 GHz. J. Biol. Phys. (18), 281–295.

Johnston, S., Boehm, S., Healy, D., Goebel, R., Linden, D., 2010. Neurofeedback: a promising tool for the self-regulation of emotion networks. NeuroImage 49 (1), 1066–1072. Available from: https://doi.org/10.1016/j.neuroimage.2009.07.056.

Kandalaft, M.R., Didehbani, N., Krawczyk, D.C., Allen, T.T., Chapman, S.B., 2013. Virtual reality social cognition training for young adults with high-functioning autism. J. Autism Dev. Disord. 43 (1), 34–44. Available from: https://doi.org/10.1007/s10803-012-1544-6.

Kessler, S.K., Minhas, P., Woods, A.J., Rosen, A., Gorman, C., Bikson, M., 2013. Dosage considerations for transcranial direct current stimulation in children: a computational modeling study. PLoS One 8 (9), e76112.

Khan, S.A., Faraone, S.V., 2006. The genetics of ADHD: a literature review of 2005. Curr. Psychiatry Rep. 8 (5), 393–397.

Kishi, T., Kafantaris, V., Sunday, S., Sheridan, E.M., Correll, C.U., 2012. Are antipsychotics effective for the treatment of anorexia nervosa? Results from a systematic review and meta-analysis. J. Clin. Psychiatry 73 (6), e757–766. Available from: https://doi.org/10.4088/JCP.12r07691.

Konrad, K., Eickhoff, S.B., 2010. Is the ADHD brain wired differently? A review on structural and functional connectivity in attention deficit hyperactivity disorder. Hum. Brain. Mapp. 31 (6), 904–916. Available from: https://doi.org/10.1002/hbm.21058.

Kouijzer, M., de Moor, J., Gerrits, B., Buitelaar, J., van Schie, H., 2009a. Long-term effects of neurofeedback treatment in autism. Res. Autism Spectr. Disord. 3, 496–501.

Kouijzer, M., de Moor, J., Gerrits, B., Congedo, M., van Schie, H., 2009b. Neurofeedback improves executive functioning in children with autism spectrum disorders. Res. Autism Spectr. Disord. 3, 145–162.

Kutcher, S., Aman, M., Brooks, S.J., Buitelaar, J., van Daalen, E., Fegert, J., et al., 2004. International consensus statement on attention-deficit/hyperactivity disorder (ADHD) and disruptive behaviour disorders (DBDs): clinical implications and treatment practice suggestions. Eur. Neuropsychopharmacol. 14 (1), 11–28.

Kuther, T.L., 2003. Medical decision-making and minors: issues of consent and assent. Adolescence 38 (150), 343–358.

Loe, I.M., Feldman, H.M., 2007. Academic and educational outcomes of children with ADHD. J. Pediatr. Psychol. 32 (6), 643–654. Available from: https://doi.org/10.1093/jpepsy/jsl054.

Lubar, J.F., Shouse, M.N., 1976. EEG and behavioral changes in a hyperkinetic child concurrent with training of the sensorimotor rhythm (SMR): a preliminary report. Biofeedback Self Regul. 1 (3), 293–306.

Maeda, F., Keenan, J.P., Tormos, J.M., Topka, H., Pascual-Leone, A., 2000. Modulation of corticospinal excitability by repetitive transcranial magnetic stimulation. Clin. Neurophysiol. 111 (5), 800–805.

Mall, V., Berweck, S., Fietzek, U.M., Glocker, F.X., Oberhuber, U., Walther, M., et al., 2004. Low level of intracortical inhibition in children shown by transcranial magnetic stimulation. Neuropediatrics 35 (02), 120–125.

Maslen, H., Douglas, T., Cohen Kadosh, R., Levy, N., Savulescu, J., 2014a. The Regulation of cognitive enhancement devices: extending the medical model. J. Law Biosci. 1, 68–93.

Maslen, H., Earp, B.D., Cohen Kadosh, R., Savulescu, J., 2014b. Brain stimulation for treatment and enhancement in children: an ethical analysis. Front. Hum. Neurosci. 8, 953. Available from: https://doi.org/10.3389/fnhum.2014.00953.

Maslen, H., Cheeran, B., Pugh, J., Pycroft, L., Boccard, S., Prangnell, S., et al., 2018. Unexpected complications of novel deep brain stimulation treatments: ethical issues and clinical recommendations. Neuromodulation 21 (2), 135–143. Available from: https://doi.org/10.1111/ner.12613.

Mayberg, H.S., Lozano, A.M., Voon, V., McNeely, H.E., Seminowicz, D., Hamani, C., et al., 2005. Deep brain stimulation for treatment-resistant depression. Neuron 45 (5), 651–660. Available from: https://doi.org/10.1016/j.neuron.2005.02.014.

Merikangas, K.R., Nakamura, E.F., Kessler, R.C., 2009. Epidemiology of mental disorders in children and adolescents. Dialogues Clin. Neurosci. 11 (1), 7–20.

Misra, M., Aggarwal, A., Miller, K.K., Almazan, C., Worley, M., Soyka, L.A., et al., 2004. Effects of anorexia nervosa on clinical, hematologic, biochemical, and bone density parameters in community-dwelling adolescent girls. Pediatrics 114 (6), 1574–1583. Available from: https://doi.org/10.1542/peds.2004-0540.

NICE, 2017. Eating Disorders: Recognition and Treatment. NICE Guideline NG69. National Institute for Health and Care Excellence.

Nitsche, M.A., Fricke, K., Henschke, U., Schlitterlau, A., Liebetanz, D., Lang, N., et al., 2003. Pharmacological modulation of cortical excitability shifts induced by transcranial direct current stimulation in humans. J. Physiol. 553 (1), 293–301.

Nye, C., Brice, A., 2005. Combined vitamin B6–magnesium treatment in autism spectrum disorder. Cochrane Database Syst. Rev. (4), CD003497. Available from: https://doi.org/10.1002/14651858.CD003497.pub2.

Oberman, L.M., Rotenberg, A., Pascual-Leone, A., 2015. Use of transcranial magnetic stimulation in autism spectrum disorders. J. Autism Dev. Disord. 45 (2), 524–536. Available from: https://doi.org/10.1007/s10803-013-1960-2.

Oberman, L.M., Enticott, P.G., Casanova, M.F., Rotenberg, A., Pascual-Leone, A., McCracken, J.T., et al., 2016. Transcranial magnetic stimulation in autism spectrum disorder: challenges, promise, and roadmap for future research. Autism Res. 9 (2), 184–203. Available from: https://doi.org/10.1002/aur.1567.

Oliviero, A., Mordillo-Mateos, L., Arias, P., Panyavin, I., Foffani, G., Aguilar, J., 2011. Transcranial static magnetic field stimulation of the human motor cortex. J. Physiol. 589 (20), 4949–4958.

Packer, R.J., Gurney, J.G., Punyko, J.A., Donaldson, S.S., Inskip, P.D., Stovall, M., et al., 2003. Long-term neurologic and neurosensory sequelae in adult survivors of a childhood brain tumor: childhood cancer survivor study. J. Clin. Oncol. 21 (17), 3255–3261. Available from: https://doi.org/10.1200/JCO.2003.01.202.

Pattison, S., Harris, B., 2006. Counselling children and young people: a review of the evidence for its effectiveness. Couns. Psychother. Res. 6, 233–237.

Peterson, F., Kennelly, A., 1892. Physiological experiments with magnets at the Edison laboratory. N.Y. Med. J. 56, 729–732.

Pierce, T., Tincani, M., 2007. Beyond consumer advocacy: autism spectrum disorders, effective instruction, and public schools. Interv. Sch. Clin. 43, 47–51.

Polanczyk, G., de Lima, M.S., Horta, B.L., Biederman, J., Rohde, L.A., 2007. The worldwide prevalence of ADHD: a systematic review and metaregression analysis. Am. J. Psychiatry 164 (6), 942–948. Available from: https://doi.org/10.1176/ajp.2007.164.6.942.

Posey, D.J., Stigler, K.A., Erickson, C.A., McDougle, C.J., 2008. Antipsychotics in the treatment of autism. J. Clin. Invest. 118 (1), 6–14. Available from: https://doi.org/10.1172/JCI32483.

Qiu, M.G., Ye, Z., Li, Q.Y., Liu, G.J., Xie, B., Wang, J., 2011. Changes of brain structure and function in ADHD children. Brain Topogr. 24 (3–4), 243–252. Available from: https://doi.org/10.1007/s10548-010-0168-4.

Quattrocki, E., Friston, K., 2014. Autism, oxytocin and interoception. Neurosci. Biobehav. Rev. 47, 410–430. Available from: https://doi.org/10.1016/j.neubiorev.2014.09.012.

Roberts, W., Weaver, L., Brian, J., Bryson, S., Emelianova, S., Griffiths, A.M., et al., 2001. Repeated doses of porcine secretin in the treatment of autism: a randomized, placebo-controlled trial. Pediatrics 107 (5), E71.

Roberts, B.W., Kuncel, N.R., Shiner, R., Caspi, A., Goldberg, L.R., 2007. The power of personality: the comparative validity of personality traits, socioeconomic status, and cognitive ability for predicting important life outcomes. Perspect. Psychol. Sci. 2 (4), 313–345. Available from: https://doi.org/10.1111/j.1745-6916.2007.00047.x.

Rotter, J.B., 1966. Generalized expectancies for internal versus external control of reinforcement. Psychol. Monogr. 80 (1), 1–28.

Rutter, M., 2005. Incidence of autism spectrum disorders: changes over time and their meaning. Acta Paediatr. 94 (1), 2–15.

Schrijvers, G., van Hoorn, A., Huiskes, N., 2012. The care pathway: concepts and theories: an introduction. Int. J. Integr. Care 12 (Spec Ed Integrated Care Pathways), e192.

Shouse, M.N., Lubar, J.F., 1979. Operant conditioning of EEG rhythms and ritalin in the treatment of hyperkinesis. Biofeedback Self Regul. 4 (4), 299–312.

Smink, F.R., van Hoeken, D., Hoek, H.W., 2012. Epidemiology of eating disorders: incidence, prevalence and mortality rates. Curr. Psychiatry Rep. 14 (4), 406–414. Available from: https://doi.org/10.1007/s11920-012-0282-y.

Soltaninejad, Z., Nejati, V., Ekhtiari, H., 2015. Effect of anodal and cathodal transcranial direct current stimulation on DLPFC on modulation of inhibitory control in ADHD. J. Atten. Disord. . Available from: https://doi.org/10.1177/1087054715618792.

Surgenor, L., Horn, J., Plumridge, E., Hudson, S., 2002. Anorexia nervosa and psychological control: a reexamination of selected theoretical accounts. Eur. Eat. Disord. Rev. 10, 85–101.

Telch, C.F., Stice, E., 1998. Psychiatric comorbidity in women with binge eating disorder: prevalence rates from a non-treatment-seeking sample. J. Consult. Clin. Psychol. 66 (5), 768.

Truong, D., Magerowski, G., Blackburn, G., Bikson, M., Alonso-Alonso, M., 2013. Computational modeling of transcranial direct current stimulation (tDCS) in obesity: impact of head fat and dose guidelines. NeuroImage: Clin. 2, 759–766.

Turner, S., Nunn, A.J., Fielding, K., Choonara, I., 1999. Adverse drug reactions to unlicensed and off-label drugs on paediatric wards: a prospective study. Acta Paediatr. 88 (9), 965–968.
Twenge, J., 2014. Time period and birth cohort differences in depressive symptoms in the U.S., 1982–2013. Soc. Indic. Res. 121, 437–454.
van Son, G.E., van Hoeken, D., Bartelds, A.I., van Furth, E.F., Hoek, H.W., 2006. Time trends in the incidence of eating disorders: a primary care study in the Netherlands. Int. J. Eat. Disord. 39 (7), 565–569. Available from: https://doi.org/10.1002/eat.20316.
Vincenti, F., 2003. Immunosuppression minimization: current and future trends in transplant immunosuppression. J. Am. Soc. Nephrol. 14 (7), 1940–1948.
Virués-Ortega, J., 2010. Applied behavior analytic intervention for autism in early childhood: meta-analysis, meta-regression and dose–response meta-analysis of multiple outcomes. Clin. Psychol. Rev. 30 (4), 387–399. Available from: https://doi.org/10.1016/j.cpr.2010.01.008.
Vitousek, K., Watson, S., Wilson, G.T., 1998. Enhancing motivation for change in treatment-resistant eating disorders. Clin. Psychol. Rev. 18 (4), 391–420.
Vöröslakos, M., Takeuchi, Y., Brinyiczki, K., Zombori, T., Oliva, A., Fernández-Ruiz, A., et al., 2018. Direct effects of transcranial electric stimulation on brain circuits in rats and humans. Nat. Commun. 9 (1), 483. Available from: https://doi.org/10.1038/s41467-018-02928-3.
Walker, J.E., Kozlowski, G.P., 2005. Neurofeedback treatment of epilepsy. Child Adolesc. Psychiatr. Clin. N. Am. 14 (1), 163–176. Available from: https://doi.org/10.1016/j.chc.2004.07.009. viii.
Weaver, L., Rostain, A.L., Mace, W., Akhtar, U., Moss, E., O'Reardon, J.P., 2012. Transcranial magnetic stimulation (TMS) in the treatment of attention-deficit/hyperactivity disorder in adolescents and young adults: a pilot study. J. ECT. 28 (2), 98–103. Available from: https://doi.org/10.1097/YCT.0b013e31824532c8.
Weil, L.G., Fleming, S.M., Dumontheil, I., Kilford, E.J., Weil, R.S., Rees, G., et al., 2013. The development of metacognitive ability in adolescence. Conscious. Cogn. 22 (1), 264–271. Available from: https://doi.org/10.1016/j.concog.2013.01.004.
Weinstein, N., Przybylski, A.K., Murayama, K., 2017. A prospective study of the motivational and health dynamics of Internet Gaming Disorder. Peer J 5, e3838. Available from: https://doi.org/10.7717/peerj.3838.
Werling, D.M., Geschwind, D.H., 2013. Sex differences in autism spectrum disorders. Curr. Opin. Neurol. 26 (2), 146–153. Available from: https://doi.org/10.1097/WCO.0b013e32835ee548.
Widdows, K., Davis, N., 2014. Ethical considerations in using brain stimulation to treat eating disorders. Front. Behav. Neurosci. 8, 351.
Wiggs, K.K., Chang, Z., Quinn, P.D., Hur, K., Gibbons, R., Dunn, D., et al., 2018. Attention-deficit/hyperactivity disorder medication and seizures. Neurology 90 (13), e1104–e1110. Available from: https://doi.org/10.1212/WNL.0000000000005213.
Zhu, Y., Hu, X., Wang, J., Chen, J., Guo, Q., Li, C., et al., 2012. Processing of food, body and emotional stimuli in anorexia nervosa: a systematic review and meta-analysis of functional magnetic resonance imaging studies. Eur. Eat. Disord. Rev. 20 (6), 439–450. Available from: https://doi.org/10.1002/erv.2197.

Further Reading

Hong, Y.H., Wu, S.W., Pedapati, E.V., Horn, P.S., Huddleston, D.A., Laue, C.S., et al., 2015. Safety and tolerability of theta burst stimulation vs. single and paired pulse transcranial magnetic stimulation: a comparative study of 165 pediatric subjects. Front. Hum. Neurosci. 9, 29.

CHAPTER

5

Transcranial Magnetic Stimulation in Autism Spectrum Disorder

Peter G. Enticott[1], *Melissa Kirkovski*[1] *and Lindsay M. Oberman*[2]

[1]Cognitive Neuroscience Unit, School of Psychology, Deakin University, Geelong, VIC, Australia [2]Center for Neuroscience and Regenerative Medicine, Henry M. Jackson Foundation for the Advancement of Military Medicine, Rockville, MD, United States

OUTLINE

Neurotechnological Pediatric Neuropsych
DOI: https://doi.org/10.1016/B978-0-12-812777-3.00005-2

ABBREVIATIONS

μV microvolts
ASD autism spectrum disorder
CPT continuous performance test
cTBS continuous theta burst stimulation
DLPFC dorsolateral prefrontal cortex
EEG electroencephalography
EMG electromyography
GABA gamma-amino-butyric-acid
ICF intracortical facilitation
ISP interhemispheric signal propagation
iTBS intermittent theta burst stimulation
NMDA *N*-methyl-D-aspartate
LICI long-interval cortical inhibition
M1 primary motor cortex
MEP motor-evoked potential
MSO maximum stimulator output
mV millivolts
NIBS noninvasive brain stimulation
RMT resting motor threshold
rTMS repetitive transcranial magnetic stimulation
SICI short-interval cortical inhibition
SMA supplementary motor area
TBS theta burst stimulation
TMS transcranial magnetic stimulation
TPJ temporoparietal junction
WCST Wisconsin card sorting test

INTRODUCTION

Among the various pediatric neurodevelopmental disorders for which there has been noninvasive brain stimulation (NIBS) research, it is perhaps autism spectrum disorder (ASD) that has generated the greatest interest. There are several reasons for this; ASD is a highly prevalent disorder for which there are few effective biomedical treatment options, but it is also a disorder that is well supported by philanthropic initiatives and high-profile advocates that increase its visibility in the broader community. Thus, potential new treatments, such as transcranial magnetic stimulation (TMS, the most widely used NIBS technique), are likely to be highly sought after by parents, and the level of interest ensures relatively broad media exposure (e.g., *New York Times* and *Newsweek*).

An interest in TMS for ASD has also been prompted by the release of John Elder Robison's 2016 memoir, *Switched On* (published by *Spiegel & Grau*), in which the author details his experiences associated with participating in TMS research at Harvard Medical School. Mr. Robison is an adult male who was diagnosed with ASD at the age of 40, and he is very well known and highly respected in the autism community as a writer, speaker, and advocate.

Switched On is a fascinating and highly affecting work, but from an empirical perspective, it is one man's subjective experience. The therapeutic use of TMS in ASD will be determined by safety and efficacy results from clinical trials, the earliest of which are reviewed in this chapter. TMS, however, has already had a number of very useful applications in ASD that have allowed us to better understand the neuropathophysiology of this frequently confounding neurodevelopmental condition. Prior to reviewing the use of TMS in ASD, both investigative and clinical, we begin this chapter by providing a brief overview of ASD.

AUTISM SPECTRUM DISORDER

Clinical Description

ASD is a relatively new diagnostic category, having been first included in the most recent revision of the American Psychiatric Association's Diagnostic and Statistical Manual of Mental Disorders (i.e., DSM-5) (American Psychiatric Association, 2013). The history of ASD dates back to the 1940s, with the concurrent yet unaffiliated work of US-based Dr. Leo Kanner (Johns Hopkins, Baltimore) and Austria-based Dr. Hans Asperger (University of Vienna). Both published a description of a group of child patients that displayed impairments in relating to other people, in addition to a variety of unusual behaviors and interests.

The conceptual notion of autism has a somewhat varied history and has undergone numerous changes in response to both clinical experience and empirical research. Following Kanner's work, autism was sometimes referred to as childhood schizophrenia, and erroneously thought to be a precursor to a later psychotic disorder. In 1980, the release of the DSM-III saw the inclusion of the diagnostic category "infantile autism," and the broader class of "pervasive developmental disorders" (PDDs). The 1987 revision (DSM-III-R) changed this category to "autistic disorder" and recognized a triad of impairments: social relating, communication, and restricted interests. DSM-IV, which was released in 1994 (and followed up with a text revision, DSM-IV-TR, in 2000), included separate diagnostic categories for autistic disorder (i.e., autism) and Asperger's disorder (in addition to seemingly related PDDs, such as Rett's disorder and childhood disintegrative disorder). For autistic and Asperger's disorders, diagnostic criteria for social and behavioral domains were essentially identical. Autistic disorder, however, included the additional "communication" domain that outlined impairments in verbal (i.e., language) and nonverbal communication. For instance, children with autistic disorder (but not Asperger's disorder) typically displayed delays in early language acquisition. There were also documented differences in cognitive function; for instance, individuals with Asperger's disorder typically displayed an intelligence quotient (IQ) in the average or above average range, while a sizeable proportion of individuals with autistic disorder also has an intellectual disability (ID) (i.e., measure IQ of <70). Those with autistic disorder who had an IQ of 70 or great were often labeled as having "high-functioning autism" (HFA), and there was extensive debate surrounding whether HFA and Asperger's disorder comprised separate disorders. It was ultimately decided that this distinction could not be justified from a clinical or empirical perspective, and DSM authors settled on the global "ASD" diagnostic entry. In any event, and in recognition of the difficulty in differentiating autism from Asperger's disorder, the researchers had been using the ASD category in their studies well before DSM-5 (for an excellent review of the history of autism, see Volkmar and McPartland, 2014).

As it stands, DSM-5 ASD is diagnosed based on clinical symptoms in two discrete domains: social communication and repetitive behaviors/restricted interests. Unlike previous versions of the DSM, the latter cluster now includes sensory abnormalities (e.g., hyper- or hyposensitivities), which are very commonly seen in ASD, and can affect all conventional sensory modalities. There was some concern that individuals diagnosed with a DSM-IV-TR PDD would not meet DSM-5 ASD criteria. DSM-5, however, states that anyone meeting diagnostic for autistic disorder or Asperger's disorder is considered to meet criteria for ASD under DSM-5.

In addition to the core symptom domains, there are a host of associated clinical features that are commonly present in ASD. These include motor impairments, ID or unusual cognitive profile, neurological illness (e.g., epilepsy), sleep disorder, and behavioral impairments (e.g., aggression, irritability). Comorbid neurodevelopmental disorder and/or psychiatric illness are also very commonly, particularly attention deficit hyperactivity disorder (ADHD), Tourette's disorder, mood disorder (depression), anxiety, and obsessive–compulsive disorder.

Epidemiology and Etiology

ASD and the previous incarnations such as autistic disorder were once considered quite rare. For instance, in the mid–late 1990s, autism was thought to affect 1 in 1000 children. Recent figures from the US-based Centers for Disease Control now suggest that 1 in 68 children have ASD, up from an estimated 1 in 150 in 2007 (Zablotsky et al., 2015). The reason for this increase is fiercely debated, but in large, it likely reflects increased awareness, a broadening of diagnostic criteria, and "diagnostic substitution."

The etiology of ASD is enormously complex, and much remains unknown. There is a clear genetic component to many instances of ASD. However, there have been hundreds of different genes implicated, and epigenetic effects (arising from gene by environment interaction) also contribute (An and Claudianos, 2016). It has been argued that divergent genetic underpinnings to autism might actually converge on similar synaptic dysfunction (Garber, 2007). A number of environment contributors have also been identified (e.g., increased parental age at conception, use of psychotropic medication during pregnancy, reduced folic acid intake pre- and during pregnancy) (Sealey et al., 2016), but these are typically associated with small effect sizes, and even in individual cases are unlikely to be sole determinants.

Neurocognition/Neuropsychology

As noted, individuals with ASD often also present with ID, which means that they have an IQ of less than 70 points on a standardized cognitive assessment (e.g., Wechsler Intelligence Scale for Children). Where IQ is within or above "normal" limits, there is typically an uneven profile of cognitive abilities, and the nature of this profile can differ from person to person. Many individuals with ASD show reduced performance on verbal comprehension, working memory, and processing speed domains.

From a neuropsychological perspective, specific social cognitive deficits, including reduced performance on theory of mind or "mentalizing" tasks (Baron-Cohen et al., 1985), are widely reported, as are deficits in facial emotion recognition (Harms et al., 2010). There are also well document impairments in executive function, such as working memory (Wang et al., 2017). Dominant theoretical models over recent decades have emphasized impaired theory of mind (Baron-Cohen et al., 1985), executive dysfunction (Ozonoff et al., 2005), and weak central coherence (i.e., a failure in information processing that impairs one's ability to perceive the environment in a global, integrated manner, instead involving "detail-oriented" processing) (Happe, 2005). A more recent theoretical account is based on Bayesian predictive coding accounts and suggests that perception is characterized by an increased reliance on sensory input and/or an attenuated balancing of knowledge or prior experience (Palmer et al., 2015).

Of note here is the sheer heterogeneity and variability in neurocognitive performance in ASD. There is no one measure that is reliably impaired across all individuals with ASD, and a formal cognitive and/or neuropsychological assessment is required to know an individual with ASD's neurocognitive strengths and weaknesses.

Neurobiology

Providing a review of neurobiological deficits in autism is well beyond the scope of this chapter, but there are some consistent findings that are worth mentioning. Increased brain volume in early childhood is widely reported and well replicated (Courchesne et al., 2011). This may reflect one or more processes, such as increased cell proliferation in utero or reduced apoptosis. Another important finding across the literature is that of reduced neural connectivity in ASD, in particular reduced "long-range" connectivity (indeed, there is some evidence for enhanced "short-range" connectivity) (O'Reilly et al., 2017).

Of particular relevance to NIBS, ASD has also been associated with abnormal activity within several specific brain regions. This includes densely connected cortical "hubs," such as the posterior superior temporal sulcus (pSTS), temporoparietal cortex [temporoparietal junction (TPJ)], dorsolateral prefrontal cortex (DLPFC), and supplementary motor area (SMA) (Stigler et al., 2011). The cerebellum has also been extensively implicated in autism, with abnormal activation and neuropathology (Fatemi et al., 2012). Conveniently, most of these regions are highly accessible when using TMS, as they are placed reasonably close to the scalp. Furthermore, there is extensive evidence for impairment in particularly neurochemical systems in autism, such as gamma-aminobutyric-acid (GABA), with a recent model suggesting an excitation/inhibition imbalance (Nelson and Valakh, 2015). Again, this is relevant because rTMS has been shown to preferentially act upon GABAergic systems, including $GABA_A$ and $GABA_B$ receptors.

TRANSCRANIAL MAGNETIC STIMULATION

TMS is a noninvasive technique that involves the administration of a strong magnetic pulse to the scalp, delivered via a plastic coated metallic coil through which a time-varying electrical current is run. The resultant magnetic pulse travels largely unimpeded through the scalp to the brain tissue, where it induces a small amount of electrical current. This current can cause activation of cortical neurons and interneurons, both inhibitory and excitatory. If applied at a sufficient intensity, TMS can trigger depolarization of excitatory pyramidal neurons. This best exemplified when stimulating the primary motor cortex (M1), where a single TMS pulse can produce brief activation peripheral muscle (e.g., hand).

TMS is a unique technique in the clinical neurosciences; in that, it can be used to both *investigate* the brain and *modulate* brain activity. The latter typically arises from the repeated administration of TMS pulses, termed repetitive transcranial magnetic stimulation (rTMS). While both modes of use have the capacity to advance our understanding of in vivo neurochemistry and brain–behavior or brain–cognition relationships, it is rTMS that provides an opportunity for intervention and treatment. Most notably, rTMS has been developed as a safe and effective intervention for treatment-resistant major depressive disorder (TR-MDD). At the time of writing, eight TMS manufacturers are registered with the US Food and Drug Administration (FDA) for antidepressant treatment, and rTMS treatments for TR-MDD are in widespread clinical use throughout America, Europe, Asia, and Australasia.

Although the field of NIBS in autism is far less advanced, both modes of TMS, investigative and intervention, have been used in individuals with ASD, with the earliest investigative publication coming in 2005 (Théoret et al., 2005), the earliest intervention study published in 2009 (Sokhadze et al., 2009), and first double-blind, sham-controlled clinical trial published in 2014 (Enticott et al., 2014). In this section, we will provide a review of the extant investigative studies using TMS in ASD, grouping studies by broad theme, and explore their contribution to our understanding of the neurobiology of ASD. In the following section, we will consider intervention studies using rTMS in ASD.

Cortical Excitability and Cortical Inhibition

The first use of TMS involved the stimulation of the M1, which produces a response in peripheral muscle that was recorded via EMG (Barker et al., 1985). This response is termed as motor-evoked potential (MEP) (see Fig. 5.1) and is thought to provide a broad measure of corticospinal excitability (CSE). The TMS intensity that is used to stimulate the brain is typically guided by an individual's "motor threshold," which is the minimum stimulator intensity required to produce a minimum MEP amplitude response over a certain number of consecutive trials (typically 50 μV amplitude in at least 5 out of 10 consecutive trials). Various "paired pulse" TMS paradigms have also been established, which provide measures of cortical inhibition or cortical facilitation. The most commonly used investigatory TMS paradigms are as follows:

- Short-interval cortical inhibition (SICI): This involves the administration of a subthreshold conditioning pulse (incapable of generating a MEP, but capable of activating inhibitory interneurons), followed 2–5 ms later by a suprathreshold test pulse. This typically produces a smaller MEP amplitude when compared with single pulse TMS, and pharmacological studies have attributed this inhibition to $GABA_A$ receptor activity (Ziemann, 2003).
- Long-interval cortical inhibition (LICI): This consists of two equivalent suprathreshold TMS pulses, separated by 100–150 ms. While a standard MEP is produced in response to the first pulse, the MEP amplitude following the second pulse is typically reduced. This is thought to reflect activity at slower acting $GABA_B$ receptors (Ziemann, 2003).
- Intracortical facilitation (ICF): When the above-mentioned SICI protocol involves a longer latency (10–15 ms), this typically produces a facilitated or increased MEP amplitude compared with single pulse TMS. From a neurobiological perspective, this has been attributed to activity within glutamatergic NMDA receptors (Ziemann, 2003).
- Cortical silent period (CSP): This consists of a single suprathreshold pulse, while the participant is actively contracting the target muscle. While a MEP is produced, the CSP refers to a period of EMG suppression immediately after the MEP. Depending on the intensity of the TMS pulse, the CSP has been shown to reflect both $GABA_A$ and $GABA_B$ receptor activity (Ziemann, 2003).

These paradigms are presented graphically in Fig. 5.1.

The motor cortical response to TMS has been used to provide an index of both cortical excitation and cortical inhibition in ASD. The latter is particularly relevant; as noted, it is believed to reflect activity of

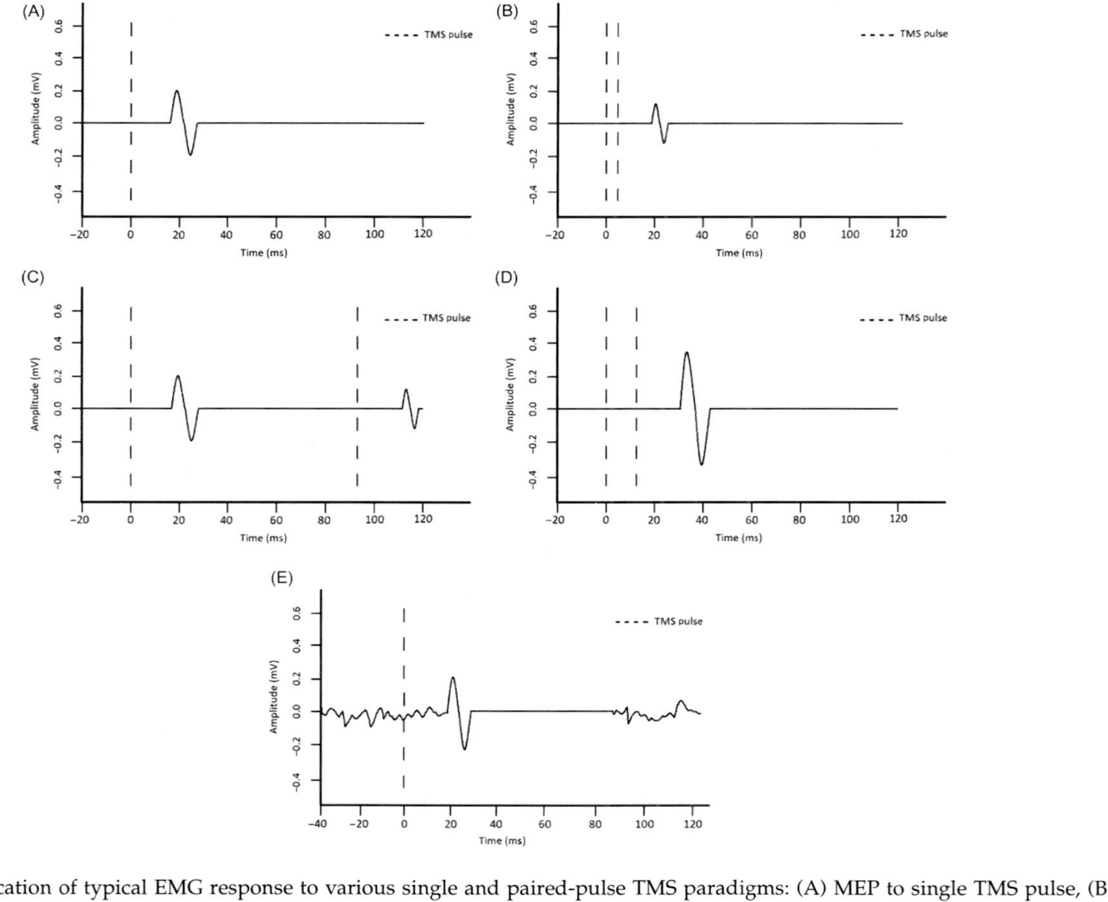

FIGURE 5.1 Stylized indication of typical EMG response to various single and paired-pulse TMS paradigms: (A) MEP to single TMS pulse, (B) reduced MEP following SICI, (C) reduced second MEP following LICI, (D) increased MEP following ICF, and (E) CSP following single pulse MEP during active contraction. *EMG*, electromyography; *TMS*, transcranial magnetic stimulation; *MEP*, motor-evoked potential; *SICI*, short-interval cortical inhibition; *LICI*, long-interval cortical inhibition; *ICF*, intracortical facilitation; *CSP*, cortical silent period.

specific GABA receptors, and GABAergic dysfunction is widely reported in ASD across genetic, postmortem, and magnetic resonance spectroscopy studies (Coghlan et al., 2012). There has also been recent preliminary evidence for clinical benefits in the use of bumetanide, a chloride-importer agonist that can promote GABAergic inhibition, in children with ASD (Lemonnier et al., 2012, 2017).

In the first published TMS study of individuals with ASD (see that paper's supplementary material), Théoret et al. (2005) found no differences between ten adults with high-functioning ASD and matched controls on resting motor threshold (RMT), SICI, ICF, or input–output curves [i.e., plotted curve assessing % maximum stimulator output (MSO) and resultant average MEP amplitude (mV)]. There was a trend toward a group difference in CSP, with the ASD group demonstrating a mean CSP 25 ms less than controls, potentially indicative of $GABA_B$ receptor deficits.

A later study by Enticott et al. (2010) examined CSE in adolescents and young adults with ASD and also found no group differences in motor threshold (resting and active) or single pulse MEP amplitude, again indicating comparable CSE. As with Théoret et al. (2005), both clinical and control groups displayed the expected facilitation with ICF. Individuals characterized as diagnosed with DSM-IV-TR autistic disorder, however, failed to show evidence for SICI in either cerebral hemisphere, potentially reflective of $GABA_A$ receptor impairment. This did not extend to participants diagnosed with DSM-IV-TR Asperger's disorder, where SICI was comparable to that seen in controls.

A follow-up study by Enticott et al. (2013) examined 36 adolescents and young adults with ASD (13 medicated) and found no evidence for differences in left hemisphere motor threshold (resting and active), but there was an increase in motor threshold (resting and active) in the right hemisphere, which could indicate reduced right-sided CSE in ASD. This is somewhat difficult to interpret, as there are many determinants of motor threshold, including scalp-to-cortex distance, and it is not necessarily indicative of a neurochemical or synaptic impairment. There were no group differences in amplitude for either resting or active MEPs, and no differences in CSP duration, cortical excitability, or cortical facilitation. When separating the ASD group according to the presence of absence of early language delay (typically present in DSM-IV-TR autistic disorder but not DSM-IV-TR Asperger's disorder), it was found that those with ASD without an early language delay appeared to account for right hemisphere increases in motor threshold. As with the previous study, those with ASD who experienced early language delay displayed reduced SICI, but this was restricted to the left hemisphere (and left-sided GABAergic dysfunction might explain these language impairments).

Thus, there is some evidence for GABAergic deficits in ASD when using TMS to probe motor CSE, although this may only affect certain ASD subtypes. Given the evidence for GABAergic impairments in ASD, it is perhaps surprising the more robust impairments in cortical inhibition have not been detected in ASD. It should be noted that putative GABAergic measures using TMS are likely to be determined by a number of other factors, and one should be cautious when attempting a purely GABAergic interpretation. These findings are also limited to a small region of M1, and GABA concentrations are not uniform across the brain nor are they particularly well correlated among brain regions at an individual level (Greenhouse et al., 2016). Nevertheless, these results are not entirely inconsistent with GABAergic models of the neurobiology of ASD (which may also help to explain increased risk of epilepsy in ASD) and lend some further preliminary support to the use of pharmacological agents that have direct or indirect effects on GABAergic function. As will be discussed later, they also lend support to targeted attempts at modulating GABAergic function via NIBS.

Mirror Neuron System

A controversial theory of the neurobiology of ASD, particular with respect to social relating, concerns the mirror neuron system (MNS). "Mirror neurons" are cells that fire during both the execution and observation of motor behavior and were originally discovered via depth electrode recordings in macaque monkeys (di Pellegrino et al., 1992). An analogous system has since been established in humans and has been extended to show mirroring for tactile sensation, emotion, and pain (Keysers and Gazzola, 2009). From a theoretical perspective, the MNS has been suggested to be involved in simulation of other peoples' minds, and thus subserving various aspects of social cognition and empathy (e.g., *direct matching hypothesis*; Keysers and Gazzola, 2009). Perhaps inevitably, it was subsequently suggested that social impairments in ASD might stem, at least in part, from a failure to adequately develop mirror neurons, or a breakdown in the broader mirror system (Williams et al., 2001; Oberman et al., 2005).

There are a number of different techniques for measuring human MNS activity, including electroencephalography (EEG) and functional magnetic resonance imaging (fMRI). Where TMS is concerned, MNS activity is inferred from the response to motor cortical stimulation, measured via MEP amplitude, during the observation of behavior that involves use of the muscle activated by TMS. This is typically a hand muscle, such as first dorsal interosseous (FDI) or abductor pollicis brevis. In general, motor cortical TMS during action observation produces a facilitated, or larger, MEP amplitude when compared with MEP amplitude during a baseline condition, such as observation of a static body part or fixation cross. From a network perspective, action observation is thought to produce mirror neuron activity within *posterior inferior frontal gyrus* (IFG, more specifically *pars opercularis*) and *ventral premotor cortex*. This produces premotor input to M1, which enhances M1 excitability. When a TMS pulse is delivered to a more excitable brain region, it produces a facilitated response to stimulation, in this case a larger MEP. Thus, the degree of MEP facilitation during action observation, relative to MEP amplitude in a baseline condition, is considered a putative measure of human MNS activity. In noting the difficulty in extrapolating from this response to the activity of individual mirror neurons, this is often spoken of in terms of "motor resonance" or "interpersonal motor resonance." We are then effectively examining the degree to which action observation engages one's own motor system, but the underlying mechanism is still assumed to be some analogue of the mirror neuron response detected via depth electrode recordings in macaques.

Indeed, the first published TMS investigations of ASD were designed to test the "mirror neuron hypothesis" of autism. Théoret et al. (2005) administered TMS to M1 during action observation in 10 adults with "high-functioning" ASD and matched controls. Action observation involved a video showing a hand performing intransitive movements (presented from both egocentric and allocentric perspectives). Unlike controls, individuals with ASD failed to demonstrate the expected motor resonance when viewing hand movements from an egocentric perspective; this was interpreted as potentially reflecting a MNS impairment in ASD that could also explain social symptoms and difficulties in self-other processing. These results, however, are ambiguous when considering that no differences were seen during action observation when presented from an allocentric perspective.

Although not strictly involving action observation, Minio-Paluello et al. (2009) used TMS with an "empathy for pain" protocol, which involves M1 TMS during video observation of a needle penetrating a hand (specifically, the muscle that TMS is applied to preferentially activate, such as FDI), and is thought to reflect a broader MNS. Compared with matched controls, 16 adult males with Asperger's syndrome failed to demonstrate the expected inhibitory effect on MEP amplitude. Again, this was interpreted as evidence for poor sensorimotor resonance in ASD that might underpin characteristic social deficits.

More recent studies of mirror system activity in ASD have been contradictory. Employing M1 TMS during hand action observation in 34 adolescents/young adults with high-functioning ASD and a matched control group, Enticott et al. (2012) found reduced motor resonance in ASD. Furthermore, reduced motor resonance was associated with higher social symptom ratings on the Autism Spectrum Quotient (AQ). A follow-up study from the same group, however, failed to verify these results when requiring participants to observe more complex and "interactive" hand actions. That is, individuals with ASD showed no evidence of reduced motor resonance, suggesting that, as with Théoret et al. (2005), motor resonance in ASD can be evoked with certain stimuli.

There is evidence for modulation of the human MNS via rTMS of the IFG among typically developing individuals (Mehta et al., 2015), but the use of rTMS to specifically target the hMNS in ASD seems unjustified when the apparent underlying neurophysiological deficit has not been reliably established. Admittedly, this could reflect heterogeneity, and a deficit in sensorimotor resonance may be applicable to at least some individuals with ASD, but the origin of any such deficit continues to be debated. It is also important to note that most studies do not control for sensory inputs (e.g., properties of visual gaze) in detecting the mirror neuron response, despite that fact that (1) individuals with ASD show abnormal gaze patterns when observing other people (Frazier et al., 2017) and (2) gaze variables have been associated with the measured mirror neuron response (Donaldson et al., 2015a). The latter is unsurprising when considering the broader MNS network, which begins, at a cortical level, with the visual representation in visual cortex and then projects to superior temporal cortical regions (which are critical for processing biological motion) (Iacoboni and Dapretto, 2006).

Neuroplasticity

Contemporary neurobiological accounts of ASD suggest that characterized abnormalities, such as reduced connectivity, are underpinned by aberrant brain plasticity (i.e., capacity of the brain to change the nature and strength of neural connections). Recent work has utilized modern rTMS protocols, such as theta burst stimulation (TBS), to further investigate the induction of a neuroplastic response in the brain. TBS is a high-frequency but low intensity protocol that involves "bursts" of three pulses at 50 Hz, which are repeated at 5 Hz. These frequencies were selected based on in vitro studies of neuronal firing and are applied in an attempt to "entrain" neural activity at this frequency. There are generally two modes of TBS. Continuous TBS (cTBS) involves the administration of TBS for an uninterrupted period of 40 seconds (600 pulses in total). By contrast, intermittent TBS (iTBS) involves the administration of TBS for 2 seconds, followed by an 8-second rest period, for 190 seconds (600 pulses in total). Using single pulse TMS to measure MEP outcomes, cTBS has been shown to have long-term depression (LTD)-like (suppression of cortical excitability) effects, while iTBS has been shown to have long-term potentiation (LTP)–like (facilitation of cortical excitability) effects, lasting approximately 30 minutes, in healthy individuals (Huang et al., 2005), although there is a great degree of inter-individual variability (Hamada et al., 2013). (See Chung et al., 2016 for a recent systematic review and meta-analysis.)

It has been suggested that neural plasticity might be increased, or heightened, in ASD, referred to as "hyperplasticity." Primarily through the work of Oberman et al., this has been demonstrated using both cTBS (Oberman et al., 2010a,b, 2012, 2014, 2016) and, to a lesser extent, iTBS (Oberman et al., 2010a, 2012; Pedapati et al., 2016). Specifically, the aforementioned research generally indicates that the effects of these protocols are greater (in terms of the effects on MEP amplitude) and longer lasting (up to 90 minutes) in ASD samples.

From a plasticity perspective, however, there is evidence that might be seen as contradictory. Using an alternative paradigm, namely, paired associative stimulation (PAS; whereby peripheral electrical stimulation is paired with TMS), Jung et al. (2013) report reduced LTP-like plasticity among an ASD cohort compared to typically developing controls. It should be noted, however, that the forms of plasticity assessed by TBS and PAS are very different (non-Hebbian vs Hebbian), and it appears that while plasticity is aberrant in ASD, changes are not uniform across modes of neuroplasticity (Enticott and Oberman, 2013).

Notably, age might be an important factor influencing the impact the effects of TBS on plasticity in ASD. Oberman et al. (2014) report a significant relationship between participant age and the duration of response, suggesting longer lasting responses in older individuals. Furthermore, in a child and adolescent cohort, Pedapati et al. (2016) note that while individuals with, and without, ASD show the expected facilitatory effects of iTBS, MEP amplitude was significantly decreased in the ASD group 20 minutes following stimulation. In contrast to expectations, a younger subset of participants in the Oberman et al. (2014) study showed a contradictory effect to cTBS, with facilitated MEPs being recorded.

The long-term effects of TBS plasticity protocols (i.e., metaplasticity) have also been investigated in ASD, by applying the alternate protocol 24 hours later. Using this approach, neurotypical individuals do not show any continued effects of the previous day's stimulation. Among individuals with ASD, however, the metaplastic effects of TBS appear to be longer lasting. The effects of applying the alternate protocol 24 hours after the initial session resulted in a reduced effect (Oberman et al., 2012). That is, when cTBS is applied, followed 24 hours later by iTBS, the effects of iTBS are attenuated, and vice versa. A reduced metaplastic response has also been seen in ASD when repeating cTBS on consecutive days (Oberman et al., 2016).

A common theme emerging from these studies, and linking from the previous section, is the possibility of GABAergic mechanisms underlying some of these findings. Jung et al. (2013), however, found no relationship between SICI and PAS plasticity indices.

The ability of TMS protocols to evoke measureable neuroplastic changes in brain ensures that we can measure and compare neuroplasticity across psychiatric, neurological, and neurodevelopmental disorders. It seems that an understanding of the neuroplastic effects of rTMS in ASD (i.e., what rTMS is actually doing to the brain) is a vital component of developing an rTMS-based intervention for the core symptoms of this condition. The diagnostic and prognostic potential of such protocols has also been considered (Oberman et al., 2012).

Combined Techniques: TMS-EEG

A relatively recent development in NIBS experimentation involves recording EEG during TMS (Hill et al., 2016). Combining these techniques poses a number of technical challenges, largely because the TMS pulse introduces a large amount of artifact in the electrophysiological trace. Advances in EEG technology (e.g., very high acquisition rates, 10–20 kHz), combined with hardware improvements and postprocessing strategies, means that it is now possible to resolve TMS-induced artifact in a very short period (~10 ms) and record TMS-evoked responses (or TEPs) emanating from the brain.

To date, just two studies have been published using this combined methodology in ASD. Kirkovski et al. (2016b) combined TMS and EEG among adults with high-functioning ASD, stimulating right M1, right DLPFC, and right TPJ (which has been linked to social and executive impairments in ASD) (Donaldson et al., 2015b) while recording brain activity from underlying scalp electrodes. EEG was used to investigate oscillatory power (reflective of neural activity) and phase-synchrony (reflective of neural connectivity) immediately following single pulse TMS. These responses did not differ between adults with ASD and matched control at any site. This finding remained when data were stratified by sex. There was some evidence, however, that the phase-synchrony response to TMS at M1 and at rTPJ might be related to ASD relevant trait in the clinical group. Interestingly, this relationship was evident in the female but not the male ASD subgroup, which supports the suggestion that ASD neurobiology is modulated by sex (Kirkovski et al., 2013).

Similarly, when comparing a sample of children and young adults with ASD to neurotypical controls, Jarczok et al. (2016) found no group differences in neural connectivity. Here, single pulse TMS was applied to left M1 to investigate interhemispheric signal propagation (ISP).

ISP reflects neural connectivity between the two hemispheres of the brain, reflective of corpus callosum microstructure. Moreover, ISP is known to be modulated by age, and while age-related effects of neurodevelopment have been reported in ASD, the expected age-related effects on ISP were present in both groups. Unlike Kirkovski et al. (2016b), however, ISP was not found to be related to any traits or characteristics associated with ASD nor ADHD (a common co-morbidity of ASD).

Despite these null findings, there are many TMS-EEG protocols that are yet to be explored in ASD. Research combining TMS-EEG with behavioral tasks, or to measure the response to other NIBS protocols such as SICI, LICI, or rTMS, may prove fruitful. As with the neuroplasticity work in this area, understanding the mechanisms that underlie the neurobiological response to TMS is integral for efforts to develop an rTMS-based biomedical intervention for ASD, and this work should continue. While technically demanding, there are open-source toolboxes now available for analyzing combined TMS-EEG data, such as the MATLAB-based EEGLAB extension TMS-EEG Signal Analyser (TESA) (Rogasch et al., 2017).

REPETITIVE TRANSCRANIAL MAGNETIC STIMULATION

We now turn to the use of rTMS in ASD, which is concerned with modulating brain function for treatment or therapeutic purposes. This has been examined at various levels, ranging from the underlying neuropathophysiology of ASD to neuropsychological and clinical impairments. Although we cannot advocate the clinical use of rTMS for ASD at this time, it will become apparent that there is emerging evidence to suggest that the developmental of an rTMS-based therapeutic intervention for ASD is a distinct possibility.

Neurophysiology

The first work examining the effect of rTMS in ASD was conduct by Dr. Manuel Casanova and colleagues, who primarily focused on ameliorating abnormal event-related potentials in ASD. This group has employed subthreshold low-frequency (0.5–1 Hz) stimulation of left and right DLPFC, which are also targets in rTMS depression treatments. From a mechanistic perspective, low-frequency stimulation was chosen based on Casanova's earlier work with postmortem ASD brain tissue, which demonstrated abnormalities in cortical "minicolumn" structure that are thought to result in GABAergic inhibitory deficits in the autistic brain (and may explain the oft-cited excitation/inhibition imbalance in ASD) (Casanova et al., 2015).

For instance, Sokhadze et al. (2009) administered six sessions of rTMS (two per week) to left DLPFC in adolescents and young adults with ASD, opting for a subthreshold (90%RMT) and low-frequency (0.5 Hz) paradigm. EEG was recorded during the presentation of Kanizsa figures. Although a sham control was not employed, following rTMS there was a reduced P3a amplitude in response to nontarget stimuli. Additional work by this group, which has typically involved sequential bilateral stimulation of DLPFC, has revealed EEG improvements in gamma power to target stimuli (Baruth et al., 2010), various ERP components in response to target stimuli (P50, N/P200, P3b) (Sokhadze et al., 2010, 2014a; Casanova et al., 2012), and error related negativity (Sokhadze et al., 2012). Thus, across these studies, rTMS appears to improve and somewhat "normalize" the neural response in ASD, providing a clear distinction between target and distractor (or nontarget) items (i.e., reduced ERP component amplitudes and reduced gamma power in the latter).

Following on from studies of motor dysfunction, which revealed a reduction in cortical movement-related cortical potentials (MRCPs) in ASD (Rinehart et al., 2006; Enticott et al., 2009), Enticott et al. (2012) investigated whether a single session of low-frequency rTMS could modulate motor function. Eleven adolescents/young adults with ASD completed three sessions, which involved EEG recording during the performance of a button pressing task designed to promote movement preparation, both before and after 15 minutes of low-frequency (1 Hz) rTMS. Each session was identical except for the location/mode of stimulation (SMA, left M1, and sham M1). While not bringing MRCPs within typical limits, SMA stimulation produced a significant improvement in early premovement EEG negativity (suggesting enhanced preparation at a neural level), while left M1 stimulation yielded a higher premovement amplitude. There were, however, no changes seen in motor performance (as determined by timing variables assessing movement preparation and execution during the button pressing task).

Language

To date, one study has examined rTMS in the context of language ability in ASD. Fecteau et al. (2011) administered five separate rTMS sessions (1 Hz, 70%MSO) to 10 adults with ASD (DSM-IV-TR Asperger's disorder) and 10 matched controls and examined change in performance (accuracy and reaction time) on an object naming task (Boston Naming Test). Active stimulation conditions were centered around the IFG (left and right pars triangularis, left and right pars opercularis), all determined via MRI-guided neuronavigation, while a sham condition was also included.

Individuals with ASD displayed an increase in response time (i.e., poorer performance) following stimulation of left pars opercularis, but a reduced response time (i.e., enhanced performance) following stimulation of pars triangularis. This was interpreted as supporting language network anomalies in ASD, but also pointed to a capacity for precisely localized rTMS to modulate neurocognitive performance in ASD.

Executive Function

Executive impairments are very well documented in ASD and include reduced performance in response inhibition, planning, cognitive flexibility, and working memory. Indeed, an executive dysfunction theory of ASD has been one of three dominant neurocognitive models of ASD over the past three decades. Accordingly, executive deficits have been the target of rTMS interventions in ASD.

Recently, Ni et al. (2017) investigated whether iTBS, which as noted earlier typically has an excitatory effect, could improve neuropsychological performance on measures of response inhibition [continuous performance test (CPT)] and planning [Wisconsin card sorting test (WCST)]. Participants attended three sessions, each of which involved undergoing a different stimulation procedure: (1) sequential bilateral stimulation (left then right) of DLPFC, (2) sequential bilateral stimulation of pSTS, and (3) stimulation of the inion (i.e., active control). For bilateral stimulation conditions, two courses of conventional iTBS (i.e., 600 pulses), separated by 5 minutes, were delivered to each hemisphere. Clinical assessments of autistic symptomatology (Social Responsiveness Scale) and obsessive–compulsive symptomatology (Yale–Brown obsessive compulsive scale) were also administered. Assessments were conducted before, 8 hours after, and 2 days after stimulation. CPT reaction time was improved following DLPFC stimulation, but there was trend toward increased errors for both the CPT and WCST following pSTS stimulation.

Another study, described in Ameis et al. (2017), is currently in progress but is based on a previous approach whereby high-frequency (20 Hz) rTMS to bilateral DLPFC improved working memory in schizophrenia (Barr et al., 2013). Working memory deficits are widely documented in ASD (Demetriou et al., 2017), and the amelioration of these impairments may produce clinical benefits. This approach is described further in a later section.

Repetitive Behaviors

Although not a primary target for intervention, Casanova's group has measured repetitive behaviors using the Repetitive Behavior Scale—Revised

(RBS-R) before and after rTMS across several studies. A reduction in RBS-R total score was reported in response to both left-sided (Sokhadze et al., 2009, 2010) and bilateral low-frequency (0.5–1 Hz) stimulation (Baruth et al., 2010; Casanova et al., 2012, 2014; Sokhadze et al., 2014a,b), although the extent to which these data overlap is unclear. In any event, improvements in repetitive behaviors is presumably attributable to enhanced cortical inhibition that impacts upon relevant executive abilities governed by networks involving DLPFC. Using parent-report data via the Y-BOCS, Ni et al. (2017) also reported a reduction in repetitive behaviors following pSTS iTBS, but these are very difficult to interpret given the short duration following treatment (i.e., 8 hours and 2 days) at which assessments were taken. Furthermore, no differences were seen for self-report data.

Irritability

Aforementioned clinical trials conducted by Casanova have revealed clinical improvements in addition to neurophysiological changes. Perhaps, the most consistent clinical finding across these studies is reduced irritability (as typically indexed by the aberrant behavior checklist) (Baruth et al., 2010; Casanova et al., 2012; Sokhadze et al., 2014a). It has generally been seen in those studies delivering bilateral dlPFC stimulation, rather than only left-sided stimulation (Sokhadze et al., 2009, 2010), suggesting the importance of right-sided stimulation in this clinical improvement. In some instances, this has been accompanied by reductions in hyperactivity (Sokhadze et al., 2014a).

Social Relating

Impairments in social relating, including reduced eye contact, failure to develop appropriate peer relationships, reduced joint attention, and less social/emotional reciprocity, are typically considered the hallmark of ASD. They can also lead to poor mental health and can cause difficulties with integrating in school, work, and other social environments. Social-relating impairments are notoriously difficult to treat in ASD, and most interventions are behavioral in nature and targeted toward very young children.

Social impairments are also relatively difficult to measure, particularly if one aims to gain a comprehensive, sensitive, reliable, and valid index of social relating in older children and adults that can be repeated in a longitudinal fashion. Many intervention studies instead take a "research domain criteria," or RDoC-style approach, measuring change in a specific aspect of social relating. For instance, studies of intranasal oxytocin, a neuropeptide that is implicated in maternal bonding and social trust, have used measures of affective theory of mind (i.e., facial emotion recognition) to measure social change (Guastella et al., 2010).

Unfortunately, measurement options are relatively limited when attempting to take a more comprehensive measure of social relating. This typically involves subjective ratings of social behavior and social cognition, completed by an individual with ASD and/or a family member. Standardized clinical assessments (e.g., CARS, ADOS) are sometimes used, but these typically lack the required sensitivity to detect anything other than very large clinical changes. As described above, other options include the use of neuropsychological assessments of "social cognition," but these can also be combined with neuroimaging or neurophysiology to determine any associated functional brain changes.

Enticott et al. (2014) conducted a small, sham-controlled, double blinded trial of rTMS with the intention of improving social relating in adults with ASD. Participants ($n = 28$) underwent 10 sessions of high-frequency (5 Hz) to bilateral dorsomedial prefrontal cortex (dmPFC) over 2 weeks. This was achieved using a deep rTMS "HAUT" coil (Brainsway Ltd., Israel), which enables direct stimulation of deeper brain structure (although at the expense of focality, as a broader field is induced). Self-report autism scales and neuropsychological measures of social cognition were completed before, after, and 1 month after rTMS. At the 1-month follow-up, individuals undergoing active stimulation reported a reduction (compared with "pre") in the "social-relatedness" scale of the Ritvo Autism-Asperger Diagnostic Scale (Ritvo et al., 2008).

This finding, while encouraging, is very preliminary and will require independent replication. Otherwise, there are no multi-session rTMS trials examining specific effects on social aspects of ASD. Of note, studies conducted by Casanova's group report no changes in social-relating ability following low-frequency DLPFC rTMS (e.g., Sokhadze et al., 2009, 2010, 2014a; Baruth et al., 2010; Casanova et al., 2014).

Sensorimotor Integration

To date, only one study has examined the use of rTMS for targeting sensorimotor integration in ASD. Across a series of four small studies ($n = 9$, 17, 4, 13), Panerai et al. (2014) explored the use of both high-frequency (8 Hz) and low-frequency (1 Hz) rTMS (with sham control) to the premotor cortex of adolescents with ASD and comorbid ID. Improved performance on eye–hand integration (measured via the Psychoeducational Profile—Revised) was seen following high-frequency stimulation and further improved when paired with eye–hand integration training that was delivered by a psychologist who was blinded to

TMS conditions. Importantly, this is the only published rTMS trial to date that has looked at so-called low-functioning ASD (i.e., where participants also had a formally assessed ID).

Case Studies

Finally, there have been a small number of case studies in the literature concerning the use of rTMS in individuals with ASD. While case studies are anecdotal and uncontrolled, and therefore should be considered extremely poor evidence for the safety and efficacy of rTMS in ASD, they can provide useful directions for future research and clinical trials.

Individual case studies report improvements in social abilities (Enticott et al., 2011), irritability (Niederhofer, 2012), and stereotypies (Niederhofer, 2012). Most recently, Avirame et al. (2017) reported two adult ASD cases involving deep rTMS of the dmPFC (based on the Enticott et al. protocol) each weekday for 5–6 weeks. There were reported improvements in a number of areas, including emotion recognition, and reduction in scores on the self-report AQ.

Summary of rTMS Studies in ASD

The studies reviewed above provide a preliminary evidence-base to suggest that rTMS can improve a number of aspects of ASD, from neurophysiological through to clinical. However, each of these studies has important limitations, such as small sample size and heterogeneity.

Thus, at this point, we are not in a position to recommend rTMS as a clinical treatment for ASD, but there is ample evidence for pursuing large-scale, multisite clinical trials (i.e., phases II and III clinical trials) that, if successful, would provide an evidence-based for recommending the clinical use of rTMS in ASD. We conclude by considering a number of conceptual issues in this space, many of which should be addressed before any large-scale trials commence.

SUMMARY AND FUTURE DIRECTIONS

In this chapter, we have described the varied use of TMS in ASD, which is an emerging yet promising literature. As a technique, TMS has contributed to our understanding of the neurobiology of ASD, such as altered plasticity, although studies of cortical inhibition and the MNS have not produced particularly robust findings. Combined approaches, such as TMS-EEG, represent a promising new direction in better understanding brain processes that might contribute to the symptom profile of ASD.

Clinical trial studies show perhaps the strongest potential for near-term translation into clinical practice. While promising, as it stands these studies are characterized by small and heterogeneous samples, inadequate controls, measurement limitations, and small effects sizes. Considering current technologies, which allow stimulation of a given cortical region or cortical and subcortical stimulation of a broader region, TMS is very unlikely to eliminate all negative (or unwanted) aspects of ASD for a given individual. Thus, it should not be described as a "cure" nor should it be assumed that clinical trials in this space intend to discover such a cure. Instead, it appears that the best we might hope for at this stage is a targeted treatment that might alleviate problematic aspects of ASD (whether social, cognitive, behavioral or psychiatric in nature), that it does so safely (i.e., with few or no side-effects), and that desired aspects of an individual's cognitive profile (or indeed their personality) are not adversely affected. This is uniquely challenging for those attempting to design and implement such trials.

Given the large clinical and physiological heterogeneity within ASD and the overlapping symptoms with other psychiatric and neurodevelopmental disorders, researchers need to thoroughly characterize and when possible stratify participants based not only on their overall diagnosis, but also the targeted behavioral or cognitive domain being studied. Unlike medication, NIBS has very specific spatial resolution and mechanism of action. Thus, it is prudent to focus on a single cognitive or behavioral domain that negatively impacts the quality of life of individuals with ASD, but also a domain whose network pathology in ASD is relatively well-known and has been shown to be responsive to TMS in other indications. Some suggested cognitive/behavioral targets include: depression, executive functioning, irritability, social cognition, and language processing. As reviewed above, studies are ongoing in these domains with promising preliminary results.

Other Forms of Noninvasive Brain Stimulation

Here we have reviewed the use of TMS in ASD, and it is certainly the most widely used NIBS technique employed in autism in recent years. There are other NIBS techniques that might be considered, particularly with respect to potential clinical interventions for ASD. For instance, there is a suite of transcranial electrical stimulation (TES) techniques that involve weak electrical stimulation of the brain. Techniques such as transcranial direct current stimulation (tDCS) and transcranial alternating current stimulation have a number of benefits over TMS; for instance, they are inexpensive, do not carry a seizure risk, and could potentially be used at home. There are some encouraging early findings of TES in ASD (Amatachaya et al., 2015; D'Urso et al., 2015; Schneider and Hopp, 2011; van Steenburgh et al., 2017). TES does, however, exert a weaker influence on the brain when compared to TMS and when used in other conditions (e.g., MDD) has generally shown less efficacy than TMS (Berlim et al., 2013).

Clinical Translation and Safety

Despite the enthusiasm surrounding the use of TMS in ASD, researchers and clinicians must proceed with caution. First and foremost, given the limitations of the studies to date, we do not have an established evidence base that would justify the clinical use of TMS to treat ASD. Furthermore, there are known synergistic effects of TMS with concomitant medications that may affect both the safety and efficacy of such protocols (e.g., atypical antipsychotic medications that are often prescribed for irritability and lower seizure threshold). Finally, the effects of TMS protocols on the developing brain are not well examined, and while TMS has an excellent safety profile, most of the safety data come from adult populations. Early intervention in ASD is associated with the best outcomes and may reflect the brain at its most "plastic," but it will be exceptionally difficult to establish the required evidence base for TMS in ASD in preadolescent samples. For clinicians who use NIBS, at this stage we must conclude that there is insufficient evidence for treating ASD with TMS outside of an institutionally regulated clinical trial, even in an "off-label" capacity.

Another important consideration is that ASD is frequently associated with various neurocognitive strengths (e.g., enhanced attentional capacity, better processing of local detail), and in using TMS in an attempt to improve problematic symptoms, we must ensure that strengths are preserved. This is yet another distinct challenge for the TMS research community. At the very least, detailed and frequent neuropsychological assessment should be a part of every TMS trial in ASD, and research participants should be closely monitored. A significant reduction in neuropsychological capacity may result in a participant being withdrawn from further TMS treatments and should trigger a review of the study protocol.

Recommendations

Although this is an emerging literature, we can make several recommendations that might guide future research efforts.

First, in clinical trials of TMS, it is critical to match targeted brain regions to clinical outcome. There is a clear association between the brain region targeted and the reported effects on clinical symptoms or features. This can be guided by an extensive neuroscience literature concerning the brain basis of ASD, with functional neuroimaging and electrophysiological techniques particularly informative. Brain targets for therapeutic intervention in ASD must align with the hypothesized outcomes and the outcome measures.

Another important consideration in study design is heterogeneity. Indeed, ASD is extremely well known for its heterogeneity in clinical presentation, and to date, there are no consistent biomarkers or brain indicators that consistent apply to those diagnosed. It perhaps goes without saying, but it is extremely unlikely that there will be a single brain stimulation approach that will be successful for a majority of individuals with ASD. This is the case with rTMS treatment for depression, where remission rates hover around 40%–50%. Future trials should consider recruitment approaches that serve to target justifiable patient characteristics (e.g., age, gender, clinical profile, cognitive profile), or recruit sufficiently large patient populations to enable cluster analysis or similar approaches.

Consider the above recommendations, an RDoC might be the best approach here. As outlined by the National Institutes of Health (NIH), this could conceivably involve clinical trials where admission is based not solely on a clinical diagnosis of DSM-5 ASD, but rather observable, specific deficits in certain areas (e.g., poor facial affection recognition). Based on group or individual neuroimaging (e.g., fMRI) or electrophysiological (e.g., EEG) data, a specific brain region or regions would be targeted in an attempt to improve function in the specific area of impairment, and as assessment of this impairment would serve as the primary outcome of the trial. Broader clinical symptoms would also be measured (and expected to improve), but these would serve as secondary outcomes.

A further recommendation is the investigation of combinations of TMS and psychological or behavioral interventions (e.g., cognitive behavioral therapy, social skills training). There is ample evidence for the benefits of "on-line" stimulation (e.g., enhanced effects of NIBS when engaging the stimulated region via a cognitive task) (Silvanto and Cattaneo, 2017) and from a clinical perspective that could translate to further benefits from TMS when also undergoing a psychological or behavioral intervention. While this presents additional challenges for researchers (e.g., a large number of treatment arms to ensure adequate controls and, therefore, more participants), there are some encouraging findings from combined TMS/behavioral therapies other fields (e.g., stroke) (Tsagaris et al., 2016).

Further Research in Progress

Building upon the current literature, there are currently numerous investigative and therapeutic trials being conducted with NIBS in ASD.

In Australia, Dr. Enticott and his team continue to investigate the use of rTMS to yield therapeutic benefits to social relating in ASD. They have recently completed a follow-up study of deep rTMS to DMPFC, which builds upon previous clinical trials using a range of clinical/neurocognitive measures, and positron–emission tomography to image concurrent brain changes. This group has just begun a new trial comparing TBS to two regions: DMPFC and rTPJ. The latter region seems a particularly promising target for social impairments in ASD (Kirkovski et al., 2016a) and has recently been investigated using both TBS (Ni et al., 2017) and tDCS (Donaldson et al., 2017; Esse Wilson et al., 2017).

Meanwhile, in the United States, Dr. Oberman and her team at the NIH in Bethesda, MD and with collaborators at Children's Hospital of Philadelphia, Kennedy Krieger Institute and John's Hopkins University in Baltimore, MD are beginning to develop TMS protocols to study and improve language processing, social responsivity, and sensory processing abnormalities. Autism research groups in the United States are beginning to embrace TMS as a promising research and therapeutic tool and studies are underway. As of now, studies are focused on late childhood and adolescent age-groups.

Finally, there have been significant efforts to build a coordinated, international approach to the use of rTMS in ASD. As has been demonstrated throughout this chapter, there is a growing interest in NIBS-based investigations and interventions for ASD. Of the NIBS modalities, TMS has been most widely investigated thus far. Despite its growing use as an investigative and clinical tool, in ASD and also in other conditions with neurobiological underpinnings, our understanding of TMS in ASD is limited. While a promising intervention option, our understanding of the therapeutic potential of TMS in ASD is in its infancy, and off-label clinical application is premature. The numerous studies currently published in the literature are plagued by small sample sizes and often show considerable variability in the findings.

Consequently, in 2014, an international consensus group for TMS in ASD was established, driven primarily by the Clearly Presentation Foundation (clearlypresent.org) and involving researchers from around the world. The group meets annually, and is currently working toward a protocol that can be implemented in a multisite capacity.

Call for Caution

Finally, it is important to here echo the call for caution made elsewhere (Davis, 2014) when it comes to exploring the use of NIBS in neurodevelopmental disorder, particularly where they involve pediatric samples (<18 years). While TMS generally has an excellent safety profile, this is primarily based on adult research, and there has been relatively little work in pediatric samples (Rajapakse and Kirton, 2013). Thus, the effect of TMS on the developing brain is not entirely clear. Furthermore, where clinical effects might be induced, it is important to recognize that an individual with ASD may be ill-equipped to deal with the resultant life changes. This is eloquently discussed by John Elder Robison in his "Switched On" memoir, where the enhanced emotional capacity that followed TMS led to unanticipated negative consequences in his personal relationships. This underscores the importance of clinical psychology as an adjunct to rTMS intervention, and future work in this space might be best served by a combination of brain stimulation and cognitive behavioral intervention.

In summary, TMS has been useful in attempting to better understand the neurobiology of ASD, and there is very preliminary evidence to suggest that it could have therapeutic utility. As it stands, however, rTMS should not be offered as a clinical treatment for core symptoms or associated features, and to do so in an "off-label" capacity is both unwarranted and reckless. If rTMS is to be established as a safe and efficacious treatment for ASD, it will require a large, coordinated, and international research approach, similar to that employed to ultimately gain FDA approval for rTMS treatment of MDD (O'Reardon et al., 2007).

Acknowledgment

The authors would like to thank Ms. Emily Grundy for her assistance with manuscript preparation.

References

Amatachaya, A., Jensen, M.P., Patjanasoontorn, N., Auvichayapat, N., Suphakunpinyo, C., Janjarasjitt, S., et al., 2015. The short-term effects of transcranial direct current stimulation on electroencephalography in children with autism: a randomized crossover controlled trial. Behav. Neurol. 928631.

Ameis, S.H., Daskalakis, Z.J., Blumberger, D.M., Desarkar, P., Drmic, I., Mabbott, D.J., et al., 2017. Repetitive transcranial magnetic stimulation for the treatment of executive function deficits in autism spectrum disorder: clinical trial approach. J. Child Adolesc. Psychopharmacol. 27, 413–421.

American Psychiatric Association, 2013. Diagnostic and Statistical Manual of Mental Disorders. Washington, DC.

An, J.Y., Claudianos, C., 2016. Genetic heterogeneity in autism: from single gene to a pathway perspective. Neurosci. Biobehav. Rev. 68, 442–453.

Avirame, K., Stehberg, J., Todder, D., 2017. Enhanced cognition and emotional recognition, and reduced obsessive compulsive symptoms in two adults with high-functioning autism as a result of deep transcranial magnetic stimulation (dTMS): a case report. Neurocase 23, 1–6.

Barker, A.T., Jalinous, R., Freeston, I.L., 1985. Non-invasive magnetic stimulation of human motor cortex. Lancet 325, 1106–1107.

Baron-Cohen, S., Leslie, A.M., Frith, U., 1985. Does the autistic child have a "theory of mind"? Cognition 21, 37–46.

Barr, M.S., Farzan, F., Rajji, T.K., Voineskos, A.N., Blumberger, D.M., Arenovich, T., et al., 2013. Can repetitive magnetic stimulation improve cognition in schizophrenia? Pilot data from a randomized controlled trial. Biol. Psychiatry 73, 510–517.

Baruth, J.M., Casanova, M.F., El-Baz, A., Horrell, T., Mathai, G., Sears, L., et al., 2010. Low-frequency repetitive transcranial magnetic stimulation modulates evoked-gamma frequency oscillations in autism spectrum disorder. J. Neurotherapy 14, 179–194.

Berlim, M.T., Van den Eynde, F., Daskalakis, Z.J., 2013. Clinical utility of transcranial direct current stimulation (tDCS) for treating major depression: a systematic review and meta-analysis of randomized, double-blind and sham-controlled trials. J. Psychiatr. Res. 47, 1–7.

Casanova, M.F., Baruth, J.M., El-Baz, A., Tasman, A., Sears, L., Sokhadze, E., 2012. Repetitive transcranial magnetic stimulation (rTMS) modulates event-related potential (ERP) indices of attention in autism. Transl. Neurosci. 3, 170–180.

Casanova, M.F., Hensley, M.K., Sokhadze, E.M., El-Baz, A.S., Wang, Y., Li, X., et al., 2014. Effects of weekly low-frequency rTMS on autonomic measures in children with autism spectrum disorder. Front. Hum. Neurosci. 8, 851.

Casanova, M.F., Sokhadze, E., Opris, I., Wang, Y., Li, X., 2015. Autism spectrum disorders: linking neuropathological findings to treatment with transcranial magnetic stimulation. Acta Paediatr. 104, 346–355.

Chung, S.W., Hill, A.T., Rogasch, N.C., Hoy, K.E., Fitzgerald, P.B., 2016. Use of theta-burst stimulation in changing excitability of motor cortex: a systematic review and meta-analysis. Neurosci. Biobehav. Rev. 63, 43–64.

Coghlan, S., Horder, J., Inkster, B., Mendez, M.A., Murphy, D.G., Nutt, D.J., 2012. GABA system dysfunction in autism and related disorders: from synapse to symptoms. Neurosci. Biobehav. Rev. 36, 2044–2055.

Courchesne, E., Campbell, K., Solso, S., 2011. Brain growth across the life span in autism: age-specific changes in anatomical pathology. Brain Res. 1380, 138–145.

Davis, N.J., 2014. Transcranial stimulation of the developing brain: a plea for extreme caution. Front. Hum. Neurosci. 8, 600.

Demetriou, E.A., Lampit, A., Quintana, D.S., Naismith, S.L., Song, Y.J.C., Pye, J.E., et al., 2017. Autism spectrum disorders: a meta-analysis of executive function. Mol. Psychiatry 23, 1198–1204.

Di Pellegrino, G., Fadiga, L., Fogassi, L., Gallese, V., Rizzolatti, G., 1992. Understanding motor events: a neurophysiological study. Exp. Brain Res. 91, 176–180.

Donaldson, P.H., Gurvich, C., Fielding, J., Enticott, P.G., 2015a. Exploring associations between gaze patterns and putative human mirror neuron system activity. Front. Hum. Neurosci. 9, 396.

Donaldson, P.H., Rinehart, N.J., Enticott, P.G., 2015b. Noninvasive stimulation of the temporoparietal junction: a systematic review. Neurosci. Biobehav. Rev. 55, 547–572.

Donaldson, P.H., Kirkovski, M., Rinehart, N.J., Enticott, P.G., 2017. Autism-relevant traits interact with temporoparietal junction stimulation effects on social cognition: a high-definition transcranial direct current stimulation and electroencephalography study. Eur. J. Neurosci. 47, 669–681.

D'urso, G., Bruzzese, D., Ferrucci, R., Priori, A., Pascotto, A., Galderisi, S., et al., 2015. Transcranial direct current stimulation for hyperactivity and noncompliance in autistic disorder. World J. Biol. Psychiatry 16, 361–366.

Enticott, P.G., Oberman, L.M., 2013. Synaptic plasticity and non-invasive brain stimulation in autism spectrum disorders. Dev. Med. Child Neurol. 55, 13–14.

Enticott, P.G., Bradshaw, J.L., Iansek, R., Tonge, B.J., Rinehart, N.J., 2009. Electrophysiological signs of supplementary-motor-area deficits in high-functioning autism but not Asperger syndrome: an examination of internally cued movement-related potentials. Dev. Med. Child Neurol. 51, 787–791.

Enticott, P.G., Rinehart, N.J., Tonge, B.J., Bradshaw, J.L., Fitzgerald, P.B., 2010. A preliminary transcranial magnetic stimulation study of cortical inhibition and excitability in high-functioning autism and Asperger disorder. Dev. Med. Child Neurol. 52, e179–83.

Enticott, P.G., Kennedy, H.A., Zangen, A., Fitzgerald, P.B., 2011. Deep repetitive transcranial magnetic stimulation associated with improved social functioning in a young woman with an autism spectrum disorder. J. ECT 27, 41–43.

Enticott, P.G., Rinehart, N.J., Tonge, B.J., Bradshaw, J.L., Fitzgerald, P.B., 2012. Repetitive transcranial magnetic stimulation (rTMS) improves movement-related cortical potentials in autism spectrum disorders. Brain Stimulation 5, 30–37.

Enticott, P.G., Kennedy, H.A., Rinehart, N.J., Tonge, B.J., Bradshaw, J.L., Fitzgerald, P.B., 2013. GABAergic activity in autism spectrum disorders: an investigation of cortical inhibition via transcranial magnetic stimulation. Neuropharmacology 68, 202–209.

Enticott, P.G., Fitzgibbon, B.M., Kennedy, H.A., Arnold, S.L., Elliot, D., Peachey, A., et al., 2014. A double-blind, randomized trial of deep repetitive transcranial magnetic stimulation (rTMS) for autism spectrum disorder. Brain Stimulation 7, 206–211.

Esse Wilson, J., Quinn, D.K., Wilson, J.K., Garcia, C.M., Tesche, C.D., 2017. Transcranial direct current stimulation to the right temporoparietal junction for social functioning in autism spectrum disorder: case report. J. ECT 34, e10–e13.

Fatemi, S.H., Aldinger, K.A., Ashwood, P., Bauman, M.L., Blaha, C.D., Blatt, G.J., et al., 2012. Consensus paper: pathological role of the cerebellum in autism. Cerebellum 11, 777–807.

Fecteau, S., Agosta, S., Oberman, L., Pascual-Leone, A., 2011. Brain stimulation over Broca's area differentially modulates naming skills in neurotypical adults and individuals with Asperger's syndrome. Eur. J. Neurosci. 34, 158–164.

Frazier, T.W., Strauss, M., Klingemier, E.W., Zetzer, E.E., Hardan, A.Y., Eng, C., et al., 2017. A meta-analysis of gaze differences to social and nonsocial information between individuals with and without autism. J. Am. Acad. Child Adolesc. Psychiatry 56, 546–555.

Garber, K., 2007. Autism's cause may reside in abnormalities at the synapse. Science 317, 190–191.

Greenhouse, I., Noah, S., Maddock, R.J., Ivry, R.B., 2016. Individual differences in GABA content are reliable but are not uniform across the human cortex. NeuroImage 139, 1–7.

Guastella, A.J., Einfeld, S.L., Gray, K.M., Rinehart, N.J., Tonge, B.J., Lambert, T.J., et al., 2010. Intranasal oxytocin improves emotion recognition for youth with autism spectrum disorders. Biol. Psychiatry 67, 692–694.

Hamada, M., Murase, N., Hasan, A., Balaratnam, M., Rothwell, J.C., 2013. The role of interneuron networks in driving human motor cortical plasticity. Cereb. Cortex 23, 1593–1605.

Happe, F., 2005. The weak central coherence account of autism. In: Volkmar, F.R., Paul, R., Klin, A., Cohen, D. (Eds.), Handbook of Autism and Pervasive Developmental Disorders, third ed Wiley, Hoboken, NJ.

Harms, M.B., Martin, A., Wallace, G.L., 2010. Facial emotion recognition in autism spectrum disorders: a review of behavioral and neuroimaging studies. Neuropsychol. Rev. 20, 290–322.

Hill, A.T., Rogasch, N.C., Fitzgerald, P.B., Hoy, K.E., 2016. TMS-EEG: a window into the neurophysiological effects of transcranial electrical stimulation in non-motor brain regions. Neurosci. Biobehav. Rev. 64, 175–184.

Huang, Y.Z., Edwards, M.J., Rounis, E., Bhatia, K.P., Rothwell, J.C., 2005. Theta burst stimulation of the human motor cortex. Neuron 45, 201–206.

Iacoboni, M., Dapretto, M., 2006. The mirror neuron system and the consequences of its dysfunction. Nat. Rev. Neurosci. 7, 942–951.

Jarczok, T.A., Fritsch, M., Kröger, A., Schneider, A.L., Althen, H., Siniatchkin, M., et al., 2016. Maturation of interhemispheric signal propagation in autism spectrum disorder and typically developing controls: a TMS-EEG study. J. Neural Transm. 123, 925–935.

Jung, N.H., Janzarik, W.G., Delvendahl, I., Münchau, A., Biscaldi, M., Mainberger, F., et al., 2013. Impaired induction of long-term potentiation-like plasticity in patients with high-functioning autism and Asperger syndrome. Dev. Med. Child Neurol. 55, 83–89.

Keysers, C., Gazzola, V., 2009. Expanding the mirror: vicarious activity for actions, emotions, and sensations. Curr. Opin. Neurobiol. 19, 666–671.

Kirkovski, M., Enticott, P.G., Fitzgerald, P.B., 2013. A review of the role of female gender in autism spectrum disorders. J. Autism Dev. Disord. 43, 2584–2603.

Kirkovski, M., Enticott, P.G., Hughes, M.E., Rossell, S.L., Fitzgerald, P.B., 2016a. Atypical neural activity in males but not females with autism spectrum disorder. J. Autism Dev. Disord. 46, 954–963.

Kirkovski, M., Rogasch, N.C., Saeki, T., Fitzgibbon, B.M., Enticott, P.G., Fitzgerald, P.B., 2016b. Single pulse transcranial magnetic stimulation-electroencephalogram reveals no electrophysiological abnormality in adults with high-functioning autism spectrum disorder. J. Child Adolesc. Psychopharmacol. 26, 606–616.

Lemonnier, E., Degrez, C., Phelep, M., Tyzio, R., Josse, F., Grandgeorge, M., et al., 2012. A randomised controlled trial of bumetanide in the treatment of autism in children. Transl. Psychiatry 2, e202.

Lemonnier, E., Villeneuve, N., Sonie, S., Serret, S., Rosier, A., Roue, M., et al., 2017. Effects of bumetanide on neurobehavioral function in children and adolescents with autism spectrum disorders. Transl. Psychiatry 7, e1124.

Mehta, U.M., Waghmare, A.V., Thirthalli, J., Venkatasubramanian, G., Gangadhar, B.N., 2015. Is the human mirror neuron system plastic? Evidence from a transcranial magnetic stimulation study. Asian J. Psychiatry 17, 71–77.

Minio-Paluello, I., Baron-Cohen, S., Avenanti, A., Walsh, V., Aglioti, S.M., 2009. Absence of embodied empathy during pain observation in Asperger syndrome. Biol. Psychiatry 65, 55–62.

Nelson, S.B., Valakh, V., 2015. Excitatory/Inhibitory balance and circuit homeostasis in autism spectrum disorders. Neuron 87, 684–698.

Ni, H., Hung, J., Wu, C., Wu, Y., Chang, C., Chen, R., et al., 2017. The impact of single session intermittent theta-burst stimulation over the dorsolateral prefrontal cortex and posterior superior temporal sulcus on adults with autism spectrum disorder. Front. Neurosci. 11 (255).

Niederhofer, H., 2012. Effectiveness of the repetitive transcranical magnetic stimulation (RTMS) of 1 Hz for autism. Clin. Neuropsychiatry 9, 107.

Oberman, L., Ifert-Miller, F., Najib, U., Bashir, S., Woollacott, I., Gonzalez-Heydrich, J., et al., 2010a. Transcranial magnetic stimulation provides means to assess cortical plasticity and excitability in humans with fragile X syndrome and autism spectrum disorder. Front. Synaptic Neurosci 2, 26.

Oberman, L., Eldaief, M., Fecteau, S., Ifert-Miller, F., Tormos, J.M., Pascual-Leone, A., 2012. Abnormal modulation of corticospinal excitability in adults with Asperger's syndrome. Eur. J. Neurosci. 36, 2782–2788.

Oberman, L.M., Hubbard, E.M., Mccleery, J.P., Altschuler, E.L., Ramachandran, V.S., Pineda, J.A., 2005. EEG evidence for mirror neuron dysfunction in autism spectrum disorders. Cogn. Brain Res. 24, 190–198.

Oberman, L.M., Horvath, J.C., Pascual-Leone, A., 2010b. TMS: using the theta-burst protocol to explore mechanism of plasticity in individuals with fragile X syndrome and autism. J. Visualized Exp. 2272.
Oberman, L.M., Pascual-Leone, A., Rotenberg, A., 2014. Modulation of corticospinal excitability by transcranial magnetic stimulation in children and adolescents with autism spectrum disorder. Front. Hum. Neurosci. 8.
Oberman, L.M., Ifert-Miller, F., Najib, U., Bashir, S., Heydrich, J.G., Picker, J., et al., 2016. Abnormal mechanisms of plasticity and metaplasticity in autism spectrum disorders and fragile X syndrome. J. Child Adolesc. Psychopharmacol. 26, 617–624.
O'reardon, J.P., Solvason, H.B., Janicak, P.G., Sampson, S., Isenberg, K.E., Nahas, Z., et al., 2007. Efficacy and safety of transcranial magnetic stimulation in the acute treatment of major depression: a multisite randomized controlled trial. Biol. Psychiatry 62, 1208–1216.
O'reilly, C., Lewis, J.D., Elsabbagh, M., 2017. Is functional brain connectivity atypical in autism? A systematic review of EEG and MEG studies. PLoS ONE 12, e0175870.
Ozonoff, S., South, M., Provencal, S., 2005. Executive functions. In: Volkmar, F.R., Paul, R., Klin, A., Cohen, D. (Eds.), Handbook of Autism and Pervasive Developmental Disorders, third ed. Wiley, Hoboken, NJ.
Palmer, C.J., Seth, A.K., Hohwy, J., 2015. The felt presence of other minds: predictive processing, counterfactual predictions, and mentalising in autism. Conscious Cogn. 36, 376–389.
Panerai, S., Tasca, D., Lanuzza, B., Trubia, G., Ferri, R., Musso, S., et al., 2014. Effects of repetitive transcranial magnetic stimulation in performing eye-hand integration tasks: four preliminary studies with children showing low-functioning autism. Autism 18, 638–650.
Pedapati, E.V., Gilbert, D.L., Erickson, C.A., Horn, P.S., Shaffer, R.C., Wink, L.K., et al., 2016. Abnormal cortical plasticity in youth with autism spectrum disorder: a transcranial magnetic stimulation case–control pilot study. J. Child Adolesc. Psychopharmacol. 26, 625–631.
Rajapakse, T., Kirton, A., 2013. Non-invasive brain stimulation in children: applications and future directions. Transl. Neurosci. 4, 217–233.
Rinehart, N.J., Tonge, B.J., Bradshaw, J.L., Iansek, R., Enticott, P.G., Johnson, K.A., 2006. Movement-related potentials in high-functioning autism and Asperger's disorder. Dev. Med. Child Neurol. 48, 272–277.
Ritvo, R.A., Ritvo, E.R., Guthrie, D., Yuwiler, A., Ritvo, M.J., Weisbender, L., 2008. A scale to assist the diagnosis of autism and Asperger's disorder in adults (RAADS): a pilot study. J. Autism Dev. Disord. 38, 213–223.
Rogasch, N.C., Sullivan, C., Thomson, R.H., Rose, N.S., Bailey, N.W., Fitzgerald, P.B., et al., 2017. Analysing concurrent transcranial magnetic stimulation and electroencephalographic data: a review and introduction to the open-source TESA software. NeuroImage 147, 934–951.
Schneider, H.D., Hopp, J.P., 2011. The use of the bilingual aphasia test for assessment and transcranial direct current stimulation to modulate language acquisition in minimally verbal children with autism. Clin. Ling. Phon. 25, 640–654.
Sealey, L.A., Hughes, B.W., Sriskanda, A.N., Guest, J.R., Gibson, A.D., Johnson-Williams, L., et al., 2016. Environmental factors in the development of autism spectrum disorders. Environ. Int. 88, 288–298.
Silvanto, J., Cattaneo, Z., 2017. Common framework for "virtual lesion" and state-dependent TMS: The facilitatory/suppressive range model of online TMS effects on behavior. Brain Cogn. 119, 32–38.
Sokhadze, E., Baruth, J., Tasman, A., Mansoor, M., Ramaswamy, R., Sears, L., et al., 2010. Low-frequency repetitive transcranial magnetic stimulation (rTMS) affects event-

related potential measures of novelty processing in autism. Appl. Psychophysiol. Biofeedback 35, 147–161.

Sokhadze, E.M., El-Baz, A., Baruth, J., Mathai, G., Sears, L., Casanova, M.F., 2009. Effects of low frequency repetitive transcranial magnetic stimulation (rTMS) on gamma frequency oscillations and event-related potentials during processing of illusory figures in Autism. J. Autism Dev. Disord. 39, 619–634.

Sokhadze, E.M., Baruth, J.M., Sears, L., Sokhadze, G.E., El-Baz, A.S., Casanova, M.F., 2012. Prefrontal neuromodulation using rTMS improves error monitoring and correction function in autism. Appl. Psychophysiol. Biofeedback 37, 91–102.

Sokhadze, E.M., El-Baz, A.S., Sears, L.L., Opris, I., Casanova, M.F., 2014a. rTMS neuromodulation improves electrocortical functional measures of information processing and behavioral responses in autism. Front. Syst. Neurosci. 8, 134.

Sokhadze, E.M., El-Baz, A.S., Tasman, A., Sears, L.L., Wang, Y., Lamina, E.V., et al., 2014b. Neuromodulation integrating rTMS and neurofeedback for the treatment of autism spectrum disorder: an exploratory study. Appl. Psychophysiol. Biofeedback 39, 237–257.

Stigler, K.A., Mcdonald, B.C., Anand, A., Saykin, A.J., Mcdougle, C.J., 2011. Structural and functional magnetic resonance imaging of autism spectrum disorders. Brain Res. 1380, 146–161.

Théoret, H., Halligan, E., Kobayashi, M., Fregni, F., Tager-Flusberg, H., Pascual-Leone, A., 2005. Impaired motor facilitation during action observation in individuals with autism spectrum disorder. Curr. Biol. 15, R84–5.

Tsagaris, K.Z., Labar, D.R., Edwards, D.J., 2016. A framework for combining rTMS with behavioral therapy. Front. Syst. Neurosci. 10, 82.

Van Steenburgh, J.J., Varvaris, M., Schretlen, D.J., Vannorsdall, T.D., Gordon, B., 2017. Balanced bifrontal transcranial direct current stimulation enhances working memory in adults with high-functioning autism: a sham-controlled crossover study. Mol. Autism 8, 40.

Volkmar, F.R., Mcpartland, J.C., 2014. From Kanner to DSM-5: autism as an evolving diagnostic concept. Annu. Rev. Clin. Psychol. 10, 193–212.

Wang, Y., Zhang, Y.B., Liu, L.L., Cui, J.F., Wang, J., Shum, D.H., et al., 2017. A meta-analysis of working memory impairments in autism spectrum disorders. Neuropsychol. Rev. 27, 46–61.

Williams, J.H.G., Whiten, A., Suddendorf, T., Perrett, D.I., 2001. Imitation, mirror neurons and autism. Neurosci. Biobehav. Rev. 25, 287–295.

Zablotsky, B., Black, L.I., Maenner, M.J., Schieve, L.A., Blumberg, S.J., 2015. Estimated prevalence of autism and other developmental disabilities following questionnaire changes in the 2014 National Health Interview Survey. Natl. Health Stat. Rep. 87, 1–20.

Ziemann, U., 2003. Pharmacology of TMS. Suppl. Clin. Neurophysiol. 56, 226–231.

CHAPTER

6

Transcranial Magnetic Stimulation in Attention Deficit Hyperactivity Disorder

Donald L. Gilbert

Pediatrics and Neurology, Cincinnati Children's Hospital Medical Center, Cincinnati, OH, United States

OUTLINE

Neurotechnological Pediatric Neuropsych

DOI: https://doi.org/10.1016/B978-0-12-812777-3.00006-4

INTRODUCTION—CHALLENGES OF ADHD AND RATIONALE FOR RESEARCH

"Attention deficit/hyperactivity disorder" (ADHD) refers to a *categorical diagnosis* that clinicians and researchers use to capture the severely impaired end of a *spectrum* of capacities that typically develop in childhood. As contained in the name of the diagnosis, these capacities encompass attention, hyperactivity, and impulse control (American Psychiatric Association, 2013). In an otherwise healthy, typically developing child, the capacity to devote and sustain attention is limited during the pre-school years but increases through the school age years, allowing for age-appropriate learning and completion of expected work. Similarly, the ability to physically remain in one location and to delay or inhibit actions in response to environmental cues predicting immediate reward also matures, allowing for age-appropriate, culturally and socially acceptable behaviors. At any given age in childhood, and to a lesser degree adulthood, there is a spectrum of these capacities. Inattentiveness, motoric hyperactivity, and impulsiveness can be considered "symptoms" when, relative to his peers, a child is struggling to learn or conform to the social expectations of his environment. Having ADHD in childhood portends high risk for poor adult social and occupational outcomes (Molina et al., 2009; Hinshaw et al., 2015).

The diagnosis of ADHD is clinical and subjective, which hampers research that could possibly provide directions for treatments that mitigate the risk of poor long-term outcomes for ADHD. For the purposes of diagnostic classification of ADHD, inattentive and hyperactive/impulsive symptoms are each described by nine observed items. For children, clinicians performing ADHD evaluations typically use rating scales that list these 18 items and ask if they are observed by adults and how often (Bard et al., 2013; DuPaul et al., 1998; Adams et al., 1984). Due to the lack of insight about functional interference from ADHD in childhood (Cloes et al., 2017), adult observations are used in the diagnostic process. A positive ADHD diagnosis is based on observation and reporting of at least six inattentive and/or six hyperactive/impulsive symptoms occurring in two or more settings (usually school and home). Further, observers must judge that the symptoms interfere to some degree with social or academic function. The result may be a diagnosis of *combined type*, *predominantly inattentive*, or, less commonly, *predominantly hyperactive/impulsive* ADHD. Of course, since these ADHD items are nonspecific, multiple other factors such as anxiety (Bauer et al., 2016) or learning disabilities (Langberg et al., 2010) may account for their presence or their apparent interference. While the diagnosis can be made in the presence of autism spectrum disorder, it cannot be made if another mental disorder better accounts for these symptoms. While in the DSM-IV, ADHD was categorized with disruptive behavior disorders, in the DSM-5, ADHD is classified in the "Neurodevelopmental Disorders" section (American Psychiatric Association, 2013), reflecting evolving conceptions of its relationship to brain development.

ADHD is one of the most commonly diagnosed neurodevelopmental conditions in childhood (Polanczyk et al., 2014; Rowland et al., 2015). For clinical or research purposes, if ADHD is considered as a cluster of symptoms, irrespective of causes, it encompasses a large, diverse number of diagnoses that result in cognitive or emotional difficulties, some of which also include physical impairments. In some genetic conditions where ADHD symptoms are prevalent (Lo-Castro et al., 2011), such as fragile X syndrome (Grefer et al., 2016; Sullivan et al., 2006), knowledge of the genetic etiology has propelled forward research into mechanisms of developmental disorder symptoms and into clinical trials of more rational treatments, through a variety of experimental approaches (Thapar et al., 2016). In other non-genetic conditions such as prematurity (Allotey et al., 2018), cerebral palsy (Suren et al., 2012), or traumatic brain injury (Yang et al., 2016), brain changes may be acquired which generate ADHD symptoms. This is particularly complex in postnatal brain injuries, where ADHD may be a risk factor (Biederman et al., 2015) as well as a consequence (Max et al., 2005).

The heterogeneity of clinical presentation of ADHD symptoms within and across genetic and acquired conditions poses substantial challenges for rational therapeutics. For the common, trial-and-error treatment approach, it is standard to consider ADHD to be a nonspecific condition resulting from the combined influence of multiple genetic susceptibilities with small effect sizes combined with ill-defined environmental factors. Presence and severity of symptoms are evaluated subjectively, without any biomarker. Treatment discussions do not involve neurobiology. Rather, they involve providing parents options, such as stimulant versus non-stimulant pharmacological options versus behavioral treatments, which are empirically validated in studies of heterogeneous groups of patients. Possible efficacy, side effects, and costs or health insurance coverage are reviewed at the level of the group diagnosis, not necessarily based on individual or personal data. Unfortunately, the best available pharmacological and behavioral treatments (Jensen, 1999; Arnold et al., 1997) compared rigorously but implemented in children based on the subjective categorical ADHD diagnosis, had disappointing long-term outcomes in terms of occupation, substance abuse, and other social and medical costs (Molina et al., 2009; Hinshaw et al., 2015). Given the high public health impact and costs from both prevalence and disability (Sayal et al., 2017), research into neural mechanisms remains vital.

Does neurophysiology, which is generally nonspecific, offer a path forward in neuroscience research for ADHD? This is a complex question. What if ADHD, as we diagnose it clinically, biologically includes multiple independent groups of patients with distinct pathophysiologies? Because the categorical diagnosis of ADHD *per se* involves such heterogeneity and possibility for misclassification, and because there are no independent, gold-standard, objective diagnostic biomarkers for the group or for subgroups, case—control studies that start from the premise that experimental techniques may identify differences between "the ADHD" and "the typical" child may face key obstacles. For example, researchers with novel hypotheses will generally begin by testing these ideas in small cohorts. These cases or controls may be nonrepresentative or utilize exploratory statistical techniques without correction for multiple comparisons, leading to spurious results due to low power, luck, or wishful thinking. Regarding negative results, there will likely be insufficient statistical power to detect biologically important factors with large effects present only in subgroups or with small effects present more pervasively in children with ADHD. Regarding positive results, published positive findings may also not represent replicable, biological relationships (Button et al., 2013).

A further challenge involves dealing with medications. At least in the United States, where behavioral and learning problems are commonly addressed in part by prescribing medications, a high proportion of children with ADHD will be taking daily medication. Studies of any size wishing to avoid potential confounding effects of widely prescribed medications may enroll children not taking medication. However, the result will be a biased, nonrepresentative sample of mildly affected participants or non-responders who have discontinued medications. The practice of asking medications be withheld for the purpose of the Transcranial Magnetic Stimulation (TMS) study may mitigate this to some extent, particularly for stimulant medications, given their short half-life. However, other medications may be unsafe to discontinue abruptly or may have long-acting pharmacodynamic effects. To some extent, including correlational analysis with independent ratings of symptom severity outside the lab, or with quantitative assessments of behavioral performance in the laboratory, may add confidence to findings of group differences. Longitudinal studies and replication of results are also critical. It may also be helpful to assess adults with confirmed childhood diagnoses where symptoms have or have not remitted (Szekely et al., 2017).

Despite these concerns, the search for a better understanding of the various pathways underlying ADHD remains active. This search can emphasize pathways for attention, hyperactivity, motor control, impulse control, and reward salience, using neurophysiology to complement neuroimaging techniques, and incorporating genetics. Understanding how these pathways subserve normal and atypical function and how they mature in childhood could ultimately play a vital role in improving treatment outcomes. As, to date, much of this research has involved neuroimaging, this may be a source of useful hypotheses. This is particularly true for studies that have involved large, well-characterized samples (Shaw et al., 2006, 2012; Greven et al., 2015). Larger studies may capture more factors which can underlie ADHD-related symptoms in ways that may ultimately point toward new and more effective treatment approaches.

TMS can fit into this larger scientific objective of understanding the "substrate" of ADHD. TMS-evoked potentials are quantifiable and can provide information about both trait and state. The most common "readout" for TMS-evoked potentials is motor-evoked potentials in intrinsic hand muscles. These reflect cortical excitability and, as discussed below, can be combined with functional tasks to elucidate trait, state, and trait by state interactions which may differ in typically developing versus ADHD children. However, with TMS-EEG, evoked potentials could probe neurophysiological features of prefrontal cortical regions shown through functional Magnetic Resonance Imaging (fMRI) or volumetric imaging studies to differ. Many ideas which could be pursued along these lines remain hypothetical.

The purpose of this chapter is to review the use of TMS in ADHD to date. The objective of most researchers has been primarily for the purpose of understanding possible symptom mechanisms. Unfortunately, most of the studies involve small samples. There are currently nearly as many review papers as actual studies. Treatment using repetitive TMS (rTMS) will also be addressed here. There are even fewer studies of this type to review. Adult studies will be emphasized less, given the different scientific and ethical considerations as well as the scope of this book.

The structure of this chapter will be to provide first a brief discussion of overarching principles of ethics, safety, statistics, study design, and validity with regard to use of TMS in childhood and in particular children with ADHD. Following this, the sections will be organized by measurement, with each section briefly describing the technique, results, potential pitfalls, and suggested future research directions for TMS research in children with ADHD.

ETHICS, SAFETY, STUDY DESIGN, AND STATISTICS IN ADHD STUDIES

Ethics, Minimal Risk, and Safety

There are several issues pertinent to TMS in children which deserve specific discussion. These are addressed more completely elsewhere (Rossi et al., 2009; Hong et al., 2015). The use of TMS and the dedication of scarce research resources for understanding mechanisms of developmental disorders such as ADHD in children can be justified on the basis of the lifelong suffering and reduced quality of life related to these disorders (Hinshaw et al., 2015; Conelea et al., 2013). However, as a rule, ADHD does not substantially increase the risk of death or imminent morbidity. With regard to determining appropriate levels of risk for research, standard diagnosis and management in ADHD does not require any risky diagnostic testing or treatment. Thus, to be ethically acceptable, TMS research which has no possibility for direct benefit for ADHD must involve minimal risk, informed consent with parents, risk/benefit assessment with parents, and equitable subject selection (Gilbert et al., 2004). Laboratories should have clearly written consent and assent documents as well as personnel skilled in providing accurate descriptions of the machines and procedures (Gilbert et al., 2004; Gilbert, 2008). It can be helpful to compare the TMS procedure to brain MRI scans, which are widely used in otherwise healthy children with headaches, and whose magnets have comparable strength. It is worth noting that TMS research data is only interpretable if the child is comfortable. Surface electromyography (EMG) data from a restless, anxious, or upset child will typically be contaminated by muscle artifact, making evoked motor evoked potential (MEP) amplitudes uninterpretable. So, the procedure has to be comfortable for awake children, and the laboratory should be experienced in working with impulsive, physically hyperactive, and inattentive children. Sample consents and assent language from our ADHD pediatric studies are available elsewhere (Gilbert, 2016a,b).

TMS Biomarker Research: Statistical Considerations

Biomarkers

To date, TMS applications in ADHD have primarily looked at brain-based measures that might reflect pathophysiology. The term *biomarker* refers to an objectively measurable, biologically based indicator. TMS measures can be considered biomarkers, based on a number of overlapping, formal definitions. A representative definition, from the 1998 National Institutes of Health Biomarkers Definitions Working Group, was "a characteristic that is objectively measured and evaluated as an indicator of normal biological processes, pathogenic processes, or pharmacologic responses to a therapeutic intervention."

"Abnormal" TMS Results

Clinicians, researchers, and parents may optimistically expect TMS to identify binary responses which are either normal or abnormal. For quantitative measures, this would require a discreet threshold around which positive and negative diagnoses occur at high proportions, as determined using receiver-operating characteristic analysis (Sackett et al., 1991; Connell and Koepsell, 1985). This is not a realistic expectation for TMS measures for developmental diagnoses.

Overlap With Controls—Spectrum Diagnoses and Developmental Changes

Many developmental and behavioral diagnoses lie on a spectrum with typical development. Behavioral problems may sometimes represent a leftward time shift on a developmental spectrum—like an 11-year old with the maturity and executive function of a 9-year old. Age norms may not help much because of the wide variation within a chronological age. Then, for any proposed diagnostic threshold, there will be a large trade-off between sensitivity and specificity (highly sensitive threshold values will have low specificity and vice versa).

Techniques Used for TMS Biomarker Research in Developmental Disorders

For those not familiar with TMS, it really is conceptually similar to evoked potentials in wide clinical use. There is a stimulus whose properties are set by the operator, and a "readout" which is evoked, is quantifiable, and reveals basic information about the neural system between the stimulus and the readout. So, for example, light or visual patterns through the eyes produce physiological, measurable changes in visual cortex: visual-evoked potentials. Sounds produce auditory evoked potentials. Electrical stimulation of nerves through the skin evoke nerve action potentials or motor unit potentials. TMS of the motor cortex is similar: the magnetic pulse evokes depolarization which travels to muscle where surface EMG leads detect the signal. The stronger the magnetic current, the greater net depolarization and, downstream, the larger the MEP amplitude. The more myelinated the system is, the more rapidly the signal travels. Interruptions due to delayed or damaged myelination, strokes, or other lesions can affect the latency of the MEP centrally just as demyelinating neuropathies affect latencies peripherally. For the techniques reviewed below, MEP amplitudes and latencies are the quantities of interest. These can be modified in predictable ways by experimental techniques which have been demonstrated in many healthy adults.

The vast majority of TMS research in developmental disorders involves single and paired pulse TMS administered over motor cortex. A single pulse applied over motor cortex can, if sufficiently strong, result in depolarization below the coil. When administered to motor cortex over the hand area, this results, after a latency of approximately 20 ms, in an evoked potential in contralateral hand muscles. For single pulses, the readout is usually a surface EMG tracing, whose onset and latency reflect the conduction velocity of the pathway and whose amplitude reflects the instantaneous cortical excitability.

Single pulses are used to identify depolarization thresholds at rest (resting motor threshold—RMT) and during voluntary activation [active motor thresholds (AMT)]. Various other measures involving single pulses and paired pulses are indexed to these thresholds. For example, the cortical silent period (CSP) is often measured in active muscle with a pulse intensity administered at 1.5 × AMT.

Paired pulse studies involve seven parameters or "decisions" by the investigator. First there is a conditioning pulse, which is usually, although not always, administered at a subthreshold intensity such as 0.6–0.8 × RMT. The conditioning pulse precedes the test pulse, whose intensity is the second parameter, a suprathreshold intensity such as 1.1 or 1.2 × RMT, or "an intensity which produces MEPs averaging 1 mV in amplitude." The third and fourth parameters are the interstimulus intervals, for example, 2, 3, or 4 ms, and the intertrial intervals, for example, 6 or 10 seconds. So for example for short-interval cortical inhibition (SICI), the MEP amplitudes in response to single pulses are compared to the MEP amplitudes in response to paired pulses. This can be expressed as a ratio. So a ratio of 0.25 would indicate 75% average inhibition by the conditioning pulse, and a ratio of 0.68 would indicate 32% inhibition. The fifth parameter is the number of trials per condition. Two additional choices labs make are the type of coil (round versus figure 8) and the intrinsic hand muscle [abductor pollicis brevis (APB), first dorsal interosseous, or abductor digiti minimi (ADM)]. In comparing studies measuring the same measure, for example, SICI, it is important to note that laboratories have often chosen and published data using different parameters.

Study Designs for TMS Biomarker Research in Developmental Disorders

The selection of pulse intensities and interstimulus intervals affects the magnitude of SICI (Orth et al., 2003). Investigators may wish to base stimulation parameters on those used more commonly in the literature. Given the limitations of any individual technique, designing TMS studies for biomarkers of developmental disorders should ideally involve more than one technique. For example, if there are hypotheses about deficient cortical inhibition, SICI (Kujirai et al., 1993), long-interval cortical inhibition (Sanger et al., 2001), CSP (Priori et al., 1994), and ipsilateral silent period (ISP) (Wassermann et al., 1991) could all be investigated. Moreover, while there is value to study designs that are case–control to yield estimates of group differences with unaffected peers, studies should also involve careful clinical phenotyping to allow for correlations between continuous measures and severity of core symptoms or severity of associated impairments.

Special Challenges: Getting Small Children and Behaviorally Impulsive Children to "Sit Still"

In any pediatric study, two challenges in utilizing TMS are smaller heads and more active bodies. The "active" part can cause bias—consistently greater noise in the more active, ADHD group. There is also the issue of blinding. Ideally, the TMS team is blinded to diagnosis, but this is not always realistic, because it is obvious when children are highly symptomatic.

Keeping a child in a chair should be given thought in a particular lab using a particular chair. In chairs with arms, wedging pillows consistently around the child can help. We use a standard maternal nursing pillow which wraps around the child and helps keep him in the center of the chair. A visual target is also helpful for maintaining the child's head in one direction. A picture on the lab wall is monotonous, so we have found it useful to allow children to watch quiet videos, which do not involve observation of human actions, on a computer monitor. Coil position is also an issue and worth considering when using the literature to plan experiments. The most commonly used coil, the double 70 mm figure-8 coil, requires care in children. Due to the smaller heads children have, the wings of the coil have more potential angles and may not maintain the initial tangential placement. Immobilizing the coil with a coil holder or stand can also be problematic. In our experience, children's wiggly, flexible bodies always seem to migrate away from a fixed coil. Double-coil paired pulse TMS studies are difficult if the size of the head is so small that these coils touch. Smaller coils may be needed, but these will have different intensities, so more up front work may be needed to measure thresholds. Neuronavigation systems can help greatly with this issue, but these are expensive, and the 3D localizing headbands and glasses may also move on the child's head during the experiment. Newer robotic systems with 3D head and coil localization may reduce this problem, but these systems may be too expensive for many labs. The rationale for using them may be stronger for therapeutic studies where consistent coil localization within and between TMS sessions is likely more critical.

TMS IN ADHD: PRACTICAL CONSIDERATIONS AND REPRESENTATIVE RESULTS

In this section, we will describe results from published studies using TMS to understand the pathophysiology of ADHD, predominantly in childhood. The sections are organized by measurement. Each section describes briefly the technique, the findings in ADHD, the pitfalls of the technique with regard to the above-described challenges, and the possible implications or future directions for those findings and approaches.

Depolarization Thresholds: Resting and Active Motor Thresholds (RMT and AMT)

Technique

RMT and AMT are the amount of energy required to evoke a response in M1 at rest or with activation of the target muscle, expressed as a percentage of the output of the TMS device being used. Methods for evaluation are roughly consistent and standardized (Mills and Nithi, 1997). Thresholds depend in part on ion channel conduction (Ziemann et al., 1996). Because threshold is reported as a percentage of maximum stimulator output, the literature may only be a rough guide if the laboratory is using a different stimulator, coil, or target. The muscles evaluated with surface EMG are usually either the first dorsal interosseous (FDI), APB, or ADM. Due to the large area of motor cortex controlling the hand, it is usually very simple and quick to identify the "hot spot" for best evoked responses and mark it on the scalp with a wax pencil. For stimulation outside of motor cortex, neuronavigation may be helpful. It is important to be reasonably accurate with RMT measures, because many other measures are indexed to this. Using a standard protocol (Mills and Nithi, 1997), RMT results are likely to have good test–retest consistency (Gilbert et al., 2005). Bearing in mind that longer TMS lab times create challenges for restless and inattentive children, and, further, that RMT has not been reported to differ between ADHD and healthy children, RMT may be an area where higher precision can be sacrificed for shorter measurement time. The AMT is measured in slightly activated muscle to enhance visualization and interpretation. The intertrial interval for administering TMS single pulses for measuring thresholds is usually >5 seconds. In participants who are younger than 10 years, higher thresholds are expected. In such cases, it can be helpful to measure the AMT before the RMT. If the AMT is >70% of the maximum stimulator output, the RMT may be near to or above the maximum stimulator output.

Results and Interpretations

For motor thresholds, differences between children with ADHD and healthy, typically developing controls have not been reported (Dutra et al., 2016; Rubio et al., 2016). For example, in a study by our laboratory involving 52 right-handed children with ADHD ages 8–12 years and 52 age-matched controls, using a Magstim BiStim stimulator (Magstim Company, New York, NY), the mean (SD) RMT (left M1 TMS, right hand EMG, expressed in percentage of maximum stimulator output) for ADHD children was 68.5 (17.5), and the mean for controls was 64.0 (17.5) ($P = .14$). AMTs in ADHD were 44.1 (12.8) and in controls 42.6 (10.8) ($P = .5$) (Gilbert et al., 2011). Consistent with prior studies showing that RMT and AMT decrease with age (Muller et al., 1991), in that same study, in ADHD children, the correlation of age with RMT was -0.57 ($P < .001$) and with AMT was -0.38 ($P = .008$). In controls, the correlation of age with RMT was -0.67 ($P < .001$) and with AMT was -0.59 ($P < .001$) (Gilbert et al., 2011).

Pitfalls

Laboratories lacking pediatric experience should expect higher variability than with adults and therefore conduct test–retest (Mills and Nithi, 1997) and cross-laboratory consistency (Gilbert et al., 2011) of this (and other) measures. In children, RMTs may appear spuriously low initially. This probably occurs because the cortex is not really "at rest." Anxiety may be higher at the beginning of the testing session. Anxious or active children may be "closer to moving," even if this is not detected as muscle artifact on the TMS tracing. As a result, the threshold appears to be lower than it is, and the intensities selected for other experiments may be too low. AMT can be challenging because some young children will startle and relax their muscles as soon as they hear and feel the TMS pulse instead of maintaining a steady contraction. The high threshold gives an indication whether an experiment can be completed. Some measures will not be possible in children with high thresholds. Also, coils may overheat sooner, so if a backup coil is not available then time will be required for coil cooling.

Future Directions

A relatively unexplored but potentially interesting area would be comparisons between the age trajectories of RMT and AMT in both dominant and nondominant cortex in ADHD versus health children. Studies of motor function show differences in the development of dominant and nondominant hand function in children with ADHD (Gilbert et al., 2011; Cole et al., 2008). It is not known whether thresholds correlate with these lateralizing motor functions.

Recruitment Within Motor Cortex: Input–Output (*I*–*O*) Curves

Motor cortex TMS input/output (*I*–*O*) curves demonstrate changes in MEP amplitudes as a function of increasing stimulation intensity, with a steeper slope indicating easier recruitability and a flatter curve indicating that cortical neurons are less readily recruited or perhaps are inhibited.

Input–Output Technique

This is a single-pulse TMS technique with varying stimulus intensities. First, the RMT is measured. Then, for each participant, a series of pulse intensities are used at set percentages above threshold, for example, RMT, RMT + 5%, RMT + 10%, etc. (Devanne et al., 1997). Given the high intertrial variability of MEP amplitudes, sufficient numbers of trials at each intensity must be obtained for the average to be meaningful at the level of the individual (Brown et al., 2017). Neurophysiology programs with an interface that controls both timing and pulse intensity are preferred, as these can allow for the data to be obtained with a single series of trials with randomization of intensities and likely less movement of the coil during the experiment. A steep curve means that neurons are readily recruitable, such that as the pulse intensity increases, the MEP amplitude increases greatly. One study found *I*–*O* curves to be steeper in 11 adult professional musicians, compared to eight nonmusicians (Rosenkranz et al., 2007). A flat or shallow curve means that increases in pulse intensity to do not result in as much increase in MEP amplitude.

Results and Interpretations

To our knowledge, there are no reports of differences in *I*–*O* curves in children with ADHD. A small study of adults with Tourette syndrome included four participants with ADHD—too small to draw conclusions. Of interest, the adults who had a shallower *I*–*O* curve had milder tics (Orth et al., 2008). But is this primary or secondary? That is—perhaps the flatter *I*–*O* curve in motor cortex reflects mechanisms that developed over time in response to tics, which resulted in tics being less severe. A problem with studying developmental disorders of childhood onset in adults is that cross-sectional studies can yield results that are difficult to interpret as to whether they represent a primary abnormality, a compensatory response, or an outcome of years of pharmacological treatment.

Pitfalls

This is a straightforward technique and has similar limitations to other MEP amplitude techniques in terms of intertrial variability and interpreting MEPs in persons who may be more often in a "premovement" state (Chen et al., 1998; Stinear et al., 2009).

Input/Output Future Directions

Given well-established fine motor differences during development in children with ADHD (Cole et al., 2008; Mostofsky et al., 2012), this might be a productive avenue for investigation. Children with ADHD may have a more excitable motor cortex at baseline which could be reflected in either a steeper *I*–*O* slope or, as has been demonstrated in adults with Tourette syndrome (Orth et al., 2008), a less steep slope, possibly indicating development or ongoing engagement of compensatory mechanisms. Comparing the trajectory of the *I*–*O* curves in childhood in both dominant and nondominant cortex in ADHD versus healthy children might yield interesting differences. This could also be performed before and after motor skills training, possibly indicating mechanisms of motor learning and plasticity.

Motor Cortex Inhibition: The Cortical Silent Period (CSP)

Technique

A total of 5–10 sequential, suprathreshold TMS pulses are administered at 5–10 second intervals to dominant (usually) M1 during robust voluntary contraction of the contralateral target muscle. Surface EMG shows the active muscle, the TMS pulse, an evoked MEP, a period of silence, then a return of motor activity to baseline. The primary determinant of within-individual variation in CSP duration is the intensity of the pulse, not the strength of the muscle contraction (Orth and Rothwell, 2004). The CSP is typically indexed to the AMT, for example, 150% × AMT. The readout is the duration of the EMG suppression. The onset and offset can be identified visually, although statistical process control can also be used (Garvey et al., 2001). It can be easier to identify the onset and offset in series of 5–10 rectified, superimposed tracings.

Results and Interpretations

Case–control studies have more often shown CSP differences in persons with tics or Tourette syndrome, but have not consistently demonstrated any difference in CSP duration between children with ADHD and controls (Rubio et al., 2016; Gilbert et al., 2011; Moll et al., 2001). A landmark paper in 64 children (60 males; mean age 12 years: 16 each of healthy controls, ADHD, tic disorder, and ADHD plus tic disorder) evaluated CSP with pulse intensities at 40% above AMT. Children with tic disorders had mean CSP of 138 (40) ms, irrespective of the presence of ADHD. The healthy controls had a CSP of 163 (27) ms ($P = .001$) (Moll et al., 2001). In a younger cohort studied by our laboratory, using a somewhat different technique (higher intensity, different hand muscle, possibly more trials), much larger sample size, and more detailed reported clinical phenotyping, the mean CSP was 56 (41) ms in ADHD and 73 (45) ms in controls ($P = .1$) (Gilbert et al., 2011).

Pitfalls

Imprecision in measuring the AMT, used for setting the intensity for measuring CSP, or children releasing instead of maintaining muscle contraction, so that there is no silent-period "offset," is problematic. Generally, however, this is a fairly easy measure to obtain, with consistent trial to trial latencies of onset and offset.

Interhemispheric Signaling: Interhemispheric Inhibition (IHI) and Ipsilateral Silent Period (ISP)

Interhemispheric signaling may allow resources of both hemispheres to be incorporated independently into motor planning and execution of finger movements independently (Lepage et al., 2012; Fling and Seidler, 2012; Wahl et al., 2007). The capacity to perform bimanual tasks may develop atypically in ADHD and other neurodevelopmental disorders (Baumer et al., 2010; Buchmann et al., 2003; Garvey, 2003). Imaging studies quantifying volume and connectivity of the corpus callosum suggest this may be relevant to motor control or plasticity (Lepage et al., 2012; McNally et al., 2010; Draganski et al., 2010; Fling et al., 2013; Reitz and Muller, 1998).

Techniques

Several TMS techniques can be used to probe trans-hemispheric physiology. The technique most widely published in ADHD is the ISP. The ISP, similar to the CSP, is a suppression of ongoing EMG activity of target muscle after stimulation of motor cortex. However, whereas for CSP the pulse is contralateral to the active, target hand, for ISP the pulse is ipsilateral. Also, to measure ISP, the target muscles for both hands are contracted. That way the coil is stimulating pre-activated cortex (Wassermann et al., 1991). Experiments usually involve 10 trials at 100% of stimulator output. Both latency of onset and duration of suppression are reported.

Interhemispheric inhibition (IHI) is a paired pulse measure where the conditioning pulse is administered through a figure of 8 coil over one hemisphere and then the test pulse is administered 5–7 ms later with a separate coil over the opposite hemisphere. The placement of the conditioning pulse coil has been described over motor cortex and also premotor cortex. The resulting inhibition of MEP amplitudes in the paired versus single pulse trials reflects trans-hemispheric inhibitory signaling (Ferbert et al., 1992; Liuzzi et al., 2010). For this procedure, both hands are at rest, but physiology can also be studied during movement preparation. One such study compared modulation of left motor cortex by right motor cortex and by right premotor cortex. The researchers found that during movement preparation contralateral premotor cortex interactions with the motor cortex can be measured earlier than contralateral motor cortex interactions, suggesting a hierarchical temporospatial pattern (Liuzzi et al., 2010).

Results and Interpretations

Although not as consistent as some other measures, several studies have shown either longer latencies or shorter durations of ISP in ADHD (Buchmann et al., 2006; Garvey et al., 2003, 2005; Wu et al., 2012). There are also publications that include typically developing children (Garvey et al., 2003; Wu et al., 2012). The general finding is that as children mature, their latency, the time from the TMS pulse to the onset of the ISP, decreases. This was demonstrated in a small study of 12 healthy children aged 7.5–14 years in which the ISP latency ranged from approximately 55 ms in the youngest child to 30–35 ms in the oldest children, with a tight correlation of $r = -0.92$. In a comparison group of 12 children with ADHD, all latency values were 40–50 ms, and this tight negative correlation was not seen. In the same cohorts, ISP durations were recorded ranging from 2 to 35 ms, positively correlating with age in both groups (Garvey et al., 2005). In a larger cohort of 8–12-year-old children from our laboratory, in 23 children with ADHD, the mean (SD) latency was 46 (7) ms and duration was 9 (7) ms. In 31 controls, the mean (SD) latency was 41 (6) ms and duration was 11 (6) ms. The latency was significantly longer ($P = .007$) in ADHD children, suggesting slower transcallosal signaling, possibly due to delayed maturation. Longer ISP latency also correlated with younger age ($r = -0.26$, $P = .06$ trend), higher (worse) ADHD rating scale scores ($r = 0.33$, $P = .02$), higher (worse) motor ratings ($r = 0.25$, $P = .08$, trend), and less SICI (higher SICI ratios) ($r = 0.36$, $P = .008$) (Wu et al., 2012).

While it would be reasonable to conclude that latency correlates with structure primarily, one study did report that administration of methylphenidate in a cohort of 23 children mean age 11 years, while improving clinical symptoms, also shortened ISP latency by 10%, from 43 (4) to 40 (4) ms ($P = .03$) and lengthened its duration by 45%, from 15 (3) to 22 (3) ms ($P < .001$) (Buchmann et al., 2006). Thus, transsynaptic or other network effects modulated by stimulants may play a role in latency, although as expected likely a larger role in duration. To date, there are no publications reporting two-coil, paired pulse IHI techniques in children with ADHD.

Pitfalls

For ISP, the main issue is the intensity of stimulation relative to the higher motor thresholds in children. ISP at the maximal stimulator output (100%), in adults, may be evaluated using a pulse intensity that is two to three times their AMT. The more robust, longer duration may result in part from the difference between the stimulation intensity and the AMT, and not be a reflection of callosal integrity or cortical maturation. In contrast, for children with high thresholds, 100% of stimulator output may be just 10%–40% above their AMT. ISPs in children tend to have short duration and can be difficult to identify. The shorter duration, or absence of any response, may reflect the diminished relative intensity of the pulse more than the callosal integrity. As long as there is a detectable silent period, the onset latency may be a better indicator of the maturation of the pathway through the corpus callosum. Another big issue is comfort and cooperation. Children with any tendency to startle or be anxious may be less tolerant of TMS at such high intensities, limiting the feasibility and utility of ISP testing in young children or children with developmental disorders (Wu et al., 2012). For this reason, when an experimental protocol involves obtaining multiple measures, it can be wise to leave this technique until the end. For double-coil IHI measures, the smaller size of the child's head relative to the coil size can create challenges in placing both coils without having them touch, particularly if one or both figure-8 coils are 70 mm.

IHI Future Directions

This line of inquiry lends itself well to studies combining motor function assessments, imaging modalities (fractional anisotropy, resting connectivity), and TMS (Dirlikov et al., 2014; Crocetti et al., 2014). It would be most informative if obtained across larger age groups of children, and particularly if longitudinal rather than cross-sectional measures could be obtained. Changes in coil size or field strength may help with obtaining more consistent ISPs in younger children. However, given the problems with ISP measures, it may be more productive to develop other paradigms. For example, single or paired pulse studies could be performed and compared in dominant and nondominant motor cortex during ipsilateral finger tapping or movement preparation (Liuzzi et al., 2010), and these could be combined with other techniques to evaluate finger movement, for example, with goniometers (O'Malley et al., 2012; Klotz et al., 2012; Macneil et al., 2011).

Paired Pulse TMS Studies

Techniques

A variety of methods have been developed. As previously discussed, the basic idea is that there is a *conditioning pulse* that activates a modulatory population of neurons for a period of milliseconds (ms). This is followed by a suprathreshold *test pulse* during that milliseconds time interval. By comparing the readout, usually the MEP amplitude, from the paired pulses to the single pulse, one can calculate an effect of those modulatory neurons as either inhibitory (ratio is <1.0) or excitatory (ratio is >1.0). Most of these experiments are performed with a TMS coil stimulating resting, dominant left motor cortex in right-handed individuals.

Results and Interpretations

A particular paired pulse measure, SICI, is probably the most widely published measure in neurological and psychiatric disorders (Bunse et al., 2014; Khedr et al., 2009). It is believed to reflect GABAergic tone in cortex, but also is dopamine sensitive (Kujirai et al., 1993; Di Lazzaro et al., 2000, 2005; McDonnell et al., 2006). SICI appears to be pervasively influenced, as many diseased populations in case–control studies show reduced SICI. SICI has been described with regard to both the diagnosis and treatment-induced changes in ADHD. Early, small studies (Moll et al., 2000, 2001; Buchmann et al., 2007) showed reduced SICI in ADHD, and these findings have generally been replicated in larger subsequent studies (Gilbert et al., 2011; Wu et al., 2012; Hoegl et al., 2012). The particular SICI values found in these studies really are not meaningful—an average value in one study for ADHD can be an average value for healthy controls reported from a different laboratory, using a different technique. For example, in one early study, the mean SICI ratio in children with ADHD ($n = 16$) was 0.87 (0.19) and the mean in controls ($n = 16$) was 0.70 (0.21) (Moll et al., 2001). Yet in a more recent study, the mean SICI ratio in children with ADHD was 0.68 (0.24) and the mean in controls was 0.47 (0.17). In both cases, the group differences were highly statistically significant.

In addition to the differences identified consistently using ADHD as a category, SICI may be useful when considering ADHD as a dimension. A number of studies have shown that reduced SICI correlates with greater parent- and clinician-rated symptoms of hyperactivity. That is, children who have higher ADHD severity scores using standard clinical ratings had higher SICI ratios (less inhibition), and this was particularly true for the hyperactive/impulsive subscore (Gilbert et al., 2005, 2011; Hoegl et al., 2012). In the largest study, the correlation between the

Conners' Parent Rating ADHD T Score and the SICI ratio was 0.5 ($P = .002$) (Gilbert et al., 2011).

Rationales for studying ADHD in motor cortex, rather than prefrontal areas, partly include convenience. However, this approach is broadly supported by commonly observed impairments in fine motor control in ADHD (Cole et al., 2008; Macneil et al., 2011) and neuroimaging findings in the frontal cortex and motor control systems (Mostofsky et al., 2002; Castellanos and Proal, 2012). Conceptually, MEP amplitude reflects the instantaneous balance of cortical inhibition and excitation at the cortical node which sends signals for actions. It shares patterns of developmental neuronal migration and functional neural transmission with prefrontal areas (Gaspar et al., 1992; Goldman-Rakic et al., 2000). Therefore, a motor cortex MEP "readout" may reflect disease processes elsewhere in the brain.

Pitfalls

Measurements with amplitude as the outcome of interest have an important disadvantage. Centrally evoked motor amplitudes have very large intraindividual, intertrial variability. They reflect both trait and state. Therefore, whatever paradigm is being used, a sufficient number of trials are needed to understand any disease or drug signal over the background of the intraindividual, intertrial noise. Careful attention to patient phenotype, muscle relaxation, and laboratory conditions is also required.

SICI and Medications

Techniques

There are many pharmacological studies, including studies of ADHD medications, where TMS is used pre- and post-dose to evaluate neurophysiological changes (Ziemann et al., 2015; Ziemann, 2004). These may involve placebo or may involve active comparators.

Results and Interpretations

ADHD studies in children are mostly small. They have also mostly involved physiological studies before and after a single dose of a stimulant. While findings vary, in general stimulant studies have shown that SICI in ADHD is diminished at baseline and is increased or "normalized" by the stimulant methylphenidate (Moll et al., 2000; Buchmann et al., 2007). Clinical studies of the non-stimulant atomoxetine have raised a question as to whether some populations of children with ADHD have, on average, SICI that is no different from typically developing peers. For example, in a study from our laboratory, children with ADHD had reduced SICI, and this reduction correlated with both ADHD severity and impaired motor function (Gilbert et al., 2011). However, in a distinct cohort of children with ADHD recruited for a study of the non-stimulant, norepinephrine reuptake inhibitor atomoxetine, where most children had not had a beneficial response to treatment with stimulants, the average pre-treatment SICI was comparable to the SICI in healthy controls. Moreover, clinical responders to 4 weeks of open-label atomoxetine showed a decrease in SICI, rather than the expected increase in SICI. Finally, the magnitude of the SICI decrease correlated with magnitude of ADHD symptom improvement (Chen et al., 2014). This seemingly paradoxical finding mirrored findings from a smaller study we conducted in children with combined Tourette syndrome and ADHD (Gilbert et al., 2007). Although a large fraction of children may respond reasonably well to either stimulants or atomoxetine, some children respond better to one, and others do not respond to either (Newcorn et al., 2008). These TMS studies suggest distinct underlying physiological characteristics may reflect the likelihood of beneficial responses to atomoxetine (Schulz et al., 2012).

Pitfalls

In most, the research participants are healthy adults, and therefore, the relevance to children with developmental disorders is uncertain. As previously discussed, research of this nature must be judged minimal risk. In designing a study involving medication and TMS, both parts have to be judged minimal risk by the Institutional Review Board. Single and paired pulse TMS is considered minimal risk. IRBs generally approve studies of a medication like methylphenidate can be administered for testing purposes to children for whom that medication is being used clinically or would be typically used clinically. However, a more rigorous scientific understanding would come from including unaffected children in the study. This is not possible—based on risk, IRBs will not approve giving healthy children neurological or psychiatric medications for the purposes of understanding brain function. Studies in children affected with ADHD can be placebo-controlled, but recruitment of children for placebo-controlled studies, if they involve more chronic treatment of impairing symptoms when known effective treatments are available, can be ethically and practically problematic.

Future Directions for SICI and Other Paired Pulse TMS Measures

It would be relatively easy to combine TMS SICI protocols which have shown statistically robust differences with less studied measures of cortical inhibition or neuroplasticity. For example, long-interval cortical inhibition, for which the conditioning and test pulse are both suprathreshold (Sanger et al., 2001), could be evaluated at rest as well as in the context of pharmacological or behavioral treatments. Newer computer/stimulator interfaces allow for much more flexibility in controlling pulse intensities and interpulse or intertrial intervals. This means that short-interval and long-interval cortical inhibition can be measured in the same session. Double-coil paired pulse studies may also be useful for identifying physiology underlying disinhibited behavior or poor executive function. Nonetheless, this work is fairly labor-intensive. Multicenter studies may achieve these goals more quickly. Some promise has been suggested by combining single and paired pulse TMS physiology with genetics (Gilbert et al., 2006; Menzler et al., 2014; Eichhammer et al., 2003; Langguth et al., 2009), but the pace of this research, which has primarily occurred in adults, has been slow.

Functional Short-Interval Cortical Inhibition

Techniques

A newly emerging technique has been to combine TMS with various activities which generate states (within traits). To the extent that these states differ between persons with different traits, probing motor cortex might result in important functional insights. For example, in small samples of healthy adults, investigators have probed motor cortex physiology, timing TMS pulses precisely during trials with behavioral and motivational tasks. These include multiple states or tasks which could be altered in a variety of developmental or behavioral disorders. Examples include tasks requiring response inhibition or selective stopping (Badry et al., 2009; Coxon et al., 2009; Majid et al., 2012), perception or selection of reward (Gupta and Aron, 2011; Thabit et al., 2011; Kapogiannis et al., 2008), and cognitive procedural learning (Wilkinson et al., 2017). For perceptual techniques that are not highly time sensitive, this is straightforward and could be done without a sophisticated interface. For protocols requiring precisely timed pulses during trials, a program will have to be used which presents stimuli for viewing, accepts and logs behavioral responses with a device, and signals a TMS device when to activate and administer pulses. Our laboratory uses Presentation (www.neurobs.com) for this purpose. The temporal resolution of these studies is very high.

Results and Interpretations

Although possibly creating additional challenges for optimization in hyperkinetic children, there is immense promise for functional TMS to be used for biomarker research in developmental studies. One of the few such examples published to date required children with ADHD and healthy children to complete a modified Go/No-Go task. TMS pulses were administered at precise times as children were either in stages of movement preparation or inhibiting pre-primed responses. The results suggest children with more hyperactive ADHD phenotypes have less SICI than peers both at rest and during movement preparation and, further, do not "engage" motor cortex SICI as robustly while inhibiting responses (Hoegl et al., 2012). Further research validating these findings is in progress using an alternate response inhibition technique called the Slater Hammel task (Stinear et al., 2009; Coxon et al., 2006). This has recently been modified into a more child-friendly form which may also allow for calculations of stop signal reaction times with concurrent measures of motor cortex inhibitory or excitatory physiology at precise time-points (Guthrie et al., 2016, 2018). This could generate findings that demonstrate distinct, aberrant, or inefficient processes in children, linked to task performance. This again might create possibilities for identifying meaningful subgroups within ADHD, based on brain-based biomarkers. This also creates ideal opportunities for multimodal studies, for example, combining TMS and EEG (Bender et al., 2005).

Pitfalls

Additional programing expertise is needed for this type of research, and designing and testing protocols can be quite time-consuming. Depending on the paradigm, there may be behavioral data published in the children with the condition of interest, but this behavioral data may involve several hundred trials, many more than is typically done for TMS and more than a child may tolerate with a coil on the head. We find too that the TMS pulses tend not to disrupt function in adults, but that children become slightly less efficient during trials with TMS.

Some protocols designed and tested in adults may be too boring for children. This can result in declining function over time related to reduced motivation, which can also be a between group confounder. We have found three protocols published in adults to be of interest scientifically (Coxon et al., 2006; Majid et al., 2015; Kapogiannis et al., 2011) but practically quite boring for our pediatric participants. These protocols can be redesigned for children (Guthrie et al., 2018), but this again requires some programing expertise and time.

Some behavioral paradigms are more challenging based on the speed at which stimuli are presented. This is a limitation during TMS studies because commonly used machines for single and paired pulse studies require 5 or more seconds for the capacitor to recharge. So trials may need to proceed at a slower pace, a pace at which response inhibition or other behaviors may become less impaired.

There can be additional challenges and confounders in hyperkinetic children. The process of participation may generally activate motor cortex. This activation may be greater in hyperkinetic children, and data may be invalidated by the presence of motion artifact if the child cannot relax his hand during the trials.

A common approach in TMS studies is to average the responses of individuals, calculate paired to single-pulse ratios based on averages and thereby lose information contained in the within-subject intertrial variability. In functional TMS, even more data may be lost by this averaging approach if variability differs between individuals or groups performing tasks. For this reason, a repeated measures regression within subjects may be more valid and robust.

Neuroplasticity-Inducing Studies, Using Repetitive TMS to Measure Neuroplasticity or Modulate Brain Function Transiently

Techniques

TMS has been used to measure and modulate brain function through repeated TMS pulse paradigms, primarily in adults. The use of rTMS is most widely known in clinical treatment protocols, where the idea is to activate or inhibit an area of cortex subserving a particular function and have that activation or inhibition last for a period of weeks to months, thereby improving symptoms in a patient (Gilbert, 2016a,b). This is accomplished through daily rTMS sessions over a period of one or more weeks. In children, such protocols have been used mostly to treat depression but have also been used for stroke/cerebral palsy, autism, Tourette syndrome, and epilepsy.

The capacity to "respond" to rTMS can be considered an indicator of underlying brain function. Thus, rTMS paradigms can also be employed using various "readouts" such as TMS-evoked potential over motor cortex, functional or metabolic brain changes, or EEG, as a means of quantifying susceptibility to change, or plasticity. rTMS most commonly involves pulses given at set interstimulus interval ranging from 0.5 to 10 Hz. Pulse intensities often range from 0.9 to 1.1 times the RMT. High-frequency paradigms often involve bursts interleaved with pauses. Theta burst stimulation involves subthreshold intensity stimulation at high frequencies, most commonly 50 Hz. These protocols may involve a short burst of three high-frequency pulses administered every 200 ms. The overall stimulation time is generally shorter for theta burst protocols. Finally, another widely used protocol is paired associative stimulation, where external electrical impulses over sensory nerves are paired repeatedly at approximately 25 ms intervals with motor cortex TMS pulses.

Results and Interpretations

A pioneering study of rTMS in children with ADHD used TMS-evoked N100 EEG potentials before, during, and after rTMS as a readout. Helfrich et al. (2012) administered rTMS in 25 children ages 8–14 years, with adverse events only occurring in three, who reported mild headache. The protocol administered TMS at 80% of RMT except, in the case of six participants, at 100% of maximum stimulator output, since the thresholds were too high to be measured. Children received 900 stimuli total. N100 amplitudes decreased during the first 500 pulses (considered as five 100 pulse blocks). N100 amplitudes were also significantly reduced after rTMS compared to sham RTMS. Still the authors stated it is unclear if the change results from augmentation of inhibitory, for example, gama-aminobutyric acid (GABA), transmission or reduction of excitatory physiology. TMS-evoked motor-evoked potentials were reduced after RMTS as well, in both the active and sham rTMS groups. There was not a control population studied. The authors proposed further study of the TMS-evoked N100 as a biomarker for monitoring changes in cortical excitability, which could be used to help optimize rTMS treatment protocols.

Pitfalls

rTMS pulse intensities are indexed to RMT. Thresholds are higher in children. This means that many commercial machines will overheat and shut off during pediatric protocols. In addition, there can be built in safety features which require trade-offs between high frequency and high intensity. So in order to get to the desired intensity relative to the child's RMT, the pulse frequency is required to be lower than is used in adult trials. Even when the machine will allow for a protocol as desired, children may potentially be more sensitive to the sound, for example, and may tolerate the procedure less well. Theta burst stimulation, with its lower intensities and stimulation times, can sometimes be a well-tolerated option (Hong et al., 2015). In our experience, healthy children and children with Tourette syndrome with or without comorbid ADHD tolerate this well (Hong et al., 2015). The appropriateness of a particular protocol, especially when there is a scientific rationale to include healthy children in a biomarker study, has to prioritize tolerability and demonstrated safety.

Choosing a stimulation site can be labor-intensive, time-consuming, and expensive. There are a variety of approaches for studies outside of motor cortex, ranging from subject-specific stimulation using neuronavigation systems linked to fMRI BOLD signal (Wu et al., 2014) to less individualized localization based on EEG coordinate systems (Noda et al., 2017; Gomez et al., 2014). Using MRI has the advantage of subject specificity but is much more expensive and requires hyperactive, impulsive children to have MRI scans. The EEG coordinate approach is less precise but also less time-consuming. Given that the field generated by a TMS coil is more diffuse than, say, intraoperative electrical stimulation or deep brain stimulation, this may not necessarily have major consequences for biomarker or treatment studies.

Clinical Trials

Techniques

The techniques used in TMS clinical trials are those described above for measuring and modulating neuroplasticity. rTMS may be used at low (≤1 Hz) or high (~5–10 Hz) frequency, typically at intensities indexed to and fairly close to RMT. Theta burst has also been used.

Results and Interpretations

Weaver et al. evaluated efficacy and tolerability of high-frequency rTMS, screening 24 adolescents and young adults, only 9 of whom ultimately agreed to participate due to the time requirement. Of those, only four were younger than 18 years of age. Stimulation was at the right dorsolateral prefrontal cortex: 2000 pulses, 100% of RMT, at 10 Hz in a pattern of 4 seconds on and 26 seconds off. There were 10 once-daily sessions, given over 2 weeks. After 1 week of washout (no treatment), there was a second 2-week treatment session. Participants were randomized to active first or placebo first and crossed over to the alternate treatment for the second session. There was a strong placebo effect of sham treatment in the first treatment phase. Overall, the improvement observed in ADHD did not differ between active and sham treatments (Weaver et al., 2012).

Gomez et al. evaluated tolerability of low-frequency rTMS among ten 7–12-year-old boys with ADHD described as "non-responders to conventional therapy." The number, type, dose, and duration of prior medications were not described. Stimulation was at the left dorsolateral prefrontal cortex, based on EEG F3 position: 1500 pulses, 90% of RMT, at 1 Hz. There were five once-daily sessions. Adverse events were transient—headaches and neck pain. All participants completed the study. Parents reported on a symptom checklist that both inattentive and hyperactive symptoms improved (Gomez et al., 2014). As this study was open label, and given the placebo response noted in the prior crossover study, it is difficult to interpret the reported clinical benefit.

Pitfalls

The pitfalls are those described above for measuring and modulating neuroplasticity. In addition, ADHD clinical trials are time intensive for families and for clinical research teams. The typical clinical protocol will involve daily treatment 5 days per week for 1–3 weeks. Thus, in considering feasibility and recruitment targets, it is important to compare ADHD to other conditions for which TMS treatment is used or studied. When TMS is used for refractory depression in young adults, it is generally used when multiple medications have failed and the individual is sufficiently disabled on a daily basis that they cannot function. In that context, when electroconvulsive therapy is under consideration, use of rTMS may have a favorable risk–benefit ratio (D'Agati et al., 2010). When rTMS is used for treatment of stroke or hemiparetic cerebral palsy, it can be integrated with intensive physical therapy (Kirton et al., 2016). As such there is potential for additional concurrent benefit from therapy, and the time commitment is within the realm of what routinely occurs in the care of children with motor impairments. In addition, there are no biological therapies expected to rewire the brain after stroke, whereas noninvasive brain stimulation holds that promise.

Contrast treatment considerations for depression and cerebral palsy to ADHD. When TMS is considered as a treatment for ADHD, the level of impairment and disability is typically going to be much less than is present with depression or cerebral palsy. Moreover, there continue to be new medications introduced for ADHD which require a much less extensive time commitment. Even in a clinical trial, the study visits for medications occur weekly or bimonthly, not daily. In that context, it is difficult to imagine, at least in the United States, that a multicenter study could recruit sufficient patients for large, rigorous, sham-controlled study of 10–15 sessions of TMS in a 2–3-week period for ADHD such that the FDA would clear TMS for ADHD treatment.

CONCLUDING REMARKS

The use of TMS to understand the neurobiology of ADHD possibly points toward better treatment which dates back 20 years. Despite challenges of pediatric research, there continue to be new and interesting studies, with methodologies advancing from basic, small case–control studies in modestly characterized groups to larger studies with more detailed phenotyping and advanced statistical methods. Probing motor cortex during functional tasks, measuring and modulating neuroplasticity, and moving TMS stimulation to areas outside motor cortex are more recent paradigms which may yield insights in children. Clinical trials to date have been disappointing and may not be feasible as a basis for long-term treatment. However, clinical trials may yield discoveries that could build a rational foundation for other successful types of treatment, possibly with noninvasive brain stimulation.

References

Adams, R.M., Macy, D.J., Kocsis, J.J., Sullivan, A.R., 1984. Attention deficit disorder with hyperactivity: normative data for Conners behavior rating scale. Tex. Med. 80, 58–61.

Allotey, J., Zamora, J., Cheong-See, F., et al., 2018. Cognitive, motor, behavioural and academic performances of children born preterm: a meta-analysis and systematic review involving 64 061 children. BJOG: An International Journal of Obstetrics & Gynaecology 125, 16–25.

American Psychiatric Association, 2013. Diagnostic and Statistical Manual of Mental Disorders (DSM-5®). American Psychiatric Pub, Washington DC.

Arnold, L.E., Abikoff, H.B., Cantwell, D.P., et al., 1997. National Institute of Mental Health Collaborative Multimodal Treatment Study of Children with ADHD (the MTA). Design challenges and choices. Arch. Gen. Psychiatry 54, 865–870.

Badry, R., Mima, T., Aso, T., et al., 2009. Suppression of human cortico-motoneuronal excitability during the stop-signal task. Clin. Neurophysiol. 120, 1717–1723.

Bard, D.E., Wolraich, M.L., Neas, B., Doffing, M., Beck, L., 2013. The psychometric properties of the Vanderbilt attention-deficit hyperactivity disorder diagnostic parent rating scale in a community population. J. Dev. Behav. Pediatr.: JDBP 34, 72–82.

Bauer, N.S., Yoder, R., Carroll, A.E., Downs, S.M., 2016. Racial/ethnic differences in the prevalence of anxiety using the Vanderbilt ADHD scale in a diverse community outpatient setting. J. Dev. Behav. Pediatr.: JDBP 37, 610–618.

Baumer, T., Thomalla, G., Kroeger, J., et al., 2010. Interhemispheric motor networks are abnormal in patients with Gilles de la Tourette syndrome. Mov. Disord. 25, 2828–2837.

Bender, S., Basseler, K., Sebastian, I., et al., 2005. Electroencephalographic response to transcranial magnetic stimulation in children: evidence for giant inhibitory potentials. Ann. Neurol. 58, 58–67.

Biederman, J., Feinberg, L., Chan, J., et al., 2015. Mild traumatic brain injury and attention-deficit hyperactivity disorder in young student athletes. J. Nerv. Ment. Dis. 203, 813–819.

Brown, K.E., Lohse, K.R., Mayer, I.M.S., et al., 2017. The reliability of commonly used electrophysiology measures. Brain Stimul. 10, 1102–1111.

Buchmann, J., Wolters, A., Haessler, F., Bohne, S., Nordbeck, R., Kunesch, E., 2003. Disturbed transcallosally mediated motor inhibition in children with attention deficit hyperactivity disorder (ADHD). Clin. Neurophysiol. 114, 2036–2042.

Buchmann, J., Gierow, W., Weber, S., et al., 2006. Modulation of transcallosally mediated motor inhibition in children with attention deficit hyperactivity disorder (ADHD) by medication with methylphenidate (MPH). Neurosci. Lett. 405, 14–18.

Buchmann, J., Gierow, W., Weber, S., et al., 2007. Restoration of disturbed intracortical motor inhibition and facilitation in attention deficit hyperactivity disorder children by methylphenidate. Biol. Psychiatry 62, 963–969.

Bunse, T., Wobrock, T., Strube, W., et al., 2014. Motor cortical excitability assessed by transcranial magnetic stimulation in psychiatric disorders: a systematic review. Brain Stimul. 7, 158–169.

Button, K.S., Ioannidis, J.P., Mokrysz, C., et al., 2013. Power failure: why small sample size undermines the reliability of neuroscience. Nat. Rev. Neurosci. 14, 365–376.

Castellanos, F.X., Proal, E., 2012. Large-scale brain systems in ADHD: beyond the prefrontal-striatal model. Trends Cogn. Sci. 16, 17–26.

Chen, R., Yaseen, Z., Cohen, L.G., Hallett, M., 1998. Time course of corticospinal excitability in reaction time and self-paced movements. Ann. Neurol. 44, 317–325.

Chen, T.H., Wu, S.W., Welge, J.A., et al., 2014. Reduced short interval cortical inhibition correlates with atomoxetine response in children with attention-deficit hyperactivity disorder (ADHD). J. Child Neurol.

Cloes, K.I., Barfell, K.S., Horn, P.S., et al., 2017. Preliminary evaluation of child self-rating using the Child Tourette Syndrome Impairment Scale. Dev. Med. Child Neurol. 59, 284–290.

Cole, W.R., Mostofsky, S.H., Larson, J.C., Denckla, M.B., Mahone, E.M., 2008. Age-related changes in motor subtle signs among girls and boys with ADHD. Neurology 71, 1514–1520.

Conelea, C.A., Woods, D.W., Zinner, S.H., et al., 2013. The impact of Tourette syndrome in adults: results from the Tourette syndrome impact survey. Commun. Ment. Health J. 49, 110–120.

Connell, F.A., Koepsell, T.D., 1985. Measures of gain in certainty from a diagnostic test. Am. J. Epidemiol. 121, 744–753.

Coxon, J.P., Stinear, C.M., Byblow, W.D., 2006. Intracortical inhibition during volitional inhibition of prepared action. J. Neurophysiol. 95, 3371–3383.

Coxon, J.P., Stinear, C.M., Byblow, W.D., 2009. Stop and go: the neural basis of selective movement prevention. J. Cogn. Neurosci. 21, 1193–1203.

Crocetti, D., Dirlikov, B., Peterson, D., Gilbert, D.L., Mostofsky, S.H., 2014. Interhemispheric motor inhibition is associated with callosal structural integrity in children with ADHD. In: The 20th Annual Meeting of the Organization for Human Brain Mapping, Hamburg, Germany.

D'Agati, D., Bloch, Y., Levkovitz, Y., Reti, I., 2010. rTMS for adolescents: safety and efficacy considerations. Psychiatry Res. 177, 280–285.

Devanne, H., Lavoie, B.A., Capaday, C., 1997. Input–output properties and gain changes in the human corticospinal pathway. Exp. Brain Res. 114, 329–338.

Di Lazzaro, V., Oliviero, A., Meglio, M., et al., 2000. Direct demonstration of the effect of lorazepam on the excitability of the human motor cortex. Clin. Neurophysiol. 111, 794–799.

Di Lazzaro, V., Oliviero, A., Saturno, E., et al., 2005. Effects of lorazepam on short latency afferent inhibition and short latency intracortical inhibition in humans. J. Physiol. 564, 661–668.

Dirlikov, B., Nebel, M.B., Barber, A.D., et al., 2014. A multimodal examination of interhemispheric connectivity and mirror overflow in children with ADHD. In: The 20th Annual Meeting of the Organization for Human Brain Mapping, Hamburg, Germany.

Draganski, B., Martino, D., Cavanna, A.E., et al., 2010. Multispectral brain morphometry in Tourette syndrome persisting into adulthood. Brain 133, 3661–3675.

DuPaul, G.J., Power, T.J., Anastopoulos, A.D., Reid, R., 1998. ADHD Rating Scale-IV: Checklists, Norms, and Clinical Interpretation. Guilford Press, New York, NY.

Dutra, T.G., Baltar, A., Monte-Silva, K.K., 2016. Motor cortex excitability in attention-deficit hyperactivity disorder (ADHD): a systematic review and meta-analysis. Res. Dev. Disabil. 56, 1–9.

Eichhammer, P., Langguth, B., Wiegand, R., Kharraz, A., Frick, U., Hajak, G., 2003. Allelic variation in the serotonin transporter promoter affects neuromodulatory effects of a selective serotonin transporter reuptake inhibitor (SSRI). Psychopharmacology 166, 294–297.

Ferbert, A., Priori, A., Rothwell, J.C., Day, B.L., Colebatch, J.G., Marsden, C.D., 1992. Interhemispheric inhibition of the human motor cortex. J. Physiol. 453, 525–546.

Fling, B.W., Seidler, R.D., 2012. Task-dependent effects of interhemispheric inhibition on motor control. Behav. Brain Res. 226, 211–217.

Fling, B.W., Benson, B.L., Seidler, R.D., 2013. Transcallosal sensorimotor fiber tract structure–function relationships. Hum. Brain Mapp. 34, 384–395.

Garvey, M.A., Ziemann, U., Becker, D.A., Barker, C.A., Bartko, J.J., 2001. New graphical method to measure silent periods evoked by transcranial magnetic stimulation. Clin. Neurophysiol. 112, 1451–1460.

Garvey, M.A., Ziemann, U., Bartko, J.J., Denckla, M.B., Barker, C.A., Wassermann, E.M., 2003. Cortical correlates of neuromotor development in healthy children. Clin. Neurophysiol. 114, 1662–1670.

Garvey, M.A., Barker, C.A., Bartko, J.J., et al., 2005. The ipsilateral silent period in boys with attention-deficit/hyperactivity disorder. Clin. Neurophysiol. 116, 1889–1896.

Gaspar, P., Stepniewska, I., Kaas, J.H., 1992. Topography and collateralization of the dopaminergic projections to motor and lateral prefrontal cortex in owl monkeys. J. Comp. Neurol. 325, 1–21.

Gilbert, D.L., 2008. Design and analysis of motor-evoked potential data in pediatric neurobehavioural disorder investigations. In: Wassermann, E.M., Epstein, C.M., Ziemann, U., Walsh, V., Paus, T., Lisanby, S.H. (Eds.), The Oxford Handbook of Transcranial Magnetic Stimulation. Oxford University Press, Oxford, UK, pp. 389–400.

Gilbert, D.L., 2016a. Therapeutic rTMS in children. In: Kirton, A., Gilbert, D.L. (Eds.), Pediatric Brain Stimulation: Mapping and Modulating the Developing Brain. Academic Press; Elsevier, Cambridge MA, pp. 71–83.

Gilbert, D.L., 2016b. TMS applications in ADHD and developmental disorders. In: Kirton, A., Gilbert, D.L. (Eds.), Pediatric Brain Stimulation: Mapping and Modulating the Developing Brain. Academic Press, Elsevier, Cambridge MA, London, pp. 154–182.

Gilbert, D.L., Garvey, M.A., Bansal, A.S., Lipps, T., Zhang, J., Wassermann, E.M., 2004. Should transcranial magnetic stimulation research in children be considered minimal risk? Clin. Neurophysiol. 115, 1730–1739.

Gilbert, D.L., Sallee, F.R., Zhang, J., Lipps, T.D., Wassermann, E.M., 2005. TMS-evoked cortical inhibition: a consistent marker of ADHD scores in Tourette syndrome. Biol. Psychiatry 57, 1597–1600.

Gilbert, D.L., Wang, Z.W., Sallee, F.R., et al., 2006. Dopamine transporter genotype influences the physiological response to medication in ADHD. Brain 129, 2038–2046.

Gilbert, D.L., Zhang, J., Lipps, T.D., et al., 2007. Atomoxetine treatment of ADHD in Tourette syndrome: reduction in motor cortex inhibition correlates with clinical improvement. Clin. Neurophysiol. 118, 1835–1841.

Gilbert, D.L., Isaacs, K.M., Augusta, M., Macneil, L.K., Mostofsky, S.H., 2011. Motor cortex inhibition: a marker of ADHD behavior and motor development in children. Neurology 76, 615–621.

Goldman-Rakic, P.S., Muly III, E.C., Williams, G.V., 2000. D1 receptors in prefrontal cells and circuits. Brain Res. Rev. 31, 295–301.

Gomez, L., Vidal, B., Morales, L., et al., 2014. Low frequency repetitive transcranial magnetic stimulation in children with attention deficit/hyperactivity disorder. Preliminary results. Brain Stimul. 7, 760–762.

Grefer, M., Flory, K., Cornish, K., Hatton, D., Roberts, J., 2016. The emergence and stability of attention deficit hyperactivity disorder in boys with fragile X syndrome. J. Intellect. Disabil. Res. 60, 167–178.

Greven, C.U., Bralten, J., Mennes, M., et al., 2015. Developmentally stable whole-brain volume reductions and developmentally sensitive caudate and putamen volume alterations in those with attention-deficit/hyperactivity disorder and their unaffected siblings. JAMA Psychiatry 72, 490–499.

Gupta, N., Aron, A.R., 2011. Urges for food and money spill over into motor system excitability before action is taken. Eur. J. Neurosci. 33, 183–188.

Guthrie, M., Gilbert, D., Huddleston, D., et al., 2016. Neurophysiologic correlates of motor response inhibition. In: 46th Annual Meeting of the Child Neurology Society, Kansas City, MO. Wiley-Blackwell, 111 River St, Hoboken 07030-5774, NJ, USA, pp. S414–S414.

Guthrie, M.D., Gilbert, D.L., Huddleston, D.A., et al., 2018. Online transcranial magnetic stimulation protocol for measuring cortical physiology associated with response inhibition. J. Visual Exp. 132, e56789.

Helfrich, C., Pierau, S.S., Freitag, C.M., Roeper, J., Ziemann, U., Bender, S., 2012. Monitoring cortical excitability during repetitive transcranial magnetic stimulation in children with ADHD: a single-blind, sham-controlled TMS-EEG study. PLoS ONE 7, e50073.

Hinshaw, S.P., Arnold, L.E., Group MTAC, 2015. Attention-deficit hyperactivity disorder, multimodal treatment, and longitudinal outcome: evidence, paradox, and challenge. Wiley Interdiscip. Rev. Cogn. Sci. 6, 39–52.

Hoegl, T., Heinrich, H., Barth, W., Losel, F., Moll, G.H., Kratz, O., 2012. Time course analysis of motor excitability in a response inhibition task according to the level of hyperactivity and impulsivity in children with ADHD. PLoS ONE 7, e46066.

Hong, Y.H., Wu, S.W., Pedapati, E.V., et al., 2015. Safety and tolerability of theta burst stimulation vs. single and paired pulse transcranial magnetic stimulation: a comparative study of 165 pediatric subjects. Front. Hum. Neurosci. 9, 29.

Jensen, P.S., 1999. A 14-month randomized clinical trial of treatment strategies for attention-deficit/hyperactivity disorder. The MTA Cooperative Group. Multimodal Treatment Study of Children with ADHD. Arch. Gen. Psychiatry 56, 1073–1086.

Kapogiannis, D., Campion, P., Grafman, J., Wassermann, E.M., 2008. Reward-related activity in the human motor cortex. Eur. J. Neurosci. 27, 1836–1842.

Kapogiannis, D., Mooshagian, E., Campion, P., et al., 2011. Reward processing abnormalities in Parkinson's disease. Mov. Disord. 26, 1451–1457.

Khedr, E.M., Abo-Elfetoh, N., Rothwell, J.C., 2009. Treatment of post-stroke dysphagia with repetitive transcranial magnetic stimulation. Acta Neurol. Scand. 119, 155–161.

Kirton, A., Andersen, J., Herrero, M., et al., 2016. Brain stimulation and constraint for perinatal stroke hemiparesis: The PLASTIC CHAMPS Trial. Neurology 86, 1659–1667.

Klotz, J.M., Johnson, M.D., Wu, S.W., Isaacs, K.M., Gilbert, D.L., 2012. Relationship between reaction time variability and motor skill development in ADHD. Child Neuropsychol. 18, 576–585.

Kujirai, T., Caramia, M.D., Rothwell, J.C., et al., 1993. Corticocortical inhibition in human motor cortex. J. Physiol. 471, 501–519.

Langberg, J.M., Vaughn, A.J., Brinkman, W.B., Froehlich, T., Epstein, J.N., 2010. Clinical utility of the Vanderbilt ADHD Rating Scale for ruling out comorbid learning disorders. Pediatrics 126, e1033–e1038.

Langguth, B., Sand, P., Marek, R., et al., 2009. Allelic variation in the serotonin transporter promoter modulates cortical excitability. Biol. Psychiatry 66, 283–286.

Lepage, J.F., Beaule, V., Srour, M., et al., 2012. Neurophysiological investigation of congenital mirror movements in a patient with agenesis of the corpus callosum. Brain Stimul 5, 137–140.

Liuzzi, G., Horniss, V., Hoppe, J., et al., 2010. Distinct temporospatial interhemispheric interactions in the human primary and premotor cortex during movement preparation. Cereb. Cortex 20, 1323–1331.

Lo-Castro, A., D'Agati, E., Curatolo, P., 2011. ADHD and genetic syndromes. Brain Dev. 33, 456–461.

Macneil, L.K., Xavier, P., Garvey, M.A., et al., 2011. Quantifying excessive mirror overflow in children with attention-deficit/hyperactivity disorder. Neurology 76, 622–628.

Majid, D.S., Cai, W., George, J.S., Verbruggen, F., Aron, A.R., 2012. Transcranial magnetic stimulation reveals dissociable mechanisms for global versus selective corticomotor suppression underlying the stopping of action. Cereb. Cortex 22, 363–371.

Majid, D.S., Lewis, C., Aron, A.R., 2015. Training voluntary motor suppression with real-time feedback of motor evoked potentials. J. Neurophysiol. 113, 3446–3452.

Max, J.E., Schachar, R.J., Levin, H.S., et al., 2005. Predictors of secondary attention-deficit/hyperactivity disorder in children and adolescents 6 to 24 months after traumatic brain injury. J. Am. Acad. Child Adolesc. Psychiatry 44, 1041–1049.

McDonnell, M.N., Orekhov, Y., Ziemann, U., 2006. The role of GABA(B) receptors in intracortical inhibition in the human motor cortex. Exp. Brain Res. 173, 86–93.

McNally, M.A., Crocetti, D., Mahone, E.M., Denckla, M.B., Suskauer, S.J., Mostofsky, S.H., 2010. Corpus callosum segment circumference is associated with response control in children with attention-deficit hyperactivity disorder (ADHD). J. Child Neurol. 25, 453–462.

Menzler, K., Hermsen, A., Balkenhol, K., et al., 2014. A common SCN1A splice-site polymorphism modifies the effect of carbamazepine on cortical excitability—a pharmacogenetic transcranial magnetic stimulation study. Epilepsia 55, 362–369.

Mills, K.R., Nithi, K.A., 1997. Corticomotor threshold to magnetic stimulation: normal values and repeatability. Muscle Nerve 20, 570–576.

Molina, B.S., Hinshaw, S.P., Swanson, J.M., et al., 2009. The MTA at 8 years: prospective follow-up of children treated for combined-type ADHD in a multisite study. J. Am. Acad. Child Adolesc. Psychiatry 48, 484–500.

Moll, G.H., Heinrich, H., Trott, G., Wirth, S., Rothenberger, A., 2000. Deficient intracortical inhibition in drug-naive children with attention-deficit hyperactivity disorder is enhanced by methylphenidate. Neurosci. Lett. 284, 121–125.

Moll, G.H., Heinrich, H., Trott, G.E., Wirth, S., Bock, N., Rothenberger, A., 2001. Children with comorbid attention-deficit-hyperactivity disorder and tic disorder: evidence for additive inhibitory deficits within the motor system. Ann. Neurol. 49, 393–396.

Mostofsky, S., Shiels, K., Dirlikov, B., Gilbert, D., 2012. Children with ADHD show increased variability in sequential finger movements (P07.144). Neurology 78 (P07), 144.

Mostofsky, S.H., Cooper, K.L., Kates, W.R., Denckla, M.B., Kaufmann, W.E., 2002. Smaller prefrontal and premotor volumes in boys with attention-deficit/hyperactivity disorder. Biol. Psychiatry 52, 785–794.

Muller, K., Homberg, V., Lenard, H.G., 1991. Magnetic stimulation of motor cortex and nerve roots in children. Maturation of cortico-motoneuronal projections. Electroencephalogr. Clin. Neurophysiol. 81, 63–70.

Newcorn, J.H., Kratochvil, C.J., Allen, A.J., et al., 2008. Atomoxetine and osmotically released methylphenidate for the treatment of attention deficit hyperactivity disorder: acute comparison and differential response. Am. J. Psychiatry 165, 721–730.

Noda, Y., Barr, M.S., Zomorrodi, R., et al., 2017. Evaluation of short interval cortical inhibition and intracortical facilitation from the dorsolateral prefrontal cortex in patients with schizophrenia. Sci. Rep. 7, 17106.

O'Malley, J., Wu, S., Huddleston, D., Mostofsky, S., Gilbert, D., 2012. Quantification of dysrhythmia in children with ADHD using a novel automated goniometer method (P07.143). Neurology 78 (P07), 143.

Orth, M., Rothwell, J.C., 2004. The cortical silent period: intrinsic variability and relation to the waveform of the transcranial magnetic stimulation pulse. Clin. Neurophysiol. 115, 1076–1082.

Orth, M., Snijders, A.H., Rothwell, J.C., et al., 2003. The variability of intracortical inhibition and facilitation. Clin. Neurophysiol. 114, 2362–2369.

Orth, M., Munchau, A., Rothwell, J.C., 2008. Corticospinal system excitability at rest is associated with tic severity in Tourette syndrome. Biol. Psychiatry 64, 248–251.

Polanczyk, G.V., Willcutt, E.G., Salum, G.A., Kieling, C., Rohde, L.A., 2014. ADHD prevalence estimates across three decades: an updated systematic review and meta-regression analysis. Int. J. Epidemiol. 43, 434–442.

Priori, A., Berardelli, A., Inghilleri, M., Accornero, N., Manfredi, M., 1994. Motor cortical inhibition and the dopaminergic system. Pharmacological changes in the silent period after transcranial brain stimulation in normal subjects, patients with Parkinson's disease and drug-induced parkinsonism. Brain 117 (Pt 2), 317–323.

Reitz, M., Muller, K., 1998. Differences between 'congenital mirror movements' and 'associated movements' in normal children: a neurophysiological case study. Neurosci. Lett. 256, 69–72.

Rosenkranz, K., Williamon, A., Rothwell, J.C., 2007. Motorcortical excitability and synaptic plasticity is enhanced in professional musicians. J. Neurosci. 27, 5200–5206.
Rossi, S., Hallett, M., Rossini, P.M., Pascual-Leone, A., 2009. Safety, ethical considerations, and application guidelines for the use of transcranial magnetic stimulation in clinical practice and research. Clin. Neurophysiol. 120, 2008–2039.
Rowland, A.S., Skipper, B.J., Umbach, D.M., et al., 2015. The prevalence of ADHD in a population-based sample. J. Atten. Disord. 19, 741–754.
Rubio, B., Boes, A.D., Laganiere, S., Rotenberg, A., Jeurissen, D., Pascual-Leone, A., 2016. Noninvasive brain stimulation in pediatric attention-deficit hyperactivity disorder (ADHD): a review. J. Child Neurol. 31, 784–796.
Sackett, D.L., Haynes, R.B., Guyatt, G., Tugwell, P., 1991. The interpretation of diagnostic data, Clinical Epidemiology: A Basic Science for Clinical Medicine, second ed Little, Brown and Company, Boston, MA, pp. 69–152.
Sanger, T.D., Garg, R.R., Chen, R., 2001. Interactions between two different inhibitory systems in the human motor cortex. J. Physiol. 530, 307–317.
Sayal, K., Prasad, V., Daley, D., Ford, T., Coghill, D., 2017. ADHD in children and young people: prevalence, care pathways, and service provision. Lancet Psychiatry .
Schulz, K.P., Fan, J., Bedard, A.C., et al., 2012. Common and unique therapeutic mechanisms of stimulant and nonstimulant treatments for attention-deficit/hyperactivity disorder. Arch. Gen. Psychiatry 69, 952–961.
Shaw, P., Lerch, J., Greenstein, D., et al., 2006. Longitudinal mapping of cortical thickness and clinical outcome in children and adolescents with attention-deficit/hyperactivity disorder. Arch. Gen. Psychiatry 63, 540–549.
Shaw, P., Malek, M., Watson, B., Sharp, W., Evans, A., Greenstein, D., 2012. Development of cortical surface area and gyrification in attention-deficit/hyperactivity disorder. Biol. Psychiatry 72, 191–197.
Stinear, C.M., Coxon, J.P., Byblow, W.D., 2009. Primary motor cortex and movement prevention: where Stop meets Go. Neurosci. Biobehav. Rev. 33, 662–673.
Sullivan, K., Hatton, D., Hammer, J., et al., 2006. ADHD symptoms in children with FXS. Am. J. Med. Genet. A 140, 2275–2288.
Suren, P., Bakken, I.J., Aase, H., et al., 2012. Autism spectrum disorder, ADHD, epilepsy, and cerebral palsy in Norwegian children. Pediatrics 130, e152–e158.
Szekely, E., Sudre, G.P., Sharp, W., Leibenluft, E., Shaw, P., 2017. Defining the neural substrate of the adult outcome of childhood ADHD: a multimodal neuroimaging study of response inhibition. Am. J. Psychiatry 174, 867–876.
Thabit, M.N., Nakatsuka, M., Koganemaru, S., Fawi, G., Fukuyama, H., Mima, T., 2011. Momentary reward induce changes in excitability of primary motor cortex. Clin. Neurophysiol. 122, 1764–1770.
Thapar, A., Martin, J., Mick, E., et al., 2016. Psychiatric gene discoveries shape evidence on ADHD's biology. Mol. Psychiatry 21, 1202–1207.
Wahl, M., Lauterbach-Soon, B., Hattingen, E., et al., 2007. Human motor corpus callosum: topography, somatotopy, and link between microstructure and function. J. Neurosci. 27, 12132–12138.
Wassermann, E.M., Fuhr, P., Cohen, L.G., Hallett, M., 1991. Effects of transcranial magnetic stimulation on ipsilateral muscles. Neurology 41, 1795–1799.
Weaver, L., Rostain, A.L., Mace, W., Akhtar, U., Moss, E., O'Reardon, J.P., 2012. Transcranial magnetic stimulation (TMS) in the treatment of attention-deficit/hyperactivity disorder in adolescents and young adults: a pilot study. J. ECT 28, 98–103.
Wilkinson, L., Koshy, P.J., Steel, A., Bageac, D., Schintu, S., Wassermann, E.M., 2017. Motor cortex inhibition by TMS reduces cognitive non-motor procedural learning when immediate incentives are present. Cortex 97, 70–80.

Wu, S.W., Gilbert, D.L., Shahana, N., Huddleston, D.A., Mostofsky, S.H., 2012. Transcranial magnetic stimulation measures in attention-deficit/hyperactivity disorder. Pediatr. Neurol. 47, 177–185.

Wu, S.W., Maloney, T., Gilbert, D.L., et al., 2014. Functional MRI-navigated repetitive transcranial magnetic stimulation over supplementary motor area in chronic tic disorders. Brain Stimul. 7, 212–218.

Yang, L.Y., Huang, C.C., Chiu, W.T., Huang, L.T., Lo, W.C., Wang, J.Y., 2016. Association of traumatic brain injury in childhood and attention-deficit/hyperactivity disorder: a population-based study. Pediatr. Res. 80, 356–362.

Ziemann, U., 2004. TMS and drugs. Clin. Neurophysiol. 115, 1717–1729.

Ziemann, U., Lonnecker, S., Steinhoff, B.J., Paulus, W., 1996. Effects of antiepileptic drugs on motor cortex excitability in humans: a transcranial magnetic stimulation study. Ann. Neurol. 40, 367–378.

Ziemann, U., Reis, J., Schwenkreis, P., et al., 2015. TMS and drugs revisited 2014. Clin. Neurophysiol. 126, 1847–1868.

CHAPTER

7

TMS in Child and Adolescent Major Depression

Charles P. Lewis[1], *Faranak Farzan*[2] *and Paul E. Croarkin*[1]

[1]Department of Psychiatry and Psychology, Division of Child and Adolescent Psychiatry, Mayo Clinic, Rochester, MN, United States [2]School of Mechatronic Systems Engineering, Simon Fraser University, Surrey, BC, Canada

OUTLINE

Neurotechnological Pediatric Neuropsych
DOI: https://doi.org/10.1016/B978-0-12-812777-3.00007-6

INTRODUCTION

Mood disorders are common across the lifespan, with first depressive episodes often occurring prior to adulthood. The prevalence of moderate-to-severe depressive symptoms in youth between the ages of 12 and 17 was recently estimated to be 5.7%, while members of certain populations (e.g., certain gender, ethnicity, and socioeconomic groups) experience even higher rates of depression (Pratt and Brody, 2014). In addition to the significant morbidity and functional impairments associated with major depression, depressive disorders are often present in persons who die by suicide, which is the second-leading cause of death in adolescents and young adults in the United States (Kann et al., 2016) and worldwide (World Health Organization, 2014).

Despite the substantial individual and societal burdens associated with mood disorders, treatment and mental health services available to depressed youth remain inadequate. In particular, concerns have arisen about the use of antidepressant medications, such as the commonly prescribed selective serotonin reuptake inhibitors (SSRIs) and serotonin-norepinephrine reuptake inhibitors (SNRIs), in adolescents and young adults. Effect sizes for clinical outcomes tend to be relatively small in this age group (Bridge et al., 2007; Cipriani et al., 2016), and some meta-analyses have cast doubt on whether antidepressant medications are superior to placebo in depressed youth (Cipriani et al., 2016; Locher et al., 2017). Prominent safety concerns also have been raised in the last decade, particularly regarding increased risk of suicidality in children, adolescents, and young adults treated with antidepressant pharmacotherapy. Some studies have found increased suicidal ideation and behavior (Dubicka et al., 2006; Hammad et al., 2006; Bridge et al., 2007; Stone et al., 2009; Sharma et al., 2016), as well as aggression (Sharma et al., 2016) in young persons treated with antidepressant medications compared with placebo. While the interpretation and clinical implications of these efficacy and safety data remain controversial (Dubicka et al., 2016; Stone, 2016; Walkup, 2017), substantial media attention and regulatory advisories in the United States, Canada, and Europe have dampened many clinicians' and patients' enthusiasm for antidepressant medications. The mechanisms by which antidepressant medications might address the underlying neural circuitry related to depression in children and adolescents is also not clear. Noninvasive brain stimulation techniques, such as transcranial magnetic stimulation (TMS), may have promise as therapeutic interventions for children and adolescents with mood disorders.

In this chapter, we discuss the use of TMS as both an investigational tool for understanding brain physiology in depression and as a therapeutic intervention. We review the literature of TMS-based measures of cortical excitatory and inhibitory physiology in both adults and youth, with particular consideration given to what is known about the development of these neural systems in children and adolescents. The limited literature focused on therapeutic repetitive TMS (rTMS) in young patients with depression is then reviewed, followed by a discussion of newer TMS dosing strategies and other emerging developments in the field that are likely to impact how TMS will be utilized in this population in the near future.

EXCITATORY AND INHIBITORY CORTICAL PHYSIOLOGY IN DEPRESSED CHILDREN AND ADOLESCENTS

TMS-based experimental methods have contributed not only to the understanding of how cortical functions are altered in depression but also to how these differences may be developmentally distinct in children and adolescents. The physiology of cortical circuits can be measured noninvasively by the application of magnetic pulses to the cerebral cortex while observing the evoked responses via electroencephalography (EEG) or electromyography (EMG). In TMS–EMG, the primary motor cortex is stimulated with single, paired, or triple magnetic pulses while simultaneously monitoring responses in peripheral muscles corresponding to the region of motor cortex being stimulated (Fig. 7.1A). EMG electrodes are typically placed over the superficial upper extremity muscles, such as the abductor pollicis brevis (APB), first dorsal interosseous, or abductor digiti minimi muscles. The active motor threshold (AMT) and resting motor threshold (RMT) index the minimum intensity of a single magnetic pulse necessary to induce a motor evoked potential (MEP) reliably in a contracting or relaxed muscle, respectively (Rossini et al., 1994, 2015). Motor thresholds are believed to represent the overall excitability of cortico-cortical axons (Di Lazzaro et al., 2008; Ziemann et al., 2015), which depends on both α-amino-3-hydroxy-5-methyl-4-isoxazolepropionic acid (AMPA) and *N*-methyl-D-aspartate (NMDA) ionotropic glutamatergic receptors as well as voltage-gated sodium channels (Ziemann et al., 2015). When a single suprathreshold pulse is delivered to the cortex while the corresponding muscle is engaged in voluntary contraction, a brief (up to 300 ms), transient interruption of motor tone is observed (Day et al., 1989a,b; Cantello et al., 1992). This effect, known as the cortical silent period (CSP; Fig. 7.1B), has been posited to reflect $GABA_B$ [γ-aminobutyric acid (GABA) metabotropic receptor family B] receptor-mediated inhibitory transmission (Siebner et al., 1998;

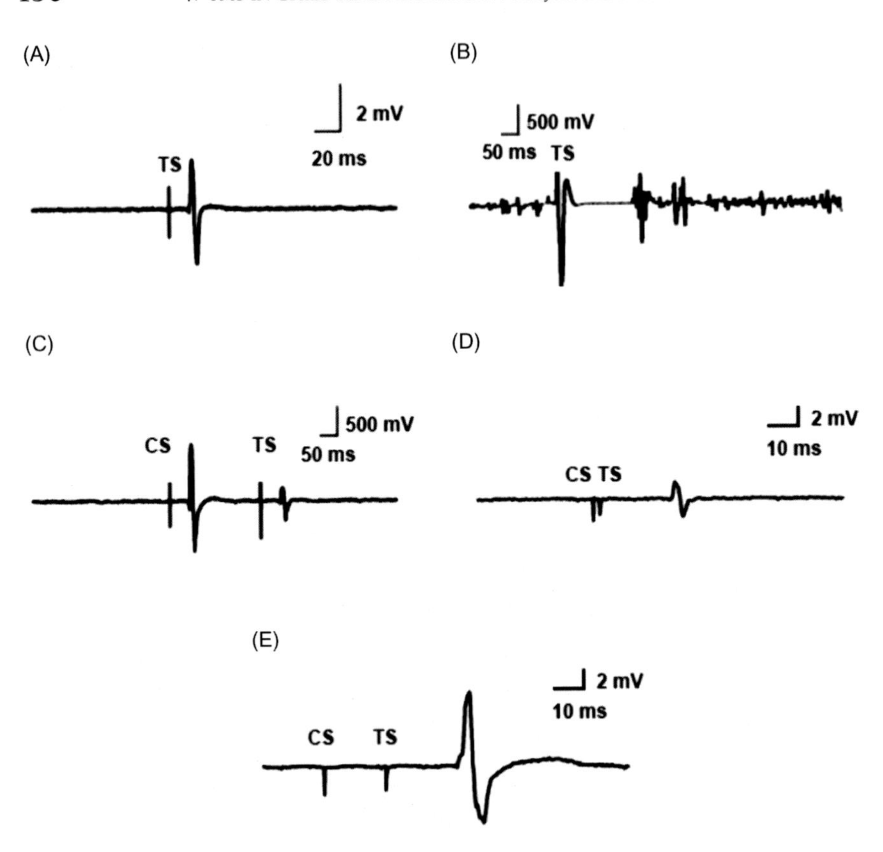

FIGURE 7.1 Electromyography recordings produced by transcranial magnetic stimulation. (A) A single test stimulus applied to the motor cortex producing an MEP. (B) The CSP starts at the onset of the MEP and ends with the return of motor activity. This is achieved by a 40% suprathreshold pulse applied to the motor cortex while the contralateral hand muscle is tonically activated. (C) LICI: A suprathreshold conditioning stimulus precedes a suprathreshold test stimulus by 100 ms, inhibiting the MEP produced by the test stimulus. (D) SICI: A subthreshold conditioning stimulus precedes a suprathreshold test stimulus by 2 ms, inhibiting the MEP produced by the test stimulus. (E) ICF: A subthreshold conditioning stimulus precedes a suprathreshold test stimulus by 20 ms, facilitating the MEP produced by the test stimulus. *CSP*, cortical silent period; *ICF*, intracortical facilitation; *LICI*, long-interval intracortical inhibition; *MEP*, motor evoked potential; *SICI*, short-interval intracortical inhibition. *Reprinted from Radhu, N., de Jesus, D.R., Ravindran, L. N., Zanjani, A., Fitzgerald, P.B., Daskalakis, Z.J., 2013. A meta-analysis of cortical inhibition and excitability using transcranial magnetic stimulation in psychiatric disorders. Clin. Neurophysiol. 124 (7), 1309–1320. © 2013, with permission from Elsevier.*

Werhahn et al., 1999; Stetkarova and Kofler, 2013) by cortical (rather than spinal) interneurons (Fuhr et al., 1991; Inghilleri et al., 1993; Roick et al., 1993; Siebner et al., 1998; Stetkarova and Kofler, 2013). However, the CSP may also be affected by $GABA_A$ (GABA ionotropic receptor family A) activity under certain circumstances (Inghilleri et al., 1996; Ziemann et al., 1996a; Kimiskidis et al., 2006).

In paired- and triple-pulse paradigms, varying the intensity of the pulses and the time intervals between stimuli results in conditioning effects that demonstrate excitation and inhibition of cortical networks. By pairing a sub-motor threshold conditioning stimulus with a subsequent suprathreshold test stimulus, the amplitude of the resultant MEP can be decreased (inhibited) or increased (facilitated). When the conditioning and test stimuli are separated by a brief interstimulus interval of 1–5 ms, the conditioned MEP amplitude is diminished, a phenomenon termed short-interval intracortical inhibition (SICI; Fig. 7.1D) (Kujirai et al., 1993; Hanajima et al., 1996; Ziemann et al., 1996b; Di Lazzaro et al., 1998; Fisher et al., 2002). SICI is believed to be related primarily to cortical $GABA_A$ receptor-mediated transmission (Hanajima et al., 1998, 2003; Di Lazzaro et al., 2000, 2005; Ilić et al., 2002). By contrast, pairing the same subthreshold conditioning stimulus and suprathreshold test stimulus, separated by a slightly longer interval of 7–20 ms, results in amplification of the conditioned MEP (relative to the amplitude of the unconditioned MEP that occurs in response to single-pulse TMS), which is known as intracortical facilitation (ICF; Fig. 7.1E) (Kujirai et al., 1993; Hanajima et al., 1996; Ziemann et al., 1996b). ICF is thought to result primarily from cortical glutamatergic receptor-mediated neurotransmission (Liepert et al., 1997; Ziemann et al., 1998; Schwenkreis et al., 1999, 2000) in a population of neurons separate from those involved in the SICI effect (Ziemann et al., 1996b; Hanajima et al., 1998). However, ICF is less well defined than other TMS–EMG paradigms, may have greater individual variation, and likely has complex mechanisms that also involve corticospinal and spinal neurons (Di Lazzaro et al., 2006). Another paired-pulse inhibitory paradigm, long-interval intracortical inhibition (LICI; Fig. 7.1C), involves two suprathreshold pulses, separated by an interval of 100–200 ms, resulting in inhibition of the conditioned MEP (again, relative to the unconditioned MEP in response to single-pulse TMS) (Valls-Solé et al., 1992; Nakamura et al., 1997; Di Lazzaro et al., 2002). LICI is posited to be mediated by the $GABA_B$ receptor (Nakamura et al., 1997; Werhahn et al., 1999; Pierantozzi et al., 2004; McDonnell et al., 2006).

A related technique, paired-associative stimulation (PAS), presents a means for assessing long-term potentiation (LTP)-related neuroplasticity. During PAS, MEPs are assessed with single-pulse TMS applied to the left motor cortex for a baseline. Subsequently, peripheral nerve stimulation is delivered to the median nerve prior to a single pulse of TMS applied to the left motor cortex with a 25-ms interstimulus interval. Single-pulse TMS MEPs are then generated and compared to baseline (Stefan et al., 2000). Recent work examined PAS changes in 34 healthy adolescents and demonstrated that the procedure was tolerable and safe. This index may have utility for the study of adolescent depression in the future (Lee et al., 2017).

The integration of TMS and EEG is an important innovation that will continue to refine our understanding of cortical neurophysiology in health and disease. This is critical both in the study of neuropsychiatric disease and in the study of neurodevelopment, as TMS–EEG methodology provides a means of examining areas of the brain beyond the motor cortex, particularly regions such as the dorsolateral prefrontal cortex (DLPFC), which undergoes substantial maturation during development and has been implicated in numerous neuropsychiatric disorders. This ongoing work has required the development of specialized amplifiers and a variety of both online and offline techniques to mitigate the effects of TMS-related artifacts during the collection of TMS-evoked potentials. However, novel work with TMS–EEG methodologies has demonstrated the ability to assess LICI (Daskalakis et al., 2008), CSP (Farzan et al., 2013), SICI (Cash et al., 2017), and ICF (Cash et al., 2017) in the DLPFC (reviewed in Farzan et al., 2016). TMS–EEG measures of LICI have demonstrated utility as biomarkers of response in brain stimulation interventions (Sun et al., 2016) and are currently under further study and development in large, multicenter trials. Similar preliminary data has been collected from adolescents and young adults with depression. Future TMS–EEG work in adolescents likely will enhance our understanding of neurodevelopment and catalyze the development of biomarkers for therapeutic brain stimulation interventions.

For a more extensive review of the animal and human literature which have helped to elucidate the mechanisms of TMS–EMG and TMS–EEG paradigms, see Ziemann et al. (2015) and Farzan et al. (2016).

Neurophysiologic Studies in Adults With Depression

Single- and paired-pulse TMS have been used to investigate cortical functions in a variety of neurologic and psychiatric conditions, including major depression. Early work in adults indicated alterations in cortical inhibition in depressed individuals. Steele et al. (2000) found that the duration of the CSP was increased in 16 adults with a current major depressive episode compared to 19 healthy controls ($P = .04$), indicating greater inhibition in the depressed patients. Depressive symptom severity, measured by the Hamilton Depression Rating Scale (HDRS), did not correlate with CSP duration. Several limitations of this study warrant consideration, however. Depressed patients could have either a unipolar or bipolar diagnosis. Whereas prior work has demonstrated that euthymic adults with bipolar disorder have shortened CSP compared to healthy controls (Levinson et al., 2007), it is uncertain whether bipolar patients in the depressive phase of illness show similar effects on the CSP relative to healthy controls and individuals with unipolar depression. In addition, all but one of the patients were taking an antidepressant medication, which may have impacted cortical inhibition; prior research has shown CSP duration to increase with citalopram (Robol et al., 2004), but not with clomipramine (Manganotti et al., 2001).

In contrast, later studies found decreased inhibition among depressed patients. Bajbouj et al. (2006) compared TMS measures of cortical excitability and inhibition between 20 right-handed, medication-free individuals with major depression and 20 right-handed controls matched for age and gender. CSP was significantly shorter in both left ($P = .003$) and right ($P = .002$) hemispheres in depressed patients than in controls, while SICI was decreased in depressed patients compared to controls (left hemisphere, $P = .001$; right hemisphere, $P = .004$). In addition, while no differences in ICF were found, the RMT was lower in depressed individuals ($P < .0001$), but only in the right hemisphere. Depression severity (HDRS score) was inversely correlated with SICI, reaching significance in the left hemisphere ($P = .0045$). In a larger sample of 35 right-handed depressed adults taking psychotropic medications and 35 right-handed controls, Lefaucheur et al. (2008) examined cortical excitability and inhibition separately in each hemisphere with similar methods. Depressed patients demonstrated increased RMT (left hemisphere, $P = .016$), reduced ICF (left hemisphere, $P = .047$), shorter CSP (left hemisphere, $P = .005$), and reduced SICI (left hemisphere, $P = .0001$; right hemisphere, $P = .0006$). The investigators also noted laterality differences, with shorter CSP and reduced SICI and ICF in the left hemisphere in depressed patients, but no left versus right differences in controls. Several correlations between indices of excitability/inhibition and clinical measures were observed.

In a landmark study, Levinson et al. (2010) utilized TMS–EMG methodology to examine cortical excitation and inhibition in the left motor cortex in four participant groups: 16 currently depressed, unmedicated patients with major depressive disorder (MDD); 19 medicated, currently euthymic patients with previous MDD; 25 patients with treatment-resistant MDD (medicated and continuing to be depressed); and 25 healthy controls. All three depressed groups had significantly shorter CSP than healthy controls (all $P \leq .001$), while there were no differences between the depressed groups in CSP duration ($P = .96$). By contrast, SICI was significantly impaired in the treatment-resistant MDD group compared to all other groups (all $P \leq .017$), while the euthymic medicated MDD, unmedicated MDD, and healthy control groups did not differ in SICI ($P = .93$). No group differences in ICF were observed, although the treatment-resistant patients had higher mean RMT than all other groups (all $P \leq .042$). In addition, no significant correlations between depression severity (HDRS score) and either CSP duration or SICI were found. In light of the observation of shortened CSP in all depressed patients but impaired SICI in only the treatment-resistant group, the authors proposed that impaired $GABA_B$ receptor-mediated inhibitory neurotransmission may be a general feature of depressive pathophysiology, whereas more severe illness, even when treated with medication, may also involve $GABA_A$ dysfunction (indexed by SICI) and general frontal hypoexcitability (indexed by RMT) (Levinson et al., 2010).

Adult findings on TMS–EMG measures of cortical excitability and inhibition were compared across studies in a meta-analysis by Radhu et al. (2013). No differences between depressed individuals and healthy controls were found for the RMT (Hedge's $g = -.043$, $P = .677$), ICF (Hedge's $g = -.062$, $P = .628$), or MEP amplitude (Hedge's $g = .162$, $P = .492$). However, in 131 depressed patients and 149 controls across studies, CSP was significantly shorter in MDD patients (Hedge's $g = -1.232$, $P = .000$). Reduced SICI was also observed among depressed adults ($n = 115$) compared to controls ($n = 130$) across studies (Hedge's $g = .641$, $P = .000$). In summary, the adult TMS–EMG literature to date has indicated fairly consistent deficits in measures of cortical inhibition (CSP and SICI), while measures of excitability (such as RMT and ICF) have demonstrated greater variability.

Cortical Excitability and Inhibition in Depressed Children and Adolescents

The literature on TMS-measured cortical physiology in children and adolescents with major depression is more limited than in adults. Utilizing similar TMS–EMG methodology, Croarkin et al. (2013) investigated excitatory (RMT, ICF) and inhibitory (CSP, SICI) measures in a sample of 46 youth (aged 9–17 years). Twenty-four children and adolescents with MDD, who were medication-naïve and were not undergoing psychotherapy, underwent TMS testing prior to enrolling in a separate antidepressant treatment study. The depressed youth were compared with 22 healthy control participants, who had no personal histories of psychiatric diagnosis or treatment and no psychiatric illness in first- or second-degree relatives. Groups did not differ in age, sex distribution, or handedness. TMS–EMG measures from both left and right hemispheres were compared between groups in a linear mixed model covarying for age, sex, and depression severity scores. In contrast to the previous adult findings, the authors found no difference in cortical inhibition between depressed and control groups. In addition, no difference in RMT was found between depressed adolescents and controls. However, significant group main effects were observed in the ICF paradigms at interstimulus intervals of 10 ms (raw $P = .01$, adjusted $P = .03$) and 15 ms (raw $P = .001$, adjusted $P = .007$), but not at 20 ms (raw $P = .47$, adjusted $P = .71$), with the MDD group demonstrating higher mean conditioned MEP amplitude (i.e., greater facilitation). When examining ICF within each hemisphere independently, depressed adolescents demonstrated significantly greater facilitation in the left hemisphere at both 10 ms (raw $P = .01$, adjusted $P = .04$) and 15 ms (raw $P = .001$, adjusted $P = .007$) and in the right hemisphere at 15 ms only (raw $P = .001$, adjusted $P = .005$).

In addition to the main group findings reported by Croarkin et al. (2013), several post hoc analyses from the same adolescent sample examined correlations between TMS–EMG measures and other clinical variables. In an exploratory analysis, Lewis et al. (2016) found negative correlations between CSP duration and depression severity in the overall sample and depressed group (with greater depression severity corresponding to shorter CSP). In the paired-pulse ICF paradigms, the conditioned MEP amplitude (relative to unconditioned MEP in single-pulse TMS) also was inversely correlated with subsyndromal depressive symptom severity in the healthy controls (at 10-, 15-, and 20-ms interstimulus intervals) and the more substantial depressive symptom severity in the depressed group (10- and 15-ms intervals). While the majority of these relationships did not retain significance after correction for the large number of comparisons, they suggest the possibility of cortical excitability and inhibition being related to important clinical characteristics, even on measures that did not differ between controls and depressed adolescents in the group analysis (such as CSP).

Other analyses with this adolescent sample explored the utility of TMS-based markers as predictors of treatment response. Among the medication-naïve depressed participants, conditioned MEP amplitude ratios in the LICI paradigms at baseline differed between participants who responded and those who failed to respond in a subsequent trial of fluoxetine, with nonresponders exhibiting more significantly impaired inhibition (100 ms: raw $P = .01$, adjusted $P = .02$; 150 ms: raw $P = .03$, adjusted $P = .03$; 200 ms: raw $P = .01$, adjusted $P = .02$) (Croarkin et al., 2014a). Future work may help to determine whether improvement in clinical symptoms with treatment is commensurate with improvement in pretreatment cortical inhibitory deficits measured with TMS.

Converging lines of evidence indicate an age-dependent excitation–inhibition imbalance in depression. For example, the pattern of increased glutamatergic cortical facilitation in children and adolescents with depression, in contrast to the adult studies that demonstrated reduced GABAergic cortical inhibition, suggests the possibility that depression has a distinct excitatory–inhibitory pathophysiology in youth. This is plausible considering the maturation that GABAergic and glutamatergic systems undergo during development, which has been investigated previously via TMS–EMG methods in healthy and depressed children and adolescents. Walther et al. (2009) obtained RMT, SICI, and ICF measures in a healthy volunteer sample of 10 children (≤10 years), 10 adolescents (11–17 years), and 10 adults (≥18 years). RMT and ICF measures did not differ significantly between age groups. However, in the SICI paradigms, adults showed significantly greater inhibition than children (1-, 3-, and 5-ms interstimulus intervals) and adolescents (5-ms interval), whereas children and adolescents did not differ. In an exploratory analysis of the 24 depressed and 22 healthy control children and adolescents in their previous study, Croarkin et al. (2014b) found a significant negative relationship between RMT and age, which was observed in both hemispheres in the overall sample as well as in depressed adolescents, replicating previous findings of an inverse age–motor threshold relationship in children. In a subset of 33 adolescents who also underwent LICI testing, age had a significant, negative relationship with conditioned MEP amplitude (relative to unconditioned single-pulse MEP) at the 200 ms interval, which occurred in both hemispheres. These relationships were observed in the overall sample and were even more pronounced in the depressed group (Croarkin et al., 2014b). These studies suggest that overall inhibitory tone increases across development in both healthy and depressed populations.

The contribution of various GABAergic and glutamatergic receptors to the excitatory–inhibitory balance is complex and dynamic across neural development. Expression and binding of GABA receptors is reduced in early life compared to later in development (Silverstein and Jensen, 2007); however, functional GABAergic synapses generally precede the development of glutamatergic transmission (Ben-Ari et al., 1997). In early postnatal development, $GABA_A$ receptors exhibit excitatory activity, causing depolarization of neurons, later shifting to their mature inhibitory function as age progresses (Ben-Ari et al., 1997; Rakhade and Jensen, 2009). Multiple mechanisms may be involved in these early paradoxical excitatory functions. The $GABA_A$ receptor is composed of various protein subunits that affect the receptor's function, and the relative expression of subunit types shifts across development (Duncan et al., 2010), and at varying rates in different brain regions, well into the early adult years (Chugani et al., 2001). Expression of potassium–chloride and sodium–potassium–chloride cotransporters that affect the overall function of the receptor's ion channel also varies with age (Rakhade and Jensen, 2009). Furthermore, the postsynaptic inhibitory function of the metabotropic $GABA_B$ receptor is absent in early life (Ben-Ari et al., 1997; Leinekugel et al., 1999). Glutamate receptors, which play an essential role in learning and neural plasticity, are also differentially expressed during early development (Silverstein and Jensen, 2007). In the adult brain, both AMPA and NMDA glutamatergic receptors interact synergistically to regulate excitatory transmission, whereas in the immature cortex, AMPA receptors are functionally inert, with NMDA receptors being relatively overexpressed and dominating excitatory function (Ben-Ari et al., 1997; Silverstein and Jensen, 2007). These considerable differences in receptor functioning, coupled with other long-term neurodevelopmental processes such as myelination, connectivity (Koerte et al., 2009), and synaptic pruning, contribute to substantial functional differences in the developing brain.

In summary, the limited investigations in children and adolescents with depression to date indicate that depressed youth may have a different pattern of aberrant cortical excitability than depressed adults. In addition to normal developmental changes in GABAergic and glutamatergic neurotransmission, children and adolescents with depression may have further alterations in excitatory and inhibitory functions that impact the development of cortical circuitry (Figs. 7.2 and 7.3). While this provides a rationale for neuromodulatory interventions in depressed youth, it simultaneously demands careful consideration of the developmental context in which disease-related changes occur.

Effects of rTMS on Cortical Excitatory and Inhibitory Systems in Depressed Children and Adolescents

rTMS is believed to exert its therapeutic effects by modulating the excitatory–inhibitory balance in neural circuits, not only in the stimulated region of cortex [typically left DLPFC (L-DLPFC)] but also through network connections to other cortical and subcortical structures (Fig. 7.4) that are related to the heterogeneous symptomatology of depression (Anderson et al., 2016). However, as noted above, the development of cortical circuits in youth is highly dynamic, and to date only a small number of studies have examined the neurobiological changes that result from rTMS in children and adolescents with depression.

The effect of rTMS on excitatory cortical physiology was reported by Croarkin et al. (2012), who assessed RMT via TMS–EMG in eight depressed adolescents undergoing an open-label trial of rTMS [10 Hz stimulus frequency, 120% motor threshold (MT) stimulus intensity, L-DLPFC coil placement, 30 treatments over 6–8 weeks, 3000 pulses per session]. RMT was measured, via a computerized algorithm and observation of APB movement, at baseline and at weeks 2, 4, and 5. The repeated-measures mixed model did not reveal an overall

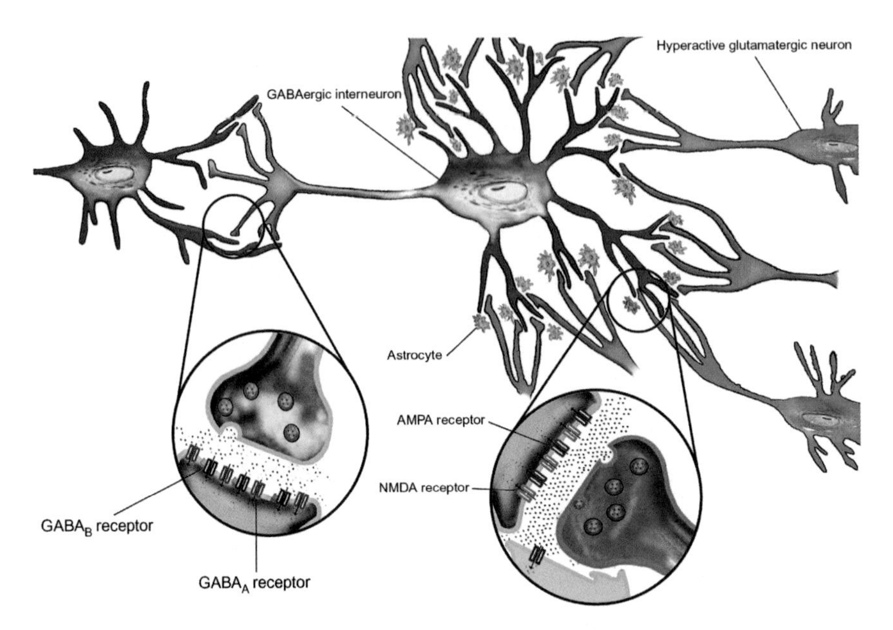

FIGURE 7.2 Increased cortical excitability in childhood depression. This illustration presents a theoretical model regarding the impact of increased glutamatergic intracortical facilitation in child and adolescent depression. Increased intracortical facilitation in depressed children and adolescents indicates increased cortical glutamatergic NMDA receptor functioning. At this stage, GABA$_A$ receptor-mediated and GABA$_B$ receptor-mediated neurotransmission are unchanged because GABAergic interneurons are functioning normally. However, increased NMDA activity is thought to play a key role in excitotoxic effects. The impact is likely widespread and could subsequently impair or damage GABAergic interneurons. *AMPA*, α-amino-3-hydroxy-5-methyl-4-isoxazolepropionic acid; *GABA$_A$*, γ-aminobutyric acid ionotropic receptor family A; *GABA$_B$*, γ-aminobutyric acid metabotropic receptor family B; *NMDA*, *N*-methyl-D-aspartate. *Reproduced with permission from Croarkin, P.E., Nakonezny, P.A., Husain, M.M., Melton, T., Buyukdura, J.S., Kennard, B.D., et al., 2013. Evidence for increased glutamatergic cortical facilitation in children and adolescents with major depressive disorder. JAMA Psychiatry 70 (3), 291–299. doi:10.1001/2013.jamapsychiatry.24.*

within-subjects effect on RMT ($P = .32$). Hence, RMT had no significant decrease across the treatment schedule. However, least squares mean change from baseline to week 5 approached significance ($P = .07$, Cohen's $d = 0.49$), which was significantly different from zero in a post hoc nonparametric test ($P = .03$). It is difficult and problematic to draw conclusions from this small sample. These results suggest that high-frequency rTMS may increase cortical excitability in depressed adolescents. Alternatively, RMT may have more individual variability in youth. Further study on the effects of rTMS on cortical excitability is warranted prior to the widespread clinical use of rTMS in adolescents with depression.

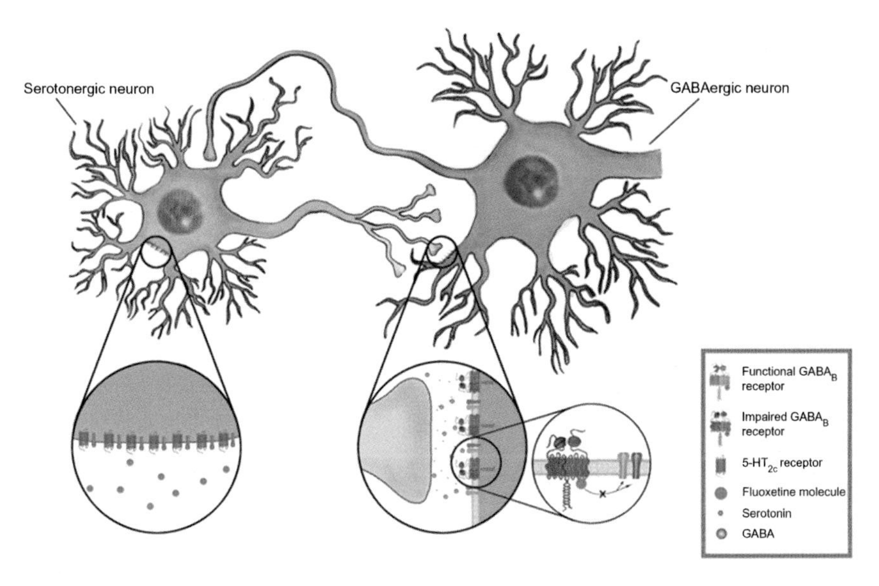

FIGURE 7.3 GABA$_B$ and 5-HT$_{2C}$ are G protein–linked receptors with putative roles in depression and widespread distribution in the cortex. Prior work suggests that presynaptic GABA$_B$ receptors regulate serotonergic neurons and that some SSRIs may potentiate GABAergic neurotransmission. Fluoxetine's mechanism of action is unique among antidepressants due to its 5-HT$_{2C}$ antagonism. This is noteworthy as 5-HT$_{2C}$ receptors are expressed on GABAergic neurons. *5-HT$_{2C}$*, 5-hydroxytryptamine (serotonin) receptor subtype 2C; *GABA$_B$*, γ-aminobutyric acid metabotropic receptor family B; SSRI, selective serotonin reuptake inhibitor. *Reprinted from Croarkin, P.E., Nakonezny, P.A., Husain, M.M., Port, J.D., Melton, T., Kennard, B.D., Emslie, G.J., Kozel, F.A., Daskalakis, Z.J., 2014. Evidence for pretreatment LICI deficits among depressed children and adolescents with nonresponse to fluoxetine. Brain Stimul. 7 (2), 243–251. © 2014, with permission from Elsevier.*

Magnetic resonance spectroscopy (MRS) is a noninvasive imaging technique that permits in vivo measurement of concentrations of neural metabolites based on their magnetic resonance spectra. Among the substances that can be assessed via MRS are amino acid neurotransmitters and their metabolites, including glutamate (Glu), glutamine (Gln), and GABA (Maddock and Buonocore, 2012) (Fig. 7.5). Several pediatric MRS studies have examined glutamatergic and GABAergic metabolites in the L-DLPFC and the anterior cingulate cortex (ACC), two cortical regions involved in the pathophysiology of depression (Kondo et al., 2011). L-DLPFC Glu concentrations did not differ between depressed adolescents and healthy controls, but significant inverse correlations were found between Glu concentration and both number of depressive episodes and duration of illness (Caetano et al., 2005). However, adolescents with MDD have demonstrated reduced concentrations of ACC Glu (Rosenberg et al., 2005) and Glu + Gln (termed Glx) (Mirza et al., 2004; Rosenberg et al., 2004) compared to healthy controls. In addition,

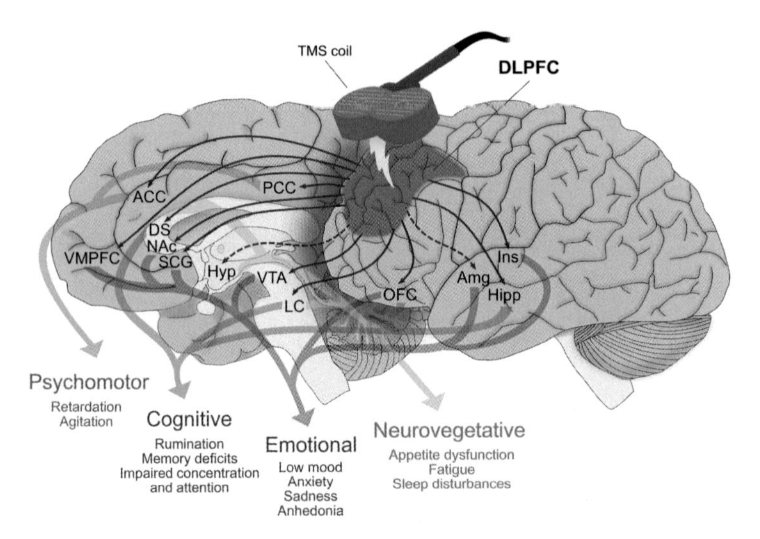

FIGURE 7.4 Anatomical connectivity of the DLPFC to regions involved in depression symptomatology. Solid arrows represent direct connectivity, dashed arrows represent sparse or indirect connectivity. *ACC*, anterior cingulate cortex; *Amg*, amygdala; *DLPFC*, dorsolateral prefrontal cortex; *DS*, dorsal striatum; *Hipp*, hippocampus; *Hyp*; hypothalamus; *Ins*, insula; *LC*, locus coeruleus; *NAc*, nucleus accumbens; *OFC*, orbitofrontal cortex; *PCC*, posterior cingulate cortex; *SCG*, subgenual cingulate gyrus; *TMS*, transcranial magnetic stimulation; *VMPFC*, ventromedial prefrontal cortex; *VTA*, ventral tegmental area. *Reprinted from Anderson, R.J., Hoy, K.E., Daskalakis, Z.J., Fitzgerald, P.B., 2016. Repetitive transcranial magnetic stimulation for treatment resistant depression: re-establishing connections. Clin. Neurophysiol. 127 (11), 3394–3405. © 2016, with permission from Elsevier.*

anterior cingulate GABA concentrations have been found to be significantly lower in depressed adolescents than in healthy youth, with ACC GABA having a particular relationship with the clinical phenotype of anhedonic depression (Gabbay et al., 2012).

Despite the increasing body of evidence for alterations of glutamate and GABA in these cortical structures implicated in pediatric depression, few studies have utilized MRS to assess the effect of rTMS on cortical glutamate concentrations in depressed youth. In a small case series, Yang et al. (2014) examined L-DLPFC glutamatergic metabolites in six patients (ages 15–21) who underwent high-frequency rTMS (10 Hz, 120% MT, L-DLPFC, 3000 pulses per session, 15 treatments over 15 consecutive workdays) for MDD. Patients who responded to rTMS ($n = 3$), defined as those who experienced a decrease of at least 50% on HDRS and Beck Depression Inventory (BDI) scores, demonstrated an 11% mean increase in L-DLPFC Glu concentration between baseline and the end of 3 weeks of treatment. rTMS nonresponders, however, exhibited higher baseline glutamate but experienced a 10% mean decrease in L-DLPFC glutamate with treatment. Croarkin et al. (2016) investigated glutamatergic metabolites in the L-DLPFC and medial ACC (Fig. 7.6) in

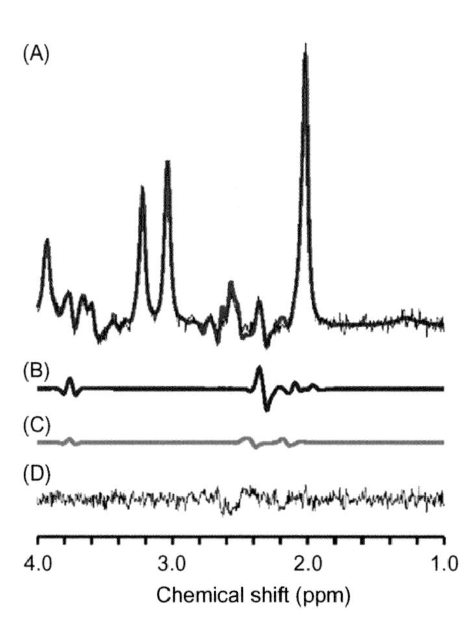

FIGURE 7.5 Representative TE-optimized PRESS spectrum obtained from the L-DLPFC. (A) Fitted LCModel spectrum (red) overlaid on the source spectrum (black). (B) Individual fit of the Glu metabolite. (C) Individual fit of the Gln metabolite. Note that the area under the curve is significantly less than for Glu due to the relatively low concentration of Gln. (D) Residual spectrum after subtracting the fitted spectrum from the original spectrum. *Gln*, glutamine; *Glu*, glutamate; *L-DLPFC*, left dorsolateral prefrontal cortex; *PRESS*, point resolved spectroscopic sequence; *TE*, echo time. *Reprinted from Croarkin, P.E., Nakonezny, P.A., Wall, C. A., Murphy, L.L., Sampson, S.M., Frye, M.A., Port, J.D., 2016. Transcranial magnetic stimulation potentiates glutamatergic neurotransmission in depressed adolescents. Psychiatry Res. Neuroimaging 247, 25–33. © 2016, with permission from Elsevier.*

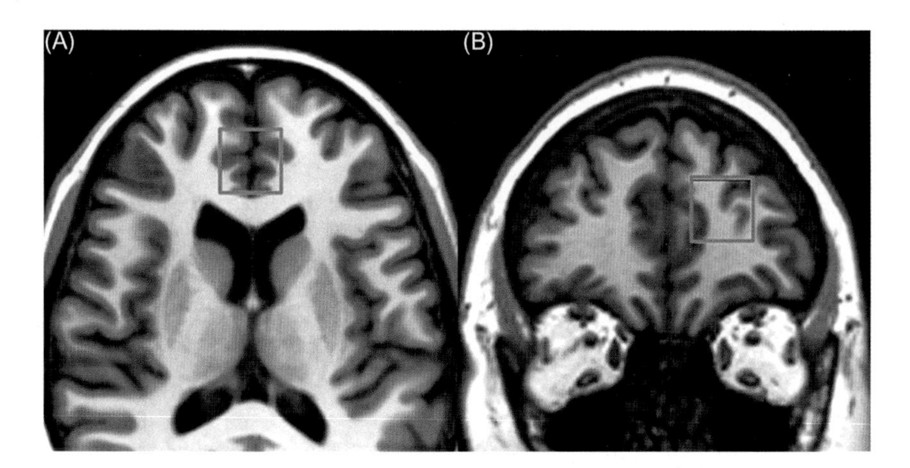

FIGURE 7.6 Locations of the two sampled MRS voxels. (A) The ACC voxel is placed in the midline to sample pregenual anterior cingulate cortex (Brodmann's areas 24a, 24b, and 32). (B) The L-DLPFC voxel is placed to sample dorsolateral prefrontal cortex and subjacent white matter (Brodmann's areas 9 and 46). *ACC*, anterior cingulate cortex, *L-DLPFC*, left dorsolateral prefrontal cortex; *MRS*, magnetic resonance spectroscopy. *Reprinted from Croarkin, P. E., Nakonezny, P.A., Wall, C.A., Murphy, L.L., Sampson, S.M., Frye, M.A., Port, J.D., 2016. Transcranial magnetic stimulation potentiates glutamatergic neurotransmission in depressed adolescents. Psychiatry Res. Neuroimaging 247, 25–33. © 2016, with permission from Elsevier.*

10 adolescents (ages 13–17) with MDD who underwent rTMS (10 Hz, 120% MT, L-DLPFC, 3000 pulses per session, up to 30 sessions over 6–8 weeks). Patients were scanned at baseline, end of treatment, and 6 months posttreatment. In a linear mixed model analysis, the Gln/Glu ratio increased in both regions during treatment and showed continued increase at 6 months; this was significant for one MRS sequence in the ACC ($P_{corrected} = .04$) and approached significance in the L-DLPFC ($P_{corrected} = .08$). The results did not change when age and baseline depression severity were included as covariates in the models. In addition, a significant inverse relationship was observed between the L-DLPFC Gln/Glu ratio and depression severity ($P_{corrected} = .01$). Although both studies were limited by small sample sizes, these preliminary data suggest that cortical glutamate is dynamic and may be altered by rTMS in depressed adolescents. Moreover, the quantitative, noninvasive measurement of excitatory neurotransmitters may have future roles in predicting which patients are most likely to benefit from stimulation and in brain-based monitoring for response to treatment with rTMS.

THERAPEUTIC APPLICATIONS OF TMS IN DEPRESSED CHILDREN AND ADOLESCENTS

In addition to its utility as a noninvasive tool for assessing the excitatory and inhibitory functions of neural circuits, TMS can be used to induce lasting changes in cortical excitability and inhibition when applied repeatedly. In general, low-frequency rTMS (≤1 Hz) is thought to decrease cortical excitability, while high-frequency rTMS (≥10 Hz) increases cortical excitability. More recently, a brief but patterned form of rTMS, theta burst stimulation (TBS), also has been shown to impact cortical inhibition and excitation. Compared to conventional rTMS approaches, TBS more closely mimics the endogenous theta/gamma rhythms of the brain and is suggested to modify cortical activity more effectively through induction of LTP-like and long-term depression (LTD)–like plasticity (Huang et al., 2005). Intermittent TBS (iTBS) produces LTP-like plasticity in the cortex, whereas continuous TBS (cTBS) produces LTD-like plasticity or inhibition in the cortex. However, these assumptions are likely overly simplistic and particularly problematic in the context of neurodevelopment (Wilson and St George, 2016). Nevertheless, the ability of repeated stimulation paradigms to induce such changes in cortical excitability and inhibition has significant potential for clinical applications in conditions that involve disruptions in the excitatory and inhibitory functioning of cortical networks. Increasing evidence indicates that several rTMS approaches are effective treatments for major depressive episodes in adults (Brunoni et al., 2017). Early, strategically timed interventions in depressed youth might hold the prospect of mitigating a lifetime of future disease burden. To date, TMS has been utilized as a therapeutic intervention in a small number of studies of depressed children and adolescents, with initial data on efficacy, adverse effects, and neurocognitive performance suggesting promising possibilities for its continued development as an antidepressant treatment in this population.

Efficacy of rTMS for Treating Depression in Children and Adolescents

Several small case series and open-label trials have examined the therapeutic efficacy of rTMS in children and adolescents with depression (Table 7.1). While these studies have shared some methodological aspects, such as enrollment of patients who had not responded to antidepressant medications, the rTMS treatment parameters have varied considerably, particularly in the earlier studies.

In the earliest published report of the use of rTMS in adolescents with depression, Walter et al. (2001) presented seven adolescent cases (three with medication-refractory unipolar major depression, one with bipolar depression, and three with schizophrenia) who had been treated with rTMS in three separate trials. The three patients with unipolar depression (all male, ages 16–17), two of whom had prior trials of electroconvulsive therapy (ECT) and none of whom were receiving psychotropic medication at the time of rTMS, received 10 Hz stimulation at 90%–110% MT to the L-DLPFC, 1600 pulses per session, for a total of 10 sessions over 2 weeks. Two of the three patients demonstrated substantial improvement in HDRS and BDI scores at the end of 2 weeks of treatment, which was maintained at 4 weeks; the third patient showed limited improvement. The single adolescent patient with bipolar depression (female, age 18) received 10 treatments of low-frequency (1 Hz) rTMS to the right DLPFC at 110% MT, 1600 pulses per session in a single train, over 2 weeks. The patient was taking gabapentin, venlafaxine, clonazepam, and nortriptyline concurrently and showed no improvement in HDRS or Clinical Global Impression (CGI) scores at the end of treatment (week 2) but substantial improvement at week 4, raising the intriguing question of whether delayed or continued improvement can occur after the cessation of stimulation sessions.

Loo et al. (2006) described the first two cases from their double-blinded, sham-controlled trial, both of whom received active stimulation. Both patients were 16 years of age and female; one was unmedicated, the other taking venlafaxine and methylphenidate at the time of treatment. The patients underwent 10 Hz rTMS to the L-DLPFC at 110% MT, 2000 pulses per session, 29–36 sessions over 6–11 weeks (one patient had a

TABLE 7.1 Case Series and Open-Label Trials of rTMS in Depressed Children and Adolescents

Study	N	Age (y)	Population	Concurrent psychotropic medication	Stimulation site	Frequency (Hz)	Stimulus intensity (%MT)	Stimuli/Train	Trains	Inter-train interval (s)	Stimuli/Session	Sessions	Duration (weeks)	Antidepressant efficacy outcomes	Adverse effects
Walter et al. (2001)[a]	3	16–17	TRD	No	L-DLPFC	10	90–110	40–80	20–40	52–56	1600	10	2	TRD: 2/3 patients demonstrated improvement on HDRS and BDI at end of treatment and 2 weeks after rTMS	Tension headache during sessions (1/3 patients)
	1	18	BD	Yes	R-DLPFC	1	110	1600	1	n/a	1600	10	2	BD: no improvement on HDRS or CGI at end of treatment but substantial improvement 2 weeks after rTMS	None
Loo et al. (2006)	2	16	MDD	Yes/no	L-DLPFC	10	110	50	40	25	2000	29–36	6–11	Improvement in MADRS, BDI, CES-DC, and CGI-S at end of treatment and 1 month after rTMS; symptomatic and functional improvement maintained at 3 and 4 months	None; no substantial changes in neuropsychological testing between baseline and end of treatment
Bloch et al. (2008)	9	16–18	TRD	Yes	L-DLPFC	10	80	20	20	58	400	14	3	Improvement in CDRS, BDI, CGI-S, and SCARD (all study stage main effects $P < .05$); BDI, CDRS, and SCARD also improved at 1 month after rTMS (all $P < .05$); 3/9 patients met criteria for clinical response (≥30% reduction in CDRS score)	Mild headache (5/9 patients); self-limited hypomanic-like symptoms requiring no additional treatment (1/9 patients); no increase in suicidal ideation (SIQ) or decrease in cognitive performance (CANTAB)

Mayer et al. (2012)[b]													No significant change in CDRS-R, BDI-II scores from 1 month post-rTMS to 3-year follow-up; 5/8 patients minimally depressed, 1/8 mildly depressed, 1/8 moderately depressed, and 1/8 severely depressed at 3 years	No decline in cognitive performance (CANTAB) at 3 years after rTMS
Wall et al. (2011)	8	14–17	TRD	Yes	L-DLPFC	10	120	40	75	26	3000	30	6–8	Improvement in CDRS-R, QIDS-A_{17}-SR, and CGI-S at end of treatment (all $P < .001$); improvement maintained at 6 months (all $P < .001$)	Scalp discomfort (3/8 patients) leading to discontinuation of rTMS (1/8 patients); no worsening of suicidal ideation; no significant decrease in neuropsychological testing performance (AMI, CAVLT-2, D-KEFS) or concern for decline in cognitive functioning; no change in auditory threshold testing
Le et al. (2013)	25	7–16	TS	Yes	Bilateral SMA	1	110	60	20	1	1200	20	4	Small but statistically significant improvement depressive symptoms (CDI) at end of treatment ($P = .001$), maintained at 3- and 6-month follow-up; anxiety and tic severity also improved	Mild somnolence (1/25 patients) during single treatment

(*Continued*)

TABLE 7.1 (Continued)

Study	*N*	Age (*y*)	Population	Concurrent psychotropic medication	Stimulation site	Frequency (Hz)	Stimulus intensity (%MT)	Stimuli/Train	Trains	Inter-train interval (s)	Stimuli/Session	Sessions	Duration (weeks)	Antidepressant efficacy outcomes	Adverse effects
Yang et al. (2014)	8	15–21	TRD	Yes	L-DLPFC	10	120	40	75	26	3000	15	3	Criteria for clinical response (≥50% reduction on HDRS) met in 4/6 patients who completed study, with mean decrease of 68% on HDRS, 84% on BDI, and 78% on HARS; nonresponders (2/6) had mean decrease of 29% on HDRS and 36% on HARS, and mean increase of 19% on BDI	Scalp discomfort and headache; 2/8 patients withdrew (one prior to rTMS, one on second day of treatment)
Wall et al. (2016)	10	13–17	TRD	Yes	L-DLPFC	10	120	40	75	26	3000	30	6–8	Improvement in CDRS-R, QIDS-A_{17}-SR, and CGI-S at end of treatment and 6 months after rTMS (all $P < .05$); 6/10 patients met criteria for clinical response (CGI-I scores of 1 or 2, with CGI-S scores ≤ 3) at end of treatment and 6 months after rTMS	Mild, transient scalp discomfort, headache, dizziness, musculoskeletal discomfort, neck stiffness, eye twitching, nausea; 3/10 patients withdrew (one due to discomfort); no significant decline in neuropsychological testing performance (CAVLT-2, D-KEFS)

[a] *Case series included three adolescents with unipolar TRD and one with BD (depressive episode), as well as three adolescents with schizophrenia.*
[b] *Eight of nine participants from Bloch et al.'s (2008) study completed follow-up evaluations 3 years after rTMS (Mayer et al., 2012).*
AMI, Autobiographical Memory Interview; *BD*, bipolar disorder; *BDI/BDI-II*, Beck Depression Inventory/–II; *CANTAB*, Cambridge Neuropsychological Test Automated Battery; *CAVLT-2*, Children's Auditory Verbal Learning Test–Second Edition; *CDI*, Children's Depression Inventory; *CDRS/CDRS-R*, Children's Depression Rating Scale/–Revised; *CES-DC*, Center for Epidemiologic Studies Depression Scale for Children; *CGI*, Clinical Global Impression; *CGI-I*, Clinical Global Impression–Improvement; *CGI-S*, Clinical Global Impression–Severity; *D-KEFS*, Delis–Kaplan Executive Function System; *HARS*, Hamilton Anxiety Rating Scale; *HDRS*, Hamilton Depression Rating Scale; *L-DLPFC*, left dorsolateral prefrontal cortex; *MADRS*, Montgomery–Åsberg Depression Rating Scale; *MDD*, major depressive disorder; *MT*, motor threshold; *QIDS-A_{17}-SR*, Quick Inventory of Depressive Symptoms–Adolescent–Self-Report; *R-DLPFC*, right dorsolateral prefrontal cortex; *rTMS*, repetitive transcranial magnetic stimulation; *SCARD*, Screen for Child Anxiety Related Disorders; *SMA*, supplemental motor area; *SIQ*, Suicidal Ideation Questionnaire; *TRD*, treatment-resistant depression; *TS*, Tourette syndrome.

13-day interruption in treatment and periodically missed treatments). Both patients demonstrated improvements on depression measures [BDI, Center for Epidemiologic Studies Depression Scale for Children (CES-DC), CGI–Severity (CGI-S), Montgomery–Åsberg Depression Rating Scale (MADRS)] at the end of treatment, which were maintained at 1 month, and continued improvement in symptoms and functioning was reported at follow-up assessments 3 and 4 months after rTMS.

More recently, several open-label trials of rTMS have been conducted in depressed adolescents. Bloch et al. (2008) treated nine adolescents (seven females, two males; ages 16–18) with treatment-resistant major depression (defined as failure of 8-week trials of fluoxetine and at least one other antidepressant as well as failure of at least one course of psychotherapy). Two patients also had previous trials of ECT resulting in partial response. The sample was diverse in terms of comorbidity (all patients carried at least one other psychiatric diagnosis, including obsessive–compulsive disorder, posttraumatic stress disorder, borderline personality disorder, attention-deficit/hyperactivity disorder, eating disorders, and substance abuse) and concurrent pharmacologic treatment (all remained on antidepressant medications during the study, while some also were prescribed atypical antipsychotic agents or methylphenidate). Patients received 10 Hz stimulation at 80% MT to the L-DLPFC (defined as 5 cm anterior to the motor cortex location determined during motor threshold testing), 400 pulses per session, in daily sessions over 14 working days. A significant within-subject effect of study stage was observed on the majority of clinical outcomes, including BDI ($P < .05$), Children's Depression Rating Scale (CDRS, $P < .01$), CGI-S ($P < .01$), and Screen for Child Anxiety Related Disorders (SCARD, $P < .01$). Improvement was observed at day 7 (BDI: $P < .05$, CDRS: $P < .05$), day 10 (BDI: $P < .05$, CDRS: $P < .05$), and at the end of treatment (CDRS: $P < .05$, CGI-S: $P < .05$, SCARD: $P < .05$), as well as 1 month after rTMS (BDI: $P < .05$, CDRS: $P < .01$, SCARD: $P < .05$). Suicidal ideation level, as assessed by the Suicidal Ideation Questionnaire score, was not significantly affected by study stage. According to the study's definition of clinical response (a reduction in CDRS score of at least 30%), three of the nine patients responded to rTMS.

Eight of the nine patients in this sample were reassessed 3 years after completion of rTMS (Mayer et al., 2012). No significant differences in depression severity scores [BDI-II, CDRS–Revised (CDRS-R)] were found between 1 month after treatment termination and 3-year follow-up. Although the patients had a diverse range of psychiatric treatments (including ECT in four patients) following the course of rTMS, overall severity of illness was improved at 3-year follow-up, with five patients classified as minimally depressed, one mildly depressed, one moderately depressed, and one severely depressed.

Later open-label trials utilized stimulation protocols consistent with typical treatment parameters used in adults with major depression, notably stimulation intensities above the motor threshold. Wall et al. (2011) investigated rTMS in a sample of eight adolescents (seven females, one male; ages 14–17) who had not responded to at least two antidepressant medication trials. Patients underwent 30 daily sessions of 10 Hz rTMS to the L-DLPFC (5 cm anterior to the hand knob of the motor cortex), at 120% MT, 3000 pulses per session, over the course of 6–8 weeks; all remained on an SSRI during rTMS. Seven patients completed the study, with one discontinuing due to scalp discomfort. In the completers, depression severity (as measured by CDRS-R score) improved compared to baseline by the 10th treatment session ($P = .016$) and continued to improve by 20 sessions ($P < .01$) and 30 sessions ($P < .0001$). Self-reported depressive symptoms [Quick Inventory of Depressive Symptoms–Adolescent (QIDS-A17)] also improved significantly by 20 and 30 treatments ($P = .011$ and $P < .001$, respectively), as did CGI-S score ($P = .016$ and $P < .001$, respectively). All depression measures remained improved 6 months after rTMS treatment ended (CDRS-R: $P < .0001$; QIDS-A17: $P < .0001$; CGI-S: $P < .001$). While three adolescents reported some degree of suicidal ideation at baseline, no worsening of suicidal ideation was observed, and only one participant had mild suicidal thoughts (passive wishes for death) at the completion of rTMS treatment.

Le et al. (2013) applied low-frequency (1 Hz) rTMS at 110% MT to bilateral supplemental motor areas (1200 pulses per session, 20 sessions over 4 weeks) in 25 children with Tourette syndrome (3 females, 22 males; ages 7–16). Significant reductions in measures of tic severity and impairment were observed at 2 and 4 weeks, which remained when reassessed at 3 and 6 months after rTMS. In addition, while the sample did not have particularly high depression severity scores at baseline, patients also had small but statistically significant reductions in Children's Depression Inventory (CDI) scores at 2 ($P = .002$) and 4 weeks ($P = .001$) compared to baseline, which also were maintained at 3- and 6-month follow-up assessments. While the authors acknowledged that improvement in depressive symptoms might be secondary to improvement in tics, the findings raise interesting questions about the potential use of low-frequency (inhibitory) rTMS for children and adolescents with depressive symptoms [see Chapter 8: Tourette Syndrome and Obsessive Compulsive Disorder (OCD) of this volume for a review of TMS in Tourette syndrome].

Yang et al. (2014) examined cerebral neurochemical changes via proton MRS (^{1}H-MRS) in young treatment-resistant depressed patients who were treated with rTMS. Patients received 10 Hz rTMS to the L-DLPFC (5 cm anterior to the motor cortex hand knob, mapped and co-registered with structural MRI image each week), at 120% MT, 3000 pulses per session, in 15 sessions over 3 weeks. One participant withdrew before beginning rTMS, while another withdrew on the second treatment day. Of the six remaining patients (four females, two males; ages 15–21), four demonstrated response (50% or greater improvement on HDRS score), while two did not meet criteria for response. Responders had mean decreases of 68% on HDRS scores and 84% on BDI scores, as well as a 78% decrease on Hamilton Anxiety Rating Scale (HARS) scores; nonresponders had a 29% mean decrease on the HDRS, a 19% increase on BDI scores, and 36% decrease on HARS scores. Responders demonstrated lower mean baseline glutamate concentrations in the L-DLPFC than nonresponders; in the responders, L-DLPFC glutamate increased by 11% with rTMS, while glutamate decreased by 10% in the nonresponders.

In another open-label study, Wall et al. (2016) treated 10 adolescents (four females, six males; ages 13–17) with MDD, who had at least one failed trial of an antidepressant medication, and who were currently taking either an SSRI or SNRI. Patients underwent 10 Hz stimulation of the L-DLPFC at 120% MT (3000 pulses per session, 30 sessions over 6–8 weeks). Unlike previous studies, the determination of the L-DLPFC location and coil positioning were MRI-guided. Seven patients completed the 30-treatment course. Compared to pretreatment, depression severity improved after 10 treatments (CDRS-R, $P = .005$; CGI-S: $P = .02$) and continued to be improved at treatment 20 (CDRS-R: $P = .001$; QIDS-A_{17}-SR: $P = .02$; CGI-S: $P = .003$) and treatment 30 (CDRS-R: $P = .002$; QIDS-A_{17}-SR: $P = .045$; CGI-S: $P = .002$), as well as at follow-up 6 months after rTMS (CDRS-R: $P = .03$; QIDS-A_{17}-SR: $P = .03$; CGI-S: $P = .002$). Six patients were considered to have responded to rTMS, with end-of-treatment Clinical Global Impression–Improvement (CGI-I) scores of 1 or 2 (corresponding to very much improved or much improved) and CGI-S scores of 1, 2, or 3 (normal/not at all ill, borderline mentally ill, or mildly ill). All six patients maintained clinical response 6 months after rTMS.

To date, no randomized, sham-controlled trials of the efficacy of rTMS for depression in children or adolescents have been published, although one trial is ongoing (NCT01804270/NCT01804296). As such, definitive data on the efficacy of rTMS in this population are lacking, although the findings from these early open-label trials are encouraging. Several important caveats should be considered in interpreting the literature to date. First, the majority of studies have involved older adolescents and young adults; it remains unclear whether rTMS is an effective treatment for depression in children or younger adolescents. This question is particularly relevant considering that rTMS is postulated to alter cortical excitatory–inhibitory physiology, which, as discussed previously, is dynamic throughout the course of neurodevelopment. Second, the majority of studies have only enrolled patients who had not responded to prior medication trials. Treatment-resistant populations may be more likely to demonstrate nonresponse to future treatments (Rush et al., 2006; Vitiello et al., 2011), and thus the findings may not be generalizable to the larger population of depressed children and adolescents. In addition, most studies to date have treated patients who remained on antidepressant pharmacotherapy (and often other psychotropic medications) while receiving rTMS. Thus, rTMS has been studied primarily as an adjunctive treatment in treatment-resistant older adolescents, and further research is necessary to determine its place in treating a broader range of ages and clinical presentations, as well as its potential as a stand-alone treatment in comparison to pharmacologic and psychotherapeutic interventions.

Adverse Effects of rTMS in Children and Adolescents

The safety and adverse effect profile of therapeutic TMS has been reviewed previously by Krishnan et al. (2015). The authors found 35 published studies involving 322 children and adolescents who had undergone rTMS for a range of neuropsychiatric conditions, including depression. The majority of studies (33) reported on the tolerability and/or adverse effects of rTMS; 15 reported no adverse effects or that treatment was "well tolerated," with no adverse effects specified. The most common adverse effect reported was headache (11.5% of total child/adolescent patients), followed by scalp discomfort (2.5%). Less than 2% of children and adolescents receiving rTMS experienced each of the following effects: mood changes, muscle twitching, pruritus, fatigue, dizziness, neck stiffness, neck pain, somnolence, or nausea (Krishnan et al., 2015). The majority of these effects were transient, subsiding within 24 hours and without medical intervention.

The occurrence of more serious adverse effects has been rare. Three cases of seizures in pediatric patients undergoing high-frequency rTMS have been reported. Hu et al. (2011) reported the case of a 15-year-old patient with depression and no prior history of seizure, neurologic disorder, or head trauma, who underwent 10 Hz rTMS at 80% MT while continuing her previous regimen of sertraline. The patient experienced a generalized tonic–clonic seizure shortly after commencing the first treatment session, which resolved after the administration of diazepam and did not recur. The patient developed hypomanic-like symptoms lasting 8–9 hours that night, which resolved without intervention. rTMS was discontinued. Chiramberro et al. (2013) described the case of a 16-year-old adolescent with MDD and no underlying neurologic disorder or head trauma, who experienced a generalized tonic–clonic seizure while undergoing her 12th daily session of 10 Hz rTMS to the L-DLPFC. The patient recovered without sequelae or recurrence. Notably, the patient had a measured blood alcohol concentration of 0.20% shortly after the seizure and was taking several psychotropic medications (sertraline, olanzapine, and hydroxyzine) as well (Chiramberro et al., 2013). As others (Wall et al., 2014; Krishnan et al., 2015) have noted, the high blood alcohol concentration and complex medication regimen may have made the patient more susceptible to seizure induction. In addition, one case of TMS-induced seizure was reported in an adolescent undergoing "deep TMS," an approach involving an H-coil designed to penetrate to deeper limbic structures. Cullen et al. (2016) administered deep TMS to an unmedicated 17-year-old female adolescent with treatment-resistant MDD and no history of seizures, structural brain anomalies, or substance use in a randomized, sham-controlled trial. The patient received sham treatment for 4 weeks, which she tolerated but showed limited clinical improvement. In the subsequent active phase of the study, the patient received daily stimulation on consecutive weekdays at 120% MT, 18 Hz pulse frequency, 55 pulse trains lasting 2 seconds each, and 1980 pulses per session. Due to initial discomfort, stimulation was started at 85% MT and titrated upward by 5% MT daily. Toward the end of the eighth treatment (first treatment at 120% MT), the patient experienced a generalized tonic–clonic seizure lasting approximately 90 seconds that resolved without medication. Postictal confusion persisted for 2–3 days and subsequently resolved. TMS was discontinued, and 6 months later, she had experienced no further seizures. The authors cautioned that the use of adult protocols without parameter adaptations, particularly in newer paradigms such as deep TMS, may be inappropriate in younger patients, highlighting the need for additional modeling and dose-finding studies in order to adapt these protocols safely to children and adolescents. This may be especially relevant to seizure risk in light of immature inhibitory networks and lesser GABAergic tone in youth.

Among other notable adverse effects, two cases of syncope in children and adolescents undergoing rTMS have been reported. Kirton et al. (2008a,b, 2010) randomized 10 children and adolescents (ages 8–20) with arterial ischemic stroke and resultant hemiparesis to active contralesional inhibitory rTMS (1 Hz, 100% RMT, 1200 pulses per session, 10 sessions over 2 weeks) or sham treatment. One 14-year-old patient with chronic subcortical stroke experienced anxiety, nausea, lightheadedness, and mild bradycardia (with mild postural hypotension and reflex tachycardia) after 5 minutes of his initial treatment, which subsided when stimulation was discontinued; no residual effects appeared present at 1 hour or 24 hours after stimulation ceased. A 16-year-old patient enrolled in the study also experienced anxiety, diaphoresis, lightheadedness, and nausea with emesis after 8 minutes of the initial treatment session. After a pause of 1 hour and a small meal (as the patient had not eaten that day), the patient and parent elected to continue; he ultimately completed the 10-session treatment course without complication. In both cases, retrospective questioning revealed that the patients had experienced syncope or presyncopal symptoms previously. The authors noted that immaturity of the autonomic system in children and adolescents may be relevant, and they recommended careful screening for prior history of presyncopal/syncopal events, preventive measures (e.g., adequate hydration and food intake, gradual escalation of stimulation intensity), and careful observation, with immediate discontinuation and available supportive care if symptoms develop (Kirton et al., 2008b).

Antidepressant treatments, including various classes of medications, can precipitate hypomanic or manic symptoms in patients with both unipolar and bipolar depression. Treatment-emergent mania has been reported in adults treated with rTMS, typically when used concurrently with antidepressant medications (Xia et al., 2008; Rachid, 2017). Consequently, concern also has arisen for potential hypomanic or manic "switching" in children and adolescents who undergo rTMS, although few reports exist to date. In one case report, described above (Hu et al., 2011), a 15-year-old adolescent undergoing high-frequency rTMS while concurrently taking sertraline experienced a generalized tonic–clonic seizure during her initial rTMS treatment, which was followed by euphoria and talkativeness that evening. Her behavior returned to baseline after 8–9 hours without treatment. One patient in Bloch et al.'s (2008) open-label trial became more talkative, exhibited excessive laughter, and was more engaged in grooming than usual after the eighth day of rTMS; this persisted for only a single day and resolved without any antimanic intervention. rTMS was suspended for 3 days before resuming, and no hypomanic or manic symptoms recurred during the remainder of her treatment course. Neither patient had a prior history of manic or hypomanic episodes. However, it is noteworthy that in both cases, hypomanic-like symptoms lasted for 1 day or less, which falls well below the duration criteria for hypomanic or manic episodes; in addition, in both cases, the symptoms resolved without any additional intervention, even while continuing antidepressant medications, and, in the latter case, did not recur with reintroduction of rTMS (Bloch et al., 2008; Hu et al., 2011). Although such cases suggest the need for caution and the careful monitoring of clinical symptoms, much larger numbers of depressed children and adolescents treated with rTMS in controlled trials will be necessary to generate an accurate, quantitative estimation of the risk of treatment-emergent hypomania and mania (particularly relative to sham) in this age group.

TMS devices generate substantial noise (typically a loud clicking sound) with each magnetic stimulus, and thus ear protection has been employed in most protocols. One study of low-frequency rTMS in children with focal epilepsy (Sun et al., 2012) reported that 2 out of 31 patients developed tinnitus with treatment at 90% RMT, while no patients in the control group ($n = 29$; received 20% RMT stimulation) developed tinnitus. However, other studies of rTMS in depressed children and adolescents (Loo et al., 2006; Bloch et al., 2008; Wall et al., 2011, 2013, 2016; Mayer et al., 2012; Le et al., 2013; Yang et al., 2014) have not reported any adverse auditory effects, such as tinnitus or hearing impairment. One study (Wall et al., 2011) conducted auditory threshold testing prior to rTMS, at treatment completion, and at 6-month follow-up, finding no significant changes from baseline for any participant.

The safety and tolerability of TBS has been investigated to a more limited extent in children and adolescents, although not specifically in depressed youth. An early study (Wu et al., 2012) examined safety and tolerability in 40 pediatric participants (17 females, 23 males; ages 8–17; 16 with Tourette syndrome, 24 healthy controls) who underwent a single session of iTBS (3 pulses at 50 Hz per burst, 5 Hz burst frequency for a 2-second cycle every 10 seconds, 600 pulses per session) to the left primary motor cortex; three participants also underwent a single session of cTBS (3 pulses at 50 Hz per burst, 5 Hz burst frequency, 600 pulses per session) to the same region. Five participants reported adverse effects in 11.6% of TBS sessions: mild headache ($n = 3$, 7%), neck stiffness ($n = 1$, 2.3%), and finger twitching sensation ($n = 1$, 2.3%). No serious adverse events occurred, and all reported adverse effects resolved on the day of stimulation. Hong et al. (2015) conducted a retrospective analysis of 76 pediatric participants (68% healthy controls, 25% with Tourette syndrome, and 7% with other motor disorders) who underwent TBS with various treatment parameters (60%–90% MT, 30–50 Hz pulse frequency, 300–600 pulses per session); these included iTBS ($n = 59$), cTBS ($n = 8$), or both ($n = 9$). On a standardized questionnaire, participants reported adverse effects in 10.5% of TBS sessions; these included headache (6.6%), arm/hand/other pain (2.6%), numbness/tingling (2.6%), other sensations (2.6%), weakness (1.3%), and "other" (1.3%). No tinnitus or other auditory adverse effects were reported by the TBS participants, and no seizures or other serious adverse events occurred. In addition, rates ($P = .71$) and severity ($P = 1.0$) of adverse effects did not differ between participants who underwent TBS and a comparator group of 89 youth who underwent single- and paired-pulse TMS protocols; rates also did not differ between active and sham TBS ($P = 1.0$) or between iTBS and cTBS ($P = 1.0$) (Hong et al., 2015).

Neurocognitive Effects in Children and Adolescents

The impact of rTMS on neural plasticity and cortical circuitry functions, particularly in light of the rapid cognitive development in children and adolescents, has raised concern for the possibility of long-term detrimental impact on cognitive functions in this population. While little longitudinal data yet exist to answer such questions definitively, a number of early open-label studies of rTMS in depressed youth have included neurocognitive assessments to assess short-term impact on cognition.

In the two patients of their initial case series, Loo et al. (2006) administered the Rey Auditory Verbal Learning Test, Wechsler digit span and digit symbol modalities tests, Trail Making Test (TMT), and Controlled Oral Word Association Test. While unable to compare statistically with two patients, no clinically relevant deterioration on any cognitive measure was observed between baseline, 4 weeks, and conclusion of treatment. In their study of nine adolescents (ages 16–18) with major depression undergoing rTMS, Bloch et al. (2008) utilized the computer-based Cambridge Neuropsychological Test Automated Battery (CANTAB) to monitor attention, working memory, planning, and motor speed. The majority of subscales failed to show any significant change with rTMS stage (measured at baseline, 7 days, 10 days, conclusion of rTMS, and 1-month follow-up); however, compared to baseline, improvement on one measure of reaction time was noted at rTMS completion and 1-month follow-up ($P < .01$), while another planning measure was improved at follow-up ($P < .05$). In the longitudinal analysis of the same sample (Mayer et al., 2012), the CANTAB was repeated in six participants at 3 years posttreatment. There were no significant effects of time on attention, working memory, planning, or motor speed subscales, although one planning measure and one reaction time measure showed sustained improvement at 3 years compared to baseline ($P < .01$ and $P < .05$, respectively). Although it would be implausible to attribute isolated cognitive improvements 3 years later directly to the effects of the initial course of rTMS, these data suggest encouraging questions about the potential cognitive effects of neuromodulatory antidepressant interventions and the restoration of depressed youth to healthy neurodevelopmental trajectories.

Wall et al. (2011) administered the Children's Auditory Verbal Learning Test–Second Edition (CAVLT-2), Delis–Kaplan Executive Function System (D-KEFS) TMT, and Autobiographical Memory Interview (AMI) at baseline, rTMS completion, and 6 months posttreatment. In the six participants who completed testing, no significant decline in performance on any measure was observed, although the authors reported a nonsignificant trend of decline in AMI score (at completion and 6 months), an effect largely driven by the scores of two patients. However, the investigators noted that patients and their parents reported no concerns about decline in cognitive function or performance during the study. With a larger sample of 18 adolescents, Wall et al. (2013) found small but statistically significant improvement on the immediate memory ($P < .0001$) and delayed recall ($P = .0319$) subscales of the CAVLT-2; immediate memory improvement remained significant ($P = .0013$) when only patients who completed all 30 treatments ($n = 14$) were analyzed. On the D-KEFS scores, small, nonsignificant improvement was observed on number sequencing and letter sequencing subscales, as well as on overall composite scores, for both the entire sample and treatment completers. Adolescents and their parents reported no subjective changes in attention, memory, or other cognitive functions.

In summary, the current literature on children and adolescents who have been treated with rTMS for depression and some other psychiatric conditions has not shown evidence of detrimental effects on measures of cognitive functioning. Several have found statistically significant improvements on certain cognitive assessments with rTMS, although not necessarily of magnitudes that are clinically or practically relevant. Few studies have examined long-term cognitive outcomes, and longitudinal data across the entire developmental spectrum are necessary to understand how stimulation during youth affects neurocognitive development. As a result, caution is warranted for the present time, and further well-designed, prospective studies will be important in adapting TMS treatments to youth in order to understand the underlying mechanisms in neurodevelopment and to examine the side effect burden more robustly. All of these factors may diverge from what is understood regarding the effects of TMS in adult populations (Davis, 2014; Geddes, 2015). Studies with animal models (Tang et al., 2017) and field modeling approaches (Deng et al., 2013) also will prove invaluable for advancing this knowledge base.

FUTURE DIRECTIONS IN TMS THERAPIES FOR DEPRESSION IN CHILDREN AND ADOLESCENTS

While the past several decades have seen rapid increases in the investigational and therapeutic applications of TMS, neuromodulation in depressed children and adolescents remains a young field. The pediatric TMS literature to date is promising, yet it also points to many unanswered questions about efficacy and how best to adapt and optimize TMS approaches in this population. Ideally, future protocol

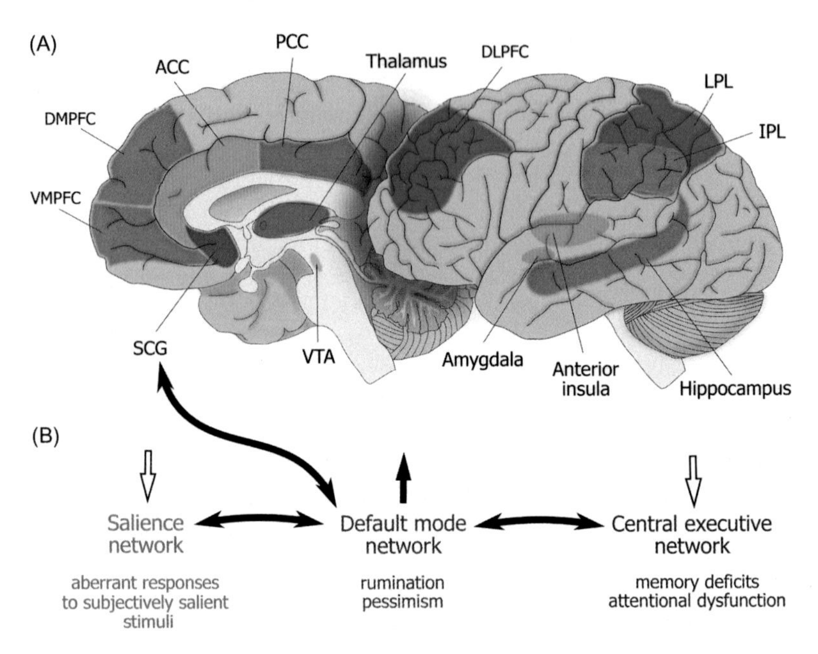

FIGURE 7.7 ICNs and depression. (A) Key nodes of the three main ICNs studied in depression; red = the default mode network, blue = the central executive network, and green = the salience network. (B) Patterns of abnormal functional connectivity within and between ICNs associated with depression and associated symptoms. Black arrows represent hyperconnectivity, and white arrows represent hypoconnectivity. *ACC*, anterior cingulate cortex; *DLPFC*, dorsolateral prefrontal cortex; *DMPFC*, dorsomedial prefrontal cortex; *ICNs*, intrinsic connectivity networks; *IPL*, inferior parietal lobe; *LPL*, lateral parietal lobe; *PCC*, posterior cingulate cortex; *SCG*, subgenual cingulate gyrus; *VMPFC*, ventromedial prefrontal cortex; *VTA*, ventral tegmental area. *Reprinted from Anderson, R.J., Hoy, K.E., Daskalakis, Z.J., Fitzgerald, P.B., 2016. Repetitive transcranial magnetic stimulation for treatment resistant depression: re-establishing connections. Clin. Neurophysiol. 127 (11), 3394–3405. © 2016, with permission from Elsevier.*

development for prospective studies of TMS for child and adolescent depression will carefully consider neurodevelopment. Prior lessons from psychopharmacological studies underscore that adult antidepressant treatments often do not have robust efficacy in child and adolescent populations (Findling et al., 1999; Emslie, 2012). Considering the differences in cortical physiology in young persons, pediatric dose-finding and effectiveness studies are critical, rather than relying on the ongoing adaptation of adult protocols. For example, in light of the evidence of excessive glutamatergic facilitation in depressed children and adolescents, 1 Hz (inhibitory) rTMS may be effective for certain presentations of depression and might present a safe and tolerable treatment option for adolescents. Unfortunately, low-frequency approaches have not been studied sufficiently in children and adolescents with depression to

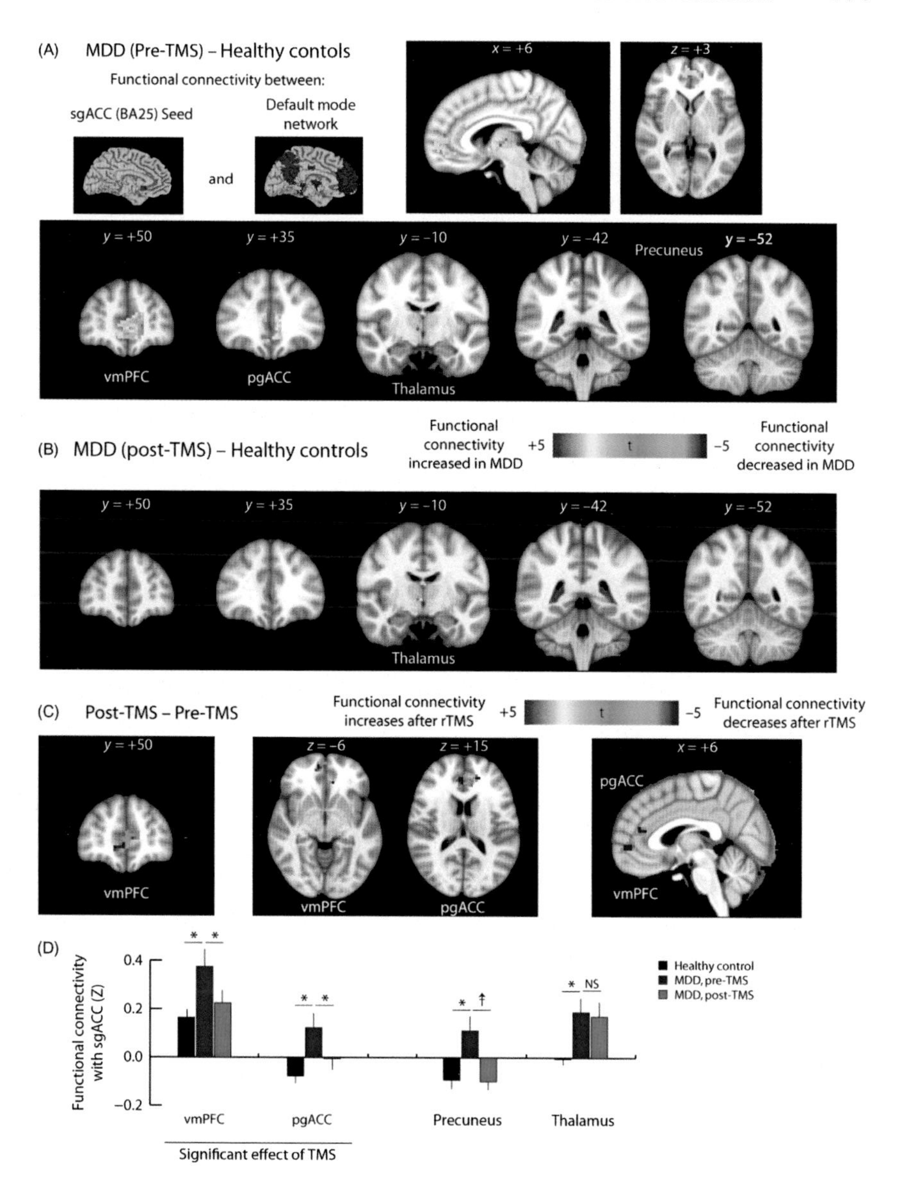

FIGURE 7.8 TMS attenuates depression-related hyperconnectivity within the default mode network. (A) Compared with healthy control subjects, depressed patients exhibited increased functional connectivity between the sgACC and multiple nodes of the default mode network, including the vmPFC, pgACC, thalamus, and precuneus. Images depict *t* statistics for the contrast of patients pretreatment versus healthy control subjects. (B) All areas of sgACC hyperconnectivity normalized after TMS, except within the thalamus. (C) Repeated measures analysis of covariance revealed significant effects of TMS on sgACC connectivity with the vmPFC and pgACC. Hyperconnectivity with the precuneus tended to normalize after treatment, but this effect did not reach significance after correcting for

date. In addition, many current rTMS protocols that involve stimulation of the DLPFC scale the treatment stimulus intensity relative to the intensity necessary to achieve thumb movement (based on visual observation) during stimulation of the primary motor cortex. Future protocols may individualize this aspect of dosing by utilizing more precise measures of excitability that are more relevant to the cortical region being targeted in treatment. Similarly, most current methods for localizing the targeted cortical structures, such as the DLPFC, fail to account for individual differences in cranial and cerebral anatomy, not to mention the rapidly developing circuitry of the young brain, which could have different optimal stimulation targets at different ages. Neuroimaging and neuro-navigated approaches may play a role in coil localization in future trials and clinical practice.

Biomarker-guided approaches to therapeutic TMS are particularly important in the context of neurodevelopment. Thoughtful approaches could concurrently optimize TMS delivery in youth and expand our understanding of neurodevelopmental mechanisms by probing the brain—via TMS, EEG, and neuroimaging techniques—during interventional trials. Increasingly, depression is viewed as a condition of cortical network dysfunction (Anderson et al., 2016) (Fig. 7.7), with distinct subtypes of network disruption (Drysdale et al., 2017). Resting-state functional magnetic resonance imaging (fMRI) studies in adults have demonstrated that TMS impacts functioning of the default mode network (DMN; Fig. 7.8) and its connectivity with the DLPFC (Liston et al., 2014). Furthermore, the antidepressant efficacy of TMS in adults may be related to resting-state functional connectivity (Fox et al., 2012), and resting-state fMRI has shown utility in predicting response to rTMS (Avissar et al., 2017; Drysdale et al., 2017). Targeting with task-based fMRI may also hold the prospect of enhancing outcomes in adults and youth (Luber et al., 2017). However, research with measures of network dysfunction in children and adolescents in the context of TMS interventions is nascent, and to date these techniques have not been used for identifying individualized cortical targets for treatment in youth.

◀ multiple comparisons. (D) Quantification of data extracted from the coordinates of the peak t statistic from each of the areas labeled in panels (A–C). For coordinates and statistics, see Table S5 in Supplement 1 of the original article. Error bars = SEM. $^{*}P < .05$, corrected for multiple comparisons. $^{\dagger}P < .01$, uncorrected, but not significant after correcting for multiple comparisons. *BA*, Brodmann area; *MDD*, major depressive disorder; *NS*, not significant; *pgACC*, pregenual anterior cingulate cortex; *rTMS*, repetitive transcranial magnetic stimulation; *sgACC*, subgenual anterior cingulate cortex; *vmPFC*, ventromedial prefrontal cortex. *Reprinted from Liston, C., Chen, A.C., Zebley, B.D., Drysdale, A.T., Gordon, R., Leuchter, B., Voss, H.U., Casey, B.J., Etkin, A., Dubin, M.J., 2014. Default mode network mechanisms of transcranial magnetic stimulation in depression. Biol. Psychiatry 76 (7), 517–526. Copyright (2014) by the Society of Biological Psychiatry, with permission from Elsevier.*

Finally, the rapid development of newer TMS paradigms and dosing strategies presents challenges and opportunities in the study of depressed youth. Current rTMS protocols are lengthy, generally consisting of daily treatment over many weeks, which limits access and feasibility for many patients. TBS and accelerated rTMS approaches (e.g., George et al., 2014) may present appealing alternatives for this population given the decreased time burden involved with these dosing approaches (Brunoni et al., 2017). However, enthusiasm for these novel stimulation techniques must be tempered with systematic study, as the safety profile and side-effect burden in youth is incompletely understood.

References

Anderson, R.J., Hoy, K.E., Daskalakis, Z.J., Fitzgerald, P.B., 2016. Repetitive transcranial magnetic stimulation for treatment resistant depression: re-establishing connections. Clin. Neurophysiol. 127 (11), 3394–3405. Available from: https://doi.org/10.1016/j.clinph.2016.08.015.

Avissar, M., Powell, F., Ilieva, I., Respino, M., Gunning, F.M., Liston, C., et al., 2017. Functional connectivity of the left DLPFC to striatum predicts treatment response of depression to TMS. Brain Stimul. 10 (5), 919–925. Available from: https://doi.org/10.1016/j.brs.2017.07.002.

Bajbouj, M., Lisanby, S.H., Lang, U.E., Danker-Hopfe, H., Heuser, I., Neu, P., 2006. Evidence for impaired cortical inhibition in patients with unipolar major depression. Biol. Psychiatry 59 (5), 395–400. Available from: https://doi.org/10.1016/j.biopsych.2005.07.036.

Ben-Ari, Y., Khazipov, R., Leinekugel, X., Caillard, O., Gaiarsa, J.L., 1997. GABAA, NMDA and AMPA receptors: a developmentally regulated 'ménage à trois'. Trends Neurosci. 20 (11), 523–529. Available from: https://doi.org/10.1016/S0166-2236(97)01147-8.

Bloch, Y., Grisaru, N., Harel, E.V., Beitler, G., Faivel, N., Ratzoni, G., et al., 2008. Repetitive transcranial magnetic stimulation in the treatment of depression in adolescents: an open-label study. J. ECT 24 (2), 156–159. Available from: https://doi.org/10.1097/YCT.0b013e318156aa49.

Bridge, J.A., Iyengar, S., Salary, C.B., Barbe, R.P., Birmaher, B., Pincus, H.A., et al., 2007. Clinical response and risk for reported suicidal ideation and suicide attempts in pediatric antidepressant treatment: a meta-analysis of randomized controlled trials. JAMA 297 (15), 1683–1696. Available from: https://doi.org/10.1001/jama.297.15.1683.

Brunoni, A.R., Chaimani, A., Moffa, A.H., Razza, L.B., Gattaz, W.F., Daskalakis, Z.J., et al., 2017. Repetitive transcranial magnetic stimulation for the acute treatment of major depressive episodes: a systematic review with network meta-analysis. JAMA Psychiatry 74 (2), 143–152. Available from: https://doi.org/10.1001/jamapsychiatry.2016.3644.

Caetano, S.C., Fonseca, M., Olvera, R.L., Nicoletti, M., Hatch, J.P., Stanley, J.A., et al., 2005. Proton spectroscopy study of the left dorsolateral prefrontal cortex in pediatric depressed patients. Neurosci. Lett. 384 (3), 321–326. Available from: https://doi.org/10.1016/j.neulet.2005.04.099.

Cantello, R., Gianelli, M., Civardi, C., Mutani, R., 1992. Magnetic brain stimulation: the silent period after the motor evoked potential. Neurology 42 (10), 1951–1959. Available from: https://doi.org/10.1212/WNL.42.10.1951.

Cash, R.F.H., Noda, Y., Zomorrodi, R., Radhu, N., Farzan, F., Rajji, T.K., et al., 2017. Characterization of glutamatergic and GABAA-mediated neurotransmission in motor and dorsolateral prefrontal cortex using paired-pulse TMS–EEG. Neuropsychopharmacology 42 (2), 502–511. Available from: https://doi.org/10.1038/npp.2016.133.

Chiramberro, M., Lindberg, N., Isometsä, E., Kähkönen, S., Appelberg, B., 2013. Repetitive transcranial magnetic stimulation induced seizures in an adolescent patient with major depression: a case report. Brain Stimul. 6 (5), 830–831. Available from: https://doi.org/10.1016/j.brs.2013.02.003.

Chugani, D.C., Muzik, O., Juhász, C., Janisse, J.J., Ager, J., Chugani, H.T., 2001. Postnatal maturation of human GABAA receptors measured with positron emission tomography. Ann. Neurol. 49 (5), 618–626. Available from: https://doi.org/10.1002/ana.1003.

Cipriani, A., Zhou, X., Del Giovane, C., Hetrick, S.E., Qin, B., Whittington, C., et al., 2016. Comparative efficacy and tolerability of antidepressants for major depressive disorder in children and adolescents: a network meta-analysis. Lancet 388 (10047), 881–890. Available from: https://doi.org/10.1016/S0140-6736(16)30385-3.

Croarkin, P.E., Wall, C.A., Nakonezny, P.A., Buyukdura, J.S., Husain, M.M., Sampson, S. M., et al., 2012. Increased cortical excitability with prefrontal high-frequency repetitive transcranial magnetic stimulation in adolescents with treatment-resistant major depressive disorder. J. Child Adolesc. Psychopharmacol. 22 (1), 56–64. Available from: https://doi.org/10.1089/cap.2011.0054.

Croarkin, P.E., Nakonezny, P.A., Husain, M.M., Melton, T., Buyukdura, J.S., Kennard, B. D., et al., 2013. Evidence for increased glutamatergic cortical facilitation in children and adolescents with major depressive disorder. JAMA Psychiatry 70 (3), 291–299. Available from: https://doi.org/10.1001/2013.jamapsychiatry.24.

Croarkin, P.E., Nakonezny, P.A., Husain, M.M., Port, J.D., Melton, T., Kennard, B.D., et al., 2014a. Evidence for pretreatment LICI deficits among depressed children and adolescents with nonresponse to fluoxetine. Brain Stimul. 7 (2), 243–251. Available from: https://doi.org/10.1016/j.brs.2013.11.006.

Croarkin, P.E., Nakonezny, P.A., Lewis, C.P., Zaccariello, M.J., Huxsahl, J.E., Husain, M. M., et al., 2014b. Developmental aspects of cortical excitability and inhibition in depressed and healthy youth: an exploratory study. Front. Hum. Neurosci. 8, 669. Available from: https://doi.org/10.3389/fnhum.2014.00669.

Croarkin, P.E., Nakonezny, P.A., Wall, C.A., Murphy, L.L., Sampson, S.M., Frye, M.A., et al., 2016. Transcranial magnetic stimulation potentiates glutamatergic neurotransmission in depressed adolescents. Psychiatry Res. Neuroimaging 247, 25–33. Available from: https://doi.org/10.1016/j.pscychresns.2015.11.005.

Cullen, K.R., Jasberg, S., Nelson, B., Klimes-Dougan, B., Lim, K.O., Croarkin, P.E., 2016. Seizure induced by deep transcranial magnetic stimulation in an adolescent with depression. J. Child Adolesc. Psychopharmacol. 26 (7), 637–641. Available from: https://doi.org/10.1089/cap.2016.0070.

Daskalakis, Z.J., Farzan, F., Barr, M.S., Maller, J.J., Chen, R., Fitzgerald, P.B., 2008. Long-interval cortical inhibition from the dorsolateral prefrontal cortex: a TMS–EEG study. Neuropsychopharmacology 33 (12), 2860–2869. Available from: https://doi.org/10.1038/npp.2008.22.

Davis, N.J., 2014. Transcranial stimulation of the developing brain: a plea for extreme caution. Front. Hum. Neurosci. 8, 600. Available from: https://doi.org/10.3389/fnhum.2014.00600.

Day, B.L., Marsden, C.D., Rothwell, J.C., Thompson, P.D., Ugawa, Y., 1989a. An investigation of the EMG silent period following stimulation of the brain in normal man. J. Physiol. 414 (Suppl), 14P. Available from: https://doi.org/10.1113/jphysiol.1989.sp017707.

Day, B.L., Rothwell, J.C., Thompson, P.D., Maertens de Noordhout, A., Nakashima, K., Shannon, K., et al., 1989b. Delay in the execution of voluntary movement by electrical or magnetic brain stimulation in intact man. Evidence for the storage of motor programs in the brain. Brain 112 (3), 649–663. Available from: https://doi.org/10.1093/brain/112.3.649.

Deng, Z.D., Lisanby, S.H., Peterchev, A.V., 2013. Electric field depth-focality tradeoff in transcranial magnetic stimulation: simulation comparison of 50 coil designs. Brain Stimul. 6 (1), 1–13. Available from: https://doi.org/10.1016/j.brs.2012.02.005.

Di Lazzaro, V., Restuccia, D., Oliviero, A., Profice, P., Ferrara, L., Insola, A., et al., 1998. Magnetic transcranial stimulation at intensities below active motor threshold activates intracortical inhibitory circuits. Exp. Brain Res. 119 (2), 265–268. Available from: https://doi.org/10.1007/s002210050341.

Di Lazzaro, V., Oliviero, A., Meglio, M., Cioni, B., Tamburrini, G., Tonali, P., et al., 2000. Direct demonstration of the effect of lorazepam on the excitability of the human motor cortex. Clin. Neurophysiol. 111 (5), 794–799. Available from: https://doi.org/10.1016/S1388-2457(99)00314-4.

Di Lazzaro, V., Oliviero, A., Mazzone, P., Pilato, F., Saturno, E., Insola, A., et al., 2002. Direct demonstration of long latency cortico-cortical inhibition in normal subjects and in a patient with vascular parkinsonism. Clin. Neurophysiol. 113 (11), 1673–1679. Available from: https://doi.org/10.1016/S1388-2457(02)00264-X.

Di Lazzaro, V., Oliviero, A., Saturno, E., Dileone, M., Pilato, F., Nardone, R., et al., 2005. Effects of lorazepam on short latency afferent inhibition and short latency intracortical inhibition in humans. J. Physiol. 564 (2), 661–668. Available from: https://doi.org/10.1113/jphysiol.2004.061747.

Di Lazzaro, V., Pilato, F., Oliviero, A., Dileone, M., Saturno, E., Mazzone, P., et al., 2006. Origin of facilitation of motor-evoked potentials after paired magnetic stimulation: direct recording of epidural activity in conscious humans. J. Neurophysiol. 96 (4), 1765–1771. Available from: https://doi.org/10.1152/jn.00360.2006.

Di Lazzaro, V., Ziemann, U., Lemon, R.N., 2008. State of the art: physiology of transcranial motor cortex stimulation. Brain Stimul. 1 (4), 345–362. Available from: https://doi.org/10.1016/j.brs.2008.07.004.

Drysdale, A.T., Grosenick, L., Downar, J., Dunlop, K., Mansouri, F., Meng, Y., et al., 2017. Resting-state connectivity biomarkers define neurophysiological subtypes of depression. Nat. Med. 23 (1), 28–38. Available from: https://doi.org/10.1038/nm.4246.

Dubicka, B., Hadley, S., Roberts, C., 2006. Suicidal behaviour in youths with depression treated with new-generation antidepressants: meta-analysis. Br. J. Psychiatry 189 (5), 393–398. Available from: https://doi.org/10.1192/bjp.bp.105.011833.

Dubicka, B., Cole-King, A., Reynolds, S., Ramchandani, P., 2016. Paper on suicidality and aggression during antidepressant treatment was flawed and the press release was misleading. BMJ 352, i911. Available from: https://doi.org/10.1136/bmj.i911.

Duncan, C.E., Webster, M.J., Rothmond, D.A., Bahn, S., Elashoff, M., Shannon Weickert, C., 2010. Prefrontal GABAA receptor α-subunit expression in normal postnatal human development and schizophrenia. J. Psychiatr. Res. 44 (10), 673–681. Available from: https://doi.org/10.1016/j.jpsychires.2009.12.007.

Emslie, G.J., 2012. Are adults just big children? Am. J. Psychiatry 169 (3), 248–250. Available from: https://doi.org/10.1176/appi.ajp.2012.12010004.

Farzan, F., Barr, M.S., Hoppenbrouwers, S.S., Fitzgerald, P.B., Chen, R., Pascual-Leone, A., et al., 2013. The EEG correlates of the TMS-induced EMG silent period in humans. NeuroImage 83, 120–134. Available from: https://doi.org/10.1016/j.neuroimage.2013.06.059.

Farzan, F., Vernet, M., Shafi, M.M.D., Rotenberg, A., Daskalakis, Z.J., Pascual-Leone, A., 2016. Characterizing and modulating brain circuitry through transcranial magnetic stimulation combined with electroencephalography. Front. Neural Circuits 10, 73. Available from: https://doi.org/10.3389/fncir.2016.00073.

Findling, R.L., Reed, M.D., Blumer, J.L., 1999. Pharmacological treatment of depression in children and adolescents. Paediatr. Drugs 1 (3), 161–182. Available from: https://doi.org/10.2165/00128072-199901030-00002.

Fisher, R.J., Nakamura, Y., Bestmann, S., Rothwell, J.C., Bostock, H., 2002. Two phases of intracortical inhibition revealed by transcranial magnetic threshold tracking. Exp. Brain Res. 143 (2), 240–248. Available from: https://doi.org/10.1007/s00221-001-0988-2.

Fox, M.D., Buckner, R.L., White, M.P., Greicius, M.D., Pascual-Leone, A., 2012. Efficacy of transcranial magnetic stimulation targets for depression is related to intrinsic functional connectivity with the subgenual cingulate. Biol. Psychiatry 72 (7), 595–603. Available from: https://doi.org/10.1016/j.biopsych.2012.04.028.

Fuhr, P., Agostino, R., Hallett, M., 1991. Spinal motor neuron excitability during the silent period after cortical stimulation. Electroencephalogr. Clin. Neurophysiol. 81 (4), 257–262. Available from: https://doi.org/10.1016/0168-5597(91)90011-L.

Gabbay, V., Mao, X., Klein, R.G., Ely, B.A., Babb, J.S., Panzer, A.M., et al., 2012. Anterior cingulate cortex γ-aminobutyric acid in depressed adolescents: relationship to anhedonia. Arch. Gen. Psychiatry 69 (2), 139–149. Available from: https://doi.org/10.1001/archgenpsychiatry.2011.131.

Geddes, L., 2015. Brain stimulation in children spurs hope—and concern. Nature 525 (7570), 436–437. Available from: https://doi.org/10.1038/525436a.

George, M.S., Raman, R., Benedek, D.M., Pelic, C.G., Grammer, G.G., Stokes, K.T., et al., 2014. A two-site pilot randomized 3 day trial of high dose left prefrontal repetitive transcranial magnetic stimulation (rTMS) for suicidal inpatients. Brain Stimul. 7 (3), 421–431. Available from: https://doi.org/10.1016/j.brs.2014.03.006.

Hammad, T.A., Laughren, T., Racoosin, J., 2006. Suicidality in pediatric patients treated with antidepressant drugs. Arch. Gen. Psychiatry 63 (3), 332–339. Available from: https://doi.org/10.1001/archpsyc.63.3.332.

Hanajima, R., Ugawa, Y., Terao, Y., Ogata, K., Kanazawa, I., 1996. Ipsilateral cortico-cortical inhibition of the motor cortex in various neurological disorders. J. Neurol. Sci. 140 (1-2), 109–116. Available from: https://doi.org/10.1016/0022-510X(96)00100-1.

Hanajima, R., Ugawa, Y., Terao, Y., Sakai, K., Furubayashi, T., Machii, K., et al., 1998. Paired-pulse magnetic stimulation of the human motor cortex: differences among I waves. J. Physiol. 509 (2), 607–618. Available from: https://doi.org/10.1111/j.1469-7793.1998.607bn.x.

Hanajima, R., Furubayashi, T., Iwata, N.K., Shiio, Y., Okabe, S., Kanazawa, I., et al., 2003. Further evidence to support different mechanisms underlying intracortical inhibition of the motor cortex. Exp. Brain Res. 151 (4), 427–434. Available from: https://doi.org/10.1007/s00221-003-1455-z.

Hong, Y.H., Wu, S.W., Pedapati, E.V., Horn, P.S., Huddleston, D.A., Laue, C.S., et al., 2015. Safety and tolerability of theta burst stimulation vs. single and paired pulse transcranial magnetic stimulation: a comparative study of 165 pediatric subjects. Front. Hum. Neurosci. 9, 29. Available from: https://doi.org/10.3389/fnhum.2015.00029.

Hu, S.H., Wang, S.S., Zhang, M.M., Wang, J.W., Hu, J.B., Huang, M.L., et al., 2011. Repetitive transcranial magnetic stimulation-induced seizure of a patient with adolescent-onset depression: a case report and literature review. J. Int. Med. Res. 39 (5), 2039–2044. Available from: https://doi.org/10.1177/147323001103900552.

Huang, Y.Z., Edwards, M.J., Rounis, E., Bhatia, K.P., Rothwell, J.C., 2005. Theta burst stimulation of the human motor cortex. Neuron 45 (2), 201–206. Available from: https://doi.org/10.1016/j.neuron.2004.12.033.

Ilić, T.V., Meintzschel, F., Cleff, U., Ruge, D., Kessler, K.R., Ziemann, U., 2002. Short-interval paired-pulse inhibition and facilitation of human motor cortex: the dimension of stimulus intensity. J. Physiol. 545 (1), 153–167. Available from: https://doi.org/10.1113/jphysiol.2002.030122.

Inghilleri, M., Berardelli, A., Cruccu, G., Manfredi, M., 1993. Silent period evoked by transcranial stimulation of the human cortex and cervicomedullary junction. J. Physiol. 466 (1), 521–534. Available from: https://doi.org/10.1113/jphysiol.1993.sp019732.

Inghilleri, M., Berardelli, A., Marchetti, P., Manfredi, M., 1996. Effects of diazepam, baclofen and thiopental on the silent period evoked by transcranial magnetic stimulation in humans. Exp. Brain Res. 109 (3), 467–472. Available from: https://doi.org/10.1007/BF00229631.

Kann, L., McManus, T., Harris, W.A., Shanklin, S.L., Flint, K.H., Hawkins, J., et al., 2016. Youth Risk Behavior Surveillance—United States, 2015. MMWR Surveill. Summ. 65 (6), 1–174. Available from: https://doi.org/10.15585/mmwr.ss6506a1.

Kimiskidis, V.K., Papagiannopoulos, S., Kazis, D.A., Sotirakoglou, K., Vasiliadis, G., Zara, F., et al., 2006. Lorazepam-induced effects on silent period and corticomotor excitability. Exp. Brain Res. 173 (4), 603–611. Available from: https://doi.org/10.1007/s00221-006-0402-1.

Kirton, A., Chen, R., Friefeld, S., Gunraj, C., Pontigon, A.M., deVeber, G., 2008a. Contralesional repetitive transcranial magnetic stimulation for chronic hemiparesis in subcortical paediatric stroke: a randomised trial. Lancet Neurol. 7 (6), 507–513. Available from: https://doi.org/10.1016/s1474-4422(08)70096-6.

Kirton, A., deVeber, G., Gunraj, C., Chen, R., 2008b. Neurocardiogenic syncope complicating pediatric transcranial magnetic stimulation. Pediatr. Neurol. 39 (3), 196–197. Available from: https://doi.org/10.1016/j.pediatrneurol.2008.06.004.

Kirton, A., deVeber, G., Gunraj, C., Chen, R., 2010. Cortical excitability and interhemispheric inhibition after subcortical pediatric stroke: plastic organization and effects of rTMS. Clin. Neurophysiol. 121 (11), 1922–1929. Available from: https://doi.org/10.1016/j.clinph.2010.04.021.

Koerte, I., Heinen, F., Fuchs, T., Laubender, R.P., Pomschar, A., Stahl, R., et al., 2009. Anisotropy of callosal motor fibers in combination with transcranial magnetic stimulation in the course of motor development. Invest. Radiol. 44 (5), 279–284. Available from: https://doi.org/10.1097/RLI.0b013e31819e9362.

Kondo, D.G., Hellem, T.L., Sung, Y.H., Kim, N., Jeong, E.K., DelMastro, K.K., et al., 2011. Review: magnetic resonance spectroscopy studies of pediatric major depressive disorder. Depress. Res. Treat. 2011, 650450. Available from: https://doi.org/10.1155/2011/650450.

Krishnan, C., Santos, L., Peterson, M.D., Ehinger, M., 2015. Safety of noninvasive brain stimulation in children and adolescents. Brain Stimul. 8 (1), 76–87. Available from: https://doi.org/10.1016/j.brs.2014.10.012.

Kujirai, T., Caramia, M.D., Rothwell, J.C., Day, B.L., Thompson, P.D., Ferbert, A., et al., 1993. Corticocortical inhibition in human motor cortex. J. Physiol. 471 (1), 501–519. Available from: https://doi.org/10.1113/jphysiol.1993.sp019912.

Le, K., Liu, L., Sun, M., Hu, L., Xiao, N., 2013. Transcranial magnetic stimulation at 1 Hertz improves clinical symptoms in children with Tourette syndrome for at least 6 months. J. Clin. Neurosci. 20 (2), 257–262. Available from: https://doi.org/10.1016/j.jocn.2012.01.049.

Lee, J.C., Croarkin, P.E., Ameis, S.H., Sun, Y., Blumberger, D.M., Rajji, T.K., et al., 2017. Paired-associative stimulation-induced long-term potentiation-like motor cortex plasticity in healthy adolescents. Front. Psychiatry 8, 95. Available from: https://doi.org/10.3389/fpsyt.2017.00095.

Lefaucheur, J.P., Lucas, B., Andraud, F., Hogrel, J.Y., Bellivier, F., Del Cul, A., et al., 2008. Inter-hemispheric asymmetry of motor corticospinal excitability in major depression studied by transcranial magnetic stimulation. J. Psychiatr. Res. 42 (5), 389–398. Available from: https://doi.org/10.1016/j.jpsychires.2007.03.001.

Leinekugel, X., Khalilov, I., McLean, H., Caillard, O., Gaiarsa, J.L., Ben-Ari, Y., et al., 1999. GABA is the principal fast-acting excitatory transmitter in the neonatal brain. In: Delgado-Escueta, A.V., Wilson, W.A., Olsen, R.W., Porter, R.J. (Eds.), Jasper's Basic Mechanisms of the Epilepsies, Third Edition: Advances in Neurology, Vol. 79. Lippincott Williams & Wilkins, Philadelphia, pp. 189–201.

Levinson, A.J., Young, L.T., Fitzgerald, P.B., Daskalakis, Z.J., 2007. Cortical inhibitory dysfunction in bipolar disorder: a study using transcranial magnetic stimulation. J. Clin. Psychopharmacol. 27 (5), 493–497. Available from: https://doi.org/10.1097/jcp.0b013e31814ce524.

Levinson, A.J., Fitzgerald, P.B., Favalli, G., Blumberger, D.M., Daigle, M., Daskalakis, Z.J., 2010. Evidence of cortical inhibitory deficits in major depressive disorder. Biol. Psychiatry 67 (5), 458–464. Available from: https://doi.org/10.1016/j.biopsych.2009.09.025.

Lewis, C.P., Nakonezny, P.A., Ameis, S.H., Vande Voort, J.L., Husain, M.M., Emslie, G.J., et al., 2016. Cortical inhibitory and excitatory correlates of depression severity in children and adolescents. J. Affect. Disord. 190, 566–575. Available from: https://doi.org/10.1016/j.jad.2015.10.020.

Liepert, J., Schwenkreis, P., Tegenthoff, M., Malin, J.P., 1997. The glutamate antagonist riluzole suppresses intracortical facilitation. J. Neural Transm. 104 (11-12), 1207–1214. Available from: https://doi.org/10.1007/BF01294721.

Liston, C., Chen, A.C., Zebley, B.D., Drysdale, A.T., Gordon, R., Leuchter, B., et al., 2014. Default mode network mechanisms of transcranial magnetic stimulation in depression. Biol. Psychiatry 76 (7), 517–526. Available from: https://doi.org/10.1016/j.biopsych.2014.01.023.

Locher, C., Koechlin, H., Zion, S.R., Werner, C., Pine, D.S., Kirsch, I., et al., 2017. Efficacy and safety of selective serotonin reuptake inhibitors, serotonin-norepinephrine reuptake inhibitors, and placebo for common psychiatric disorders among children and adolescents: a systematic review and meta-analysis. JAMA Psychiatry 74 (10), 1011–1020. Available from: https://doi.org/10.1001/jamapsychiatry.2017.2432.

Loo, C., McFarquhar, T., Walter, G., 2006. Transcranial magnetic stimulation in adolescent depression. Australas. Psychiatry 14 (1), 81–85. Available from: https://doi.org/10.1080/j.1440-1665.2006.02251.x.

Luber, B.M., Davis, S., Bernhardt, E., Neacsiu, A., Kwapil, L., Lisanby, S.H., et al., 2017. Using neuroimaging to individualize TMS treatment for depression: toward a new paradigm for imaging-guided intervention. NeuroImage 148, 1–7. Available from: https://doi.org/10.1016/j.neuroimage.2016.12.083.

Maddock, R.J., Buonocore, M.H., 2012. MR spectroscopic studies of the brain in psychiatric disorders. In: Carter, C.S., Dalley, J.W. (Eds.), Brain Imaging in Behavioral Neuroscience. Springer, Berlin, Heidelberg, pp. 199–251. Available from: https://doi.org/10.1007/7854_2011_197.

Manganotti, P., Bortolomasi, M., Zanette, G., Pawelzik, T., Giacopuzzi, M., Fiaschi, A., 2001. Intravenous clomipramine decreases excitability of human motor cortex. A study with paired magnetic stimulation. J. Neurol. Sci. 184 (1), 27–32. Available from: https://doi.org/10.1016/S0022-510X(00)00495-0.

Mayer, G., Aviram, S., Walter, G., Levkovitz, Y., Bloch, Y., 2012. Long-term follow-up of adolescents with resistant depression treated with repetitive transcranial magnetic stimulation. J. ECT 28 (2), 84–86. Available from: https://doi.org/10.1097/YCT.0b013e318238f01a.

McDonnell, M.N., Orekhov, Y., Ziemann, U., 2006. The role of GABAB receptors in intracortical inhibition in the human motor cortex. Exp. Brain Res. 173 (1), 86–93. Available from: https://doi.org/10.1007/s00221-006-0365-2.

Mirza, Y., Tang, J., Russell, A., Banerjee, S.P., Bhandari, R., Ivey, J., et al., 2004. Reduced anterior cingulate cortex glutamatergic concentrations in childhood major depression. J.

Am. Acad. Child Adolesc. Psychiatry 43 (3), 341–348. Available from: https://doi.org/10.1097/00004583-200403000-00017.

Nakamura, H., Kitagawa, H., Kawaguchi, Y., Tsuji, H., 1997. Intracortical facilitation and inhibition after transcranial magnetic stimulation in conscious humans. J. Physiol. 498 (3), 817–823. Available from: https://doi.org/10.1113/jphysiol.1997.sp021905.

Pierantozzi, M., Marciani, M.G., Palmieri, M.G., Brusa, L., Galati, S., Caramia, M.D., et al., 2004. Effect of Vigabatrin on motor responses to transcranial magnetic stimulation: an effective tool to investigate in vivo GABAergic cortical inhibition in humans. Brain Res. 1028 (1), 1–8. Available from: https://doi.org/10.1016/j.brainres.2004.06.009.

Pratt, L.A., Brody, D.J., 2014. Depression in the U.S. Household Population, 2009-2012. National Center for Health Statistics, Hyattsville, MD.

Rachid, F., 2017. Repetitive transcranial magnetic stimulation and treatment-emergent mania and hypomania: a review of the literature. J. Psychiatr. Pract. 23 (2), 150–159. Available from: https://doi.org/10.1097/pra.0000000000000219.

Radhu, N., de Jesus, D.R., Ravindran, L.N., Zanjani, A., Fitzgerald, P.B., Daskalakis, Z.J., 2013. A meta-analysis of cortical inhibition and excitability using transcranial magnetic stimulation in psychiatric disorders. Clin. Neurophysiol. 124 (7), 1309–1320. Available from: https://doi.org/10.1016/j.clinph.2013.01.014.

Rakhade, S.N., Jensen, F.E., 2009. Epileptogenesis in the immature brain: emerging mechanisms. Nat. Rev. Neurol. 5 (7), 380–391. Available from: https://doi.org/10.1038/nrneurol.2009.80.

Robol, E., Fiaschi, A., Manganotti, P., 2004. Effects of citalopram on the excitability of the human motor cortex: a paired magnetic stimulation study. J. Neurol. Sci. 221 (1-2), 41–46. Available from: https://doi.org/10.1016/j.jns.2004.03.007.

Roick, H., von Giesen, H.J., Benecke, R., 1993. On the origin of the postexcitatory inhibition seen after transcranial magnetic brain stimulation in awake human subjects. Exp. Brain Res. 94 (3), 489–498. Available from: https://doi.org/10.1007/BF00230207.

Rosenberg, D.R., Mirza, Y., Russell, A., Tang, J., Smith, J.M., Banerjee, S.P., et al., 2004. Reduced anterior cingulate glutamatergic concentrations in childhood OCD and major depression versus healthy controls. J. Am. Acad. Child Adolesc. Psychiatry 43 (9), 1146–1153. Available from: https://doi.org/10.1097/01.chi.0000132812.44664.2d.

Rosenberg, D.R., MacMaster, F.P., Mirza, Y., Smith, J.M., Easter, P.C., Banerjee, S.P., et al., 2005. Reduced anterior cingulate glutamate in pediatric major depression: a magnetic resonance spectroscopy study. Biol. Psychiatry 58 (9), 700–704. Available from: https://doi.org/10.1016/j.biopsych.2005.05.007.

Rossini, P.M., Barker, A.T., Berardelli, A., Caramia, M.D., Caruso, G., Cracco, R.Q., et al., 1994. Non-invasive electrical and magnetic stimulation of the brain, spinal cord and roots: basic principles and procedures for routine clinical application. Report of an IFCN committee. Electroencephalogr. Clin. Neurophysiol. 91 (2), 79–92. Available from: https://doi.org/10.1016/0013-4694(94)90029-9.

Rossini, P.M., Burke, D., Chen, R., Cohen, L.G., Daskalakis, Z., Di Iorio, R., et al., 2015. Non-invasive electrical and magnetic stimulation of the brain, spinal cord, roots and peripheral nerves: basic principles and procedures for routine clinical and research application. An updated report from an I.F.C.N. Committee. Clin. Neurophysiol. 126 (6), 1071–1107. Available from: https://doi.org/10.1016/j.clinph.2015.02.001.

Rush, A.J., Trivedi, M.H., Wisniewski, S.R., Nierenberg, A.A., Stewart, J.W., Warden, D., et al., 2006. Acute and longer-term outcomes in depressed outpatients requiring one or several treatment steps: a STAR*D report. Am. J. Psychiatry 163 (11), 1905–1917. Available from: https://doi.org/10.1176/ajp.2006.163.11.1905.

Schwenkreis, P., Witscher, K., Janssen, F., Addo, A., Dertwinkel, R., Zenz, M., et al., 1999. Influence of the *N*-methyl-D-aspartate antagonist memantine on human motor cortex excitability. Neurosci. Lett. 270 (3), 137–140. Available from: https://doi.org/10.1016/S0304-3940(99)00492-9.

Schwenkreis, P., Liepert, J., Witscher, K., Fischer, W., Weiller, C., Malin, J.P., et al., 2000. Riluzole suppresses motor cortex facilitation in correlation to its plasma level: a study using transcranial magnetic stimulation. Exp. Brain Res. 135 (3), 293–299. Available from: https://doi.org/10.1007/s002210000532.

Sharma, T., Guski, L.S., Freund, N., Gøtzsche, P.C., 2016. Suicidality and aggression during antidepressant treatment: systematic review and meta-analyses based on clinical study reports. BMJ 352, i65. Available from: https://doi.org/10.1136/bmj.i65.

Siebner, H.R., Dressnandt, J., Auer, C., Conrad, B., 1998. Continuous intrathecal baclofen infusions induced a marked increase of the transcranially evoked silent period in a patient with generalized dystonia. Muscle Nerve 21 (9), 1209–1212. Available from: https://doi.org/10.1002/(SICI)1097-4598(199809)21:9%3c1209::AID-MUS15%3e3.0.CO;2-M.

Silverstein, F.S., Jensen, F.E., 2007. Neonatal seizures. Ann. Neurol. 62 (2), 112–120. Available from: https://doi.org/10.1002/ana.21167.

Steele, J.D., Glabus, M.F., Shajahan, P.M., Ebmeier, K.P., 2000. Increased cortical inhibition in depression: a prolonged silent period with transcranial magnetic stimulation (TMS). Psychol. Med. 30 (3), 565–570. Available from: https://doi.org/10.1017/S0033291799002032.

Stefan, K., Kunesch, E., Cohen, L.G., Benecke, R., Classen, J., 2000. Induction of plasticity in the human motor cortex by paired associative stimulation. Brain 123 (3), 572–584. Available from: https://doi.org/10.1093/brain/123.3.572.

Stetkarova, I., Kofler, M., 2013. Differential effect of baclofen on cortical and spinal inhibitory circuits. Clin. Neurophysiol. 124 (2), 339–345. Available from: https://doi.org/10.1016/j.clinph.2012.07.005.

Stone, M., 2016. Suicidality and aggression during antidepressant treatment: authors misinterpreted earlier paper from the FDA. BMJ 352, i906. Available from: https://doi.org/10.1136/bmj.i906.

Stone, M., Laughren, T., Jones, M.L., Levenson, M., Holland, P.C., Hughes, A., et al., 2009. Risk of suicidality in clinical trials of antidepressants in adults: analysis of proprietary data submitted to US Food and Drug Administration. BMJ 339, b2880. Available from: https://doi.org/10.1136/bmj.b2880.

Sun, W., Mao, W., Meng, X., Wang, D., Qiao, L., Tao, W., et al., 2012. Low-frequency repetitive transcranial magnetic stimulation for the treatment of refractory partial epilepsy: a controlled clinical study. Epilepsia 53 (10), 1782–1789. Available from: https://doi.org/10.1111/j.1528-1167.2012.03626.x.

Sun, Y., Farzan, F., Mulsant, B.H., Rajji, T.K., Fitzgerald, P.B., Barr, M.S., et al., 2016. Indicators for remission of suicidal ideation following magnetic seizure therapy in patients with treatment-resistant depression. JAMA Psychiatry 73 (4), 337–345. Available from: https://doi.org/10.1001/jamapsychiatry.2015.3097.

Tang, A., Thickbroom, G., Rodger, J., 2017. Repetitive transcranial magnetic stimulation of the brain: mechanisms from animal and experimental models. Neuroscientist 23 (1), 82–94. Available from: https://doi.org/10.1177/1073858415618897.

Valls-Solé, J., Pascual-Leone, A., Wassermann, E.M., Hallett, M., 1992. Human motor evoked responses to paired transcranial magnetic stimuli. Electroencephalogr. Clin. Neurophysiol. 85 (6), 355–364. Available from: https://doi.org/10.1016/0168-5597(92)90048-G.

Vitiello, B., Emslie, G., Clarke, G., Wagner, K.D., Asarnow, J.R., Keller, M.B., et al., 2011. Long-term outcome of adolescent depression initially resistant to selective serotonin reuptake inhibitor treatment: a follow-up study of the TORDIA sample. J. Clin. Psychiatry 72 (3), 388–396. Available from: https://doi.org/10.4088/JCP.09m05885blu.

Walkup, J.T., 2017. Antidepressant efficacy for depression in children and adolescents: industry- and NIMH-funded studies. Am. J. Psychiatry 174 (5), 430–437. Available from: https://doi.org/10.1176/appi.ajp.2017.16091059.

Wall, C.A., Croarkin, P.E., Sim, L.A., Husain, M.M., Janicak, P.G., Kozel, F.A., et al., 2011. Adjunctive use of repetitive transcranial magnetic stimulation in depressed adolescents: a prospective, open pilot study. J. Clin. Psychiatry 72 (9), 1263–1269. Available from: https://doi.org/10.4088/JCP.11m07003.

Wall, C.A., Croarkin, P.E., McClintock, S.M., Murphy, L.L., Bandel, L.A., Sim, L.A., et al., 2013. Neurocognitive effects of repetitive transcranial magnetic stimulation in adolescents with major depressive disorder. Front. Psychiatry 4, 165. Available from: https://doi.org/10.3389/fpsyt.2013.00165.

Wall, C., Croarkin, P., Bandel, L., Schaefer, K., 2014. Response to repetitive transcranial magnetic stimulation induced seizures in an adolescent patient with major depression: a case report. Brain Stimul. 7 (2), 337–338. Available from: https://doi.org/10.1016/j.brs.2013.12.001.

Wall, C.A., Croarkin, P.E., Maroney-Smith, M.J., Haugen, L.M., Baruth, J.M., Frye, M.A., et al., 2016. Magnetic resonance imaging-guided, open-label, high-frequency repetitive transcranial magnetic stimulation for adolescents with major depressive disorder. J. Child Adolesc. Psychopharmacol. 26 (7), 582–589. Available from: https://doi.org/10.1089/cap.2015.0217.

Walter, G., Tormos, J.M., Israel, J.A., Pascual-Leone, A., 2001. Transcranial magnetic stimulation in young persons: a review of known cases. J. Child Adolesc. Psychopharmacol. 11 (1), 69–75. Available from: https://doi.org/10.1089/104454601750143483.

Walther, M., Berweck, S., Schessl, J., Linder-Lucht, M., Fietzek, U.M., Glocker, F.X., et al., 2009. Maturation of inhibitory and excitatory motor cortex pathways in children. Brain Dev. 31 (7), 562–567. Available from: https://doi.org/10.1016/j.braindev.2009.02.007.

Werhahn, K.J., Kunesch, E., Noachtar, S., Benecke, R., Classen, J., 1999. Differential effects on motorcortical inhibition induced by blockade of GABA uptake in humans. J. Physiol. 517 (2), 591–597. Available from: https://doi.org/10.1111/j.1469-7793.1999.0591t.x.

Wilson, M.T., St George, L., 2016. Repetitive transcranial magnetic stimulation: a call for better data. Front. Neural Circuits 10, 57. Available from: https://doi.org/10.3389/fncir.2016.00057.

World Health Organization, 2014. Preventing Suicide: A Global Imperative. WHO Press, Geneva.

Wu, S.W., Shahana, N., Huddleston, D.A., Lewis, A.N., Gilbert, D.L., 2012. Safety and tolerability of theta-burst transcranial magnetic stimulation in children. Dev. Med. Child Neurol. 54 (7), 636–639. Available from: https://doi.org/10.1111/j.1469-8749.2012.04300.x.

Xia, G., Gajwani, P., Muzina, D.J., Kemp, D.E., Gao, K., Ganocy, S.J., et al., 2008. Treatment-emergent mania in unipolar and bipolar depression: focus on repetitive transcranial magnetic stimulation. Int. J. Neuropsychopharmacol. 11 (1), 119–130. Available from: https://doi.org/10.1017/s1461145707007699.

Yang, X.R., Kirton, A., Wilkes, T.C., Pradhan, S., Liu, I., Jaworska, N., et al., 2014. Glutamate alterations associated with transcranial magnetic stimulation in youth depression: a case series. J. ECT 30 (3), 242–247. Available from: https://doi.org/10.1097/YCT.0000000000000094.

Ziemann, U., Chen, R., Cohen, L.G., Hallett, M., 1998. Dextromethorphan decreases the excitability of the human motor cortex. Neurology 51 (5), 1320–1324. Available from: https://doi.org/10.1212/WNL.51.5.1320.

Ziemann, U., Lönnecker, S., Steinhoff, B.J., Paulus, W., 1996a. The effect of lorazepam on the motor cortical excitability in man. Exp. Brain Res. 109 (1), 127–135. Available from: https://doi.org/10.1007/BF00228633.
Ziemann, U., Rothwell, J.C., Ridding, M.C., 1996b. Interaction between intracortical inhibition and facilitation in human motor cortex. J. Physiol. 496 (3), 873–881. Available from: https://doi.org/10.1113/jphysiol.1996.sp021734.
Ziemann, U., Reis, J., Schwenkreis, P., Rosanova, M., Strafella, A., Badawy, R., et al., 2015. TMS and drugs revisited 2014. Clin. Neurophysiol. 126 (10), 1847–1868. Available from: https://doi.org/10.1016/j.clinph.2014.08.028.

CHAPTER

8

Transcranial Magnetic Stimulation in Tourette Syndrome and Obsessive–Compulsive Disorder

Christine A. Conelea[1] *and*
Nicole C.R. McLaughlin[2,3]

[1]Department of Psychiatry, University of Minnesota, Minneapolis, MN, United States [2]Department of Psychiatry and Human Behavior, Alpert Medical School of Brown University, Providence, RI, United States
[3]Butler Hospital, Providence, RI, United States

OUTLINE

Neurotechnological Pediatric Neuropsych
DOI: https://doi.org/10.1016/B978-0-12-812777-3.00008-8

Tic disorders (TDs) and obsessive–compulsive disorder (OCD) are neurodevelopmental conditions characterized by repetitive, unwanted behaviors that are difficult to control. These disorders often co-occur in youth and are thought to have shared genetic susceptibility and overlapping neuropathology in cortico-striatal-thalamo-cortical (CSTC) circuits (Hashemiyoon et al., 2017; Hirschtritt et al., 2015). Transcranial magnetic stimulation (TMS) has been used to study the pathophysiology of pediatric TDs and OCD, and more recent research has begun testing the efficacy of TMS intervention protocols. In the current chapter, we review TMS applications in both disorders. For each disorder, we first briefly describe current models of disorder-specific neurocircuitry. We then summarize the findings from TMS studies focused on probing disorder neurophysiology, followed by studies using TMS as an intervention tool. Given the relative paucity of TMS research focused on pediatric vs adult populations in these disorders, for each TMS section we first summarize relevant adult findings before describing pediatric findings in more detail.

OVERVIEW OF TDs

Tourette's disorder and persistent motor/vocal TDs (hereafter referred to collectively as "TDs") are chronic, childhood-onset disorders characterized by the presence of tics for at least 1 year (American Psychiatric Association, 2013). Tics are repetitive, sudden, involuntary motor movements or vocalizations that typically occur at a high rate and over irregular intervals (e.g., in "bouts"). Premonitory urges are aversive somatic sensations, such as tension, pressure, or a "not just right" feeling, that can precede tic onset and are alleviated upon tic execution.

TDs occur in 1%–3% of youth and are more common in males (Centers for Disease Control and Prevention, 2009). The longitudinal course of TDs is typified by an onset around ages 5–6 years and a peak in severity between ages 10–12 years. A decrease or remission of tics into adulthood occurs in approximately two-thirds of those affected, and those who continue to have tics as adults tend to have greater symptom severity or complexity (Steinberg et al., 2010). Tic severity waxes and wanes throughout the course of the disorder across a variety of dimensions, including tic frequency, intensity, complexity, and interference (Leckman et al., 1989). About 80% of those with TDs have one or more comorbid psychiatric disorders, the most common being OCD (30%–66% comorbidity rate), attention deficit hyperactivity disorder (ADHD; 55%–60%), anxiety disorders (18%–36%), mood disorders (20%–30%), and disruptive behavior disorders (30%–40%) (Freeman et al., 2000; Hirschtritt et al., 2015).

The neurological bases of TDs are incompletely understood, but the convergence of evidence, summarized below, implicates a failure of specific CSTC circuits to inhibit motor output and premonitory urges (Wang et al., 2011). Functionally distinct pathways within CSTC are hypothesized to be involved in particular aspects of tic symptomatology (for reviews, see Worbe et al., 2015; Yael et al., 2015).

Tic generation is driven by the CSTC motor pathway. Basal ganglia abnormalities produce disinhibition of excitatory neurons in the ventral thalamus, leading to hyperexcitability of cortical motor areas (Mink, 2006). Supplementary motor area (SMA) hyperactivation in particular has been implicated in tic generation, as SMA activity has been found to be abnormally elevated in the seconds preceding and following tic execution compared to similar movements performed voluntarily (Hampson et al., 2009). Furthermore, excitatory stimulation of SMA with TMS in those without tics has been found to produce tic-like movements and urges (Finis et al., 2013). Voluntary tic suppression is thought to involve an inhibitory control network that exerts top-down control over tic generation (e.g., modulation of basal ganglia activity by inferior frontal gyrus and prefrontal, cingulate, and supplementary motor cortices). Neurotransmitter activity in these regions is also thought to play a key role in tic inhibition, such as GABAergic activity in SMA. The tendency toward reduced tic severity across the developmental course of TDs is thought to involve enhanced control over motor outputs via compensatory alterations in both short and long-range patterns of cortical connectivity (Jackson et al., 2013).

Premonitory urges are mediated by increased activity in CSTC sensorimotor loop areas, including SMA and somatosensory cortex, in conjunction with limbic and paralimbic areas, such as the amygdala, insula, and mid-cingulate cortices. Tic expression is also highly influenced by different contextual states (e.g., emotional state, concurrent behavioral activity, environmental precursors, or consequences for tics; Conelea and Woods, 2008). Dopamine and noradrenaline activity in CSTC circuits is thought to underlie this contextual-state-dependent modulation of tics (Worbe et al., 2015).

OVERVIEW OF OCD

OCD is characterized by obsessions, which are intrusive, persistent thoughts that cause distress, as well as compulsions, which are behaviors that are designed to reduce the distress induced by the compulsions (American Psychiatric Association, 2013). To meet diagnostic criteria for OCD, symptoms must interfere with function, which is often characterized by an hour or more per day of symptoms. Of note, the content and form of OCD symptoms can be quite heterogeneous across patients. Factor and cluster analysis studies have identified a number of symptom dimensions, such as contamination/washing, obsessions/checking, symmetry/ordering, and sexual/religious. However, in pediatric OCD samples, the content of identified dimensions is less thematically consistent (Conelea et al., 2012).

OCD affects 2% of the world population, and it is associated with major impairment and suffering (Ruscio et al., 2010). Its intrusive, anxiety-provoking obsessions and ritualized compulsions are distressing and can be disabling. The World Health Organization ranks OCD as one of the 10 most disabling conditions by lost income and decreased quality of life (Veale and Roberts, 2014). It typically has an early age at onset, chronic course, and low rates of remission, factors that in combination contribute to high levels of economic burden, suicidality, and premature death (Eaton et al., 2013; Pinto et al., 2006). There is a high level of psychiatric comorbidity within OCD, with 60%–80% of individuals with OCD meeting lifetime criteria for depression. Other comorbid disorders include ADHD (up to 60%), bipolar disorder (21.5%), specific phobia (22%), social anxiety disorder (18%), panic disorder (13%–56%), tics (up to 60% lifetime history), and generalized anxiety disorder (30%) (Abramovitch et al., 2015; Pallanti et al., 2011; Leonard et al., 1992).

Neuroimaging consistently implicates aberrant frontal-basal ganglia-thalamic circuitry in OCD. Structural neuroimaging studies have shown differences between participants with OCD and controls in regions within this circuit, including orbitofrontal cortex (OFC), striatum, and thalamus, both in adults and children (Gilbert et al., 2008; Jenike et al., 1996; Robinson et al., 1995; Szeszko et al., 2004). Structural connectivity studies are consistent with those looking at individual regions, pointing to abnormalities in the same frontal-subcortical circuitry. Diffusion tensor imaging studies have shown abnormalities within the cingulate, corpus callosum, and anterior limb of the internal capsule, though the direction of these abnormalities varies between children (increased connectivity) and adults (decreased connectivity; Koch et al., 2014; Piras et al., 2013). Cross-sectional resting state studies in OCD show abnormalities in frontal-subcortical connectivity in both adults and children (Fitzgerald et al., 2011; Fontenelle et al., 2009; Harrison et al., 2013). Though not entirely clear, there are early suggestions that there may be neurocircuitry differences between various OCD symptom presentations. For example, disgust appears to have a strong relationship with insular connectivity (Husted et al., 2006), and patterns of activation have been shown to be different in washing, checking, and hoarding (Mataix-Cols et al., 2004). Some research has explored treatment-related connectivity changes in OCD. In the 1990s, studies showed abnormal pretreatment positron emission tomography (PET)–fludeoxyglucose correlations among caudate, OFC, and thalamic activity that were abolished after either medication or behavior therapy (Baxter, 1992; Schwartz et al., 1996). While these PET "connectivity" methods were rudimentary, posttreatment changes in nodes of a putative OCD circuit have generally been consistent (Baxter, 1992; Saxena et al., 2009, 2002; Schwartz et al., 1996).

OVERLAP BETWEEN TDs AND OCD

TDs and OCD are often conceptualized as variants within an obsessive–compulsive spectrum of disorders due to overlapping genetic liability, symptomatology, and neuropathology. A strong genetic relationship between OCD and TDs has been replicated in multiple studies (Hirschtritt et al., 2015; Mathews and Grados, 2011). In terms of symptoms, both tics and compulsions are repetitive behaviors performed to alleviate an aversive internal experience. While tics are typically preceded by somatic premonitory urges and compulsions by specific cognitions or obsessions, some repetitive behaviors have features of both and complicate differential diagnosis (e.g., tapping repeatedly until it feels "just right"). Various terms have been coined by researchers to describe these symptoms, including "compulsive tics" and "Tourettic OCD" (Mansueto and Keuler, 2005; Palumbo and Kurlan, 2007). In the formal diagnostic nomenclature, the most recent iteration of the Diagnostic and Statistical Manual of Mental Disorders (DSM-5; American Psychiatric Association, 2013) now includes a new "tic-related" specifier for OCD diagnosis.

The co-occurrence of tics and OCD has been associated with particular clinical characteristics, including younger age of onset, higher prevalence in males, and increased occurrence of sensory phenomena, symmetry/exactness concerns, and "not just right experiences" (i.e., the urge to perform an action until it feels "just right"; Conelea et al., 2014; Dell'Osso et al., 2017). While patients with comorbid tics and OCD seem to respond equally well to cognitive behavioral therapy focused on tics (Sukhodolsky et al., 2017) and OCD (Conelea et al., 2014; March et al., 2007; Skarphedinsson et al., 2015), tics have been shown to adversely impact the outcome of medication management of pediatric OCD with sertraline (March et al., 2007; Skarphedinsson et al., 2015).

As described above, neurobiological models of both tics and OCD implicate frontal–striatal circuits. The common and distinct neural correlates of these disorders continue to be investigated. A common substrate underlying obsessive–compulsive spectrum disorders seems to be dysfunction in the sensorimotor CSTC loop (Mataix-Cols and van den Heuvel, 2006). Abnormally heightened activation in SMA and somatosensory cortex have been shown in association with both tic-related premonitory urges and sensory phenomena in OCD patients (e.g., "not just right" experiences; Subirà et al., 2015; Wang et al., 2011). Areas implicated as distinctly impaired in OCD include OFC, anterior cingulate, amygdala, and insula, although the extent to which all of these areas are relevant to the various symptom dimensions of OCD remains unknown (Mataix-Cols and van den Heuvel, 2006), and some of these regions have also been implicated in tics. In terms of TMS treatment research, the circuit commonalities between these disorders have led researchers to examine how TMS delivered over single targets impacts symptoms of both disorders. For example, SMA is emerging as a common treatment target as described below.

TMS PROTOCOLS IN TD AND OCD RESEARCH

Several single and paired-pulse TMS protocols have been used to study the pathophysiology of TD and OCD. Single-pulse TMS over motor cortex has been conducted to study corticospinal conduction by measuring motor evoked potentials (MEPs) at rest (resting motor threshold; RMT) and during active contraction of hand muscles (active motor threshold; AMT) (Orth and Rothwell, 2009). MEP parameters of interest have included amplitude, stimulation threshold, and cortical silent period (CSP) duration, which is a period of electromyography (EMG) silence that occurs between the MEP and a return to baseline activity. Paired-pulse TMS procedures allow one to examine how local circuit or afferent inputs modulate motor cortex excitability. Paired-pulse techniques involve delivering a conditioning stimulus somewhere in the brain prior to a test stimulus over motor cortex. The impact of the conditioning stimulus on motor cortex excitability is then assessed by examining changes in MEP size (Hanajima and Ugawa, 2008). In TD and OCD research, these protocols have been used to examine interhemispheric connectivity, intracortical inhibition (short-interval cortical-inhibition; SICI), and intracortical facilitation (ICF).

TMS STUDIES OF TD NEUROPHYSIOLOGY

TMS studies of TD neurophysiology have primarily focused on the tic generation network by interrogating the motor cortex using single and paired-pulse TMS. Motor cortex has been the primary target of these electrophysiology probes given that it is the main outflow tract of the motor basal ganglia and amenable to noninvasive stimulation effects that can be measured objectively with MEPs in hand muscles. Most of this research has compared adults with TDs to healthy controls (for a recent review, see Orth and Münchau, 2013). We first provide a brief summary of findings from studies conducted in adult samples, followed by more detailed discussion of child-focused studies.

Findings in adults indicate that similar stimulation threshold intensity is needed to induce MEPs at rest and in an active state, suggesting that axonal excitability of corticospinal neurons and intracortical interneurons are normal in TDs (Orth et al., 2005; Ziemann et al., 1997). Inhibitory systems seem to be differentially affected in adults with TDs. Asymmetry of interhemispheric inhibition has been shown in adult TDs (Bäumer et al., 2010). While there is evidence for reduced recruitment of synaptic inhibition in the SICI circuit (Heise et al., 2010; Orth et al., 2005), there do not appear to be notable differences in CSP duration between patients and controls when accounting for differences in absolute MEP amplitude between patients vs controls (Orth et al., 2008). Diminished SICI in TDs could reflect increased glutamatergic output from disinhibited thalamus to motor cortex or deficient $GABA_A$-mediated inhibitory capacity within cortex (Gilbert et al., 2004). Sensory afferent inhibition (SAI), a measure of sensorimotor integration, was found to be reduced in a TD group (Orth et al., 2005), which could reflect reduced efficiency of synaptic inhibition or increased access of sensory input to motor output in TDs (Orth and Münchau, 2013). Examination of motor cortex excitability in association with voluntary movement control using TMS suggests that adults with TDs start a motor task with an abnormally disinhibited level of SICI that subsequently normalizes quickly (Heise et al., 2010).

The relationship between tic severity and motor cortex excitability was assessed in a few adult TD studies. Orth and Rothwell (2009) found no significant relationship between tic severity as measured by the clinician-rated Yale Global Tic Severity Scale (YGTSS; Leckman et al., 1989) and measures of RMT, AMT, SICI, ICF, and SAI. In another study by the same group (Orth et al., 2008), YGTSS scores predicted input–output curves of MEPs for resting and AMTs and SICI but only for patients with uncomplicated TDs (i.e., no psychiatric comorbidities). Some ratings of tic frequency and tic type (e.g., head, facial, hand, etc.) derived from video recordings were related to RMT/AMT and SICI, but results again seemed to be driven by the uncomplicated patients. Thus, the precise nature of the relationship between tic severity and motor cortex electrophysiology remains unclear but is likely influenced by psychiatric comorbidities and the method used to measure tic symptomatology (i.e., clinical interview vs direct observation).

While these TMS studies in adults inform our understanding of motor cortex function in TDs, the applicability of adult TD findings to pediatric samples is difficult to ascertain without direct testing given the neurodevelopmental nature of the disorder. Although this area of the literature is small, several studies to date have used single and paired-pulse TMS paradigms to probe indices of motor cortex excitability in pediatric TDs.

Moll et al. (1999) first examined motor cortex excitability in pediatric TDs by comparing children with TDs (and no ADHD) to healthy controls. Similar to adult findings, no differences were found in RMT or AMT, again supporting the notion that axonal excitability of cortical–spinal neurons is normal in TDs. In contrast to adult findings, no differences were found for SICI and ICF. CSP was shorter in the TD group, but only at higher stimulation intensities, which could reflect hyperexcitability in motor cortex or throughout the CSTC motor loop.

A follow-up study by the same research group (Moll et al., 2001) examined the influence of comorbid ADHD on motor system excitability by comparing children with TDs only to those with ADHD only, ADHD and TD, and healthy controls. Once again, no significant group effects were found for RMT, AMT, or ICF. Children with TD (TD only or ADHD + TD) had shorter CSP than those without TD, and no difference was found in CSP between medicated and unmedicated TD patients, supporting TD-specific deficient motor inhibition in the CSTC sensorimotor circuit. SICI was reduced in children with ADHD (ADHD only or ADHD + TD), suggesting that ADHD may contribute to motor cortex inhibitory deficits observed in TD samples.

The specific contribution of ADHD severity to impaired motor cortex inhibition in TDs was examined more closely by Gilbert et al. (2004, 2005). In their first study, Gilbert et al. (2004) tested the association between cortical inhibition measures (SICI and CSP) and severity of tic, ADHD, and OCD symptoms in TD patients (aged 8–47 years). Reduced SICI was significantly associated with greater ADHD and motor tic severity. Importantly, SICI was more strongly associated with ADHD severity than tic severity. SICI was not associated with OCD severity nor was CSP duration significantly correlated with motor tic, ADHD, or OCD severity. The stability of this observed relationship between greater ADHD severity and reduced SICI in those with TDs was confirmed in a subsequent study (Gilbert et al., 2005). Taken together, these studies suggest that cortical inhibitory function in those with TDs may be more closely linked to the neuropathology underlying hyperactive/impulsive behaviors than to circuitry involved specifically in tic generation.

Motor cortex excitability prior to voluntary movement has been examined in two pediatric TD samples. Jackson et al. (2013) tested whether TMS delivered immediately prior to a movement results in reduced cortical excitability in adolescents with "pure" TD vs controls. Single TMS pulses were delivered during the premovement phase of a manual choice reaction time task. Although behavioral performance on the task did not differ between groups, the TD group had reduced MEP amplitudes in the premovement period. This parallels findings in adults (Heise et al., 2010) and supports the neurodevelopmental model of tic control, which suggests that tic control improves over time due to adaptive changes driven by active suppression of hyperexcitable motor cortex. Dual site paired-pulse TMS was also administered in this study to investigate interhemispheric connectivity between primary motor cortices, but evidence for altered interhemispheric function was not found.

In a similar study of adolescents with TDs, Draper et al. (2015) administered single-pulse TMS in conjunction with a manual Go/No-Go task to assess corticospinal excitability ahead of volitional movements. Consistent with the Jackson et al. (2013) findings, MEP amplitudes were significantly reduced in the TD group in the premovement period. This stood in contrast to the pattern observed in controls, in which an increase in MEP amplitude and decrease in MEP variability immediately preceded volitional movements. Motor tic severity was significantly negatively correlated with the rise of MEP amplitude across the movement preparation period, indicating that those with more severe motor tics were least able to modulate excitability of motor cortex. This study corroborates the notion that tic control relies on inhibition of motor cortex excitability.

The neurodevelopmental model of tic control was tested most directly in a study by Pépés et al. (2016). TMS was used to test motor cortex excitability in a sample of children, adolescents, and young adults with TD and age-matched controls. Consistent with previous adult findings, children and adolescents with TD showed reduced gain in motor excitability, as measured by TMS recruitment curves and motor excitability during preparation of voluntary movements. Other results revealed important developmental differences. In contrast to adults with TD, children showed increased RMT vs controls, and this difference normalized with age over adolescence. Variability in MEP responses was greater in children and decreased with increasing age. Increased tic severity was associated with reduced values of motor gain function (tested by systematically increasing TMS stimulation intensity from RMT and measuring MEP amplitudes during rest and during the pre-voluntary movement period). The authors concluded that these collective results support the notion that delays in the structure and development of CSTC networks contribute to tic occurrence but may normalize with age during adolescence for the majority of those affected.

Taken together, these child-focused TMS neurophysiology studies highlight the importance of interpreting findings of TD-specific abnormalities within a developmental lens and with careful consideration of co-occurring pathology. As noted above, the typical developmental course of TDs is characterized by remission or significant symptom reduction by young adulthood for about two-thirds of individuals (Steinberg et al., 2010). Therefore, adults with TDs may represent an important but distinct clinical subtype with differential underlying neuropathology (e.g., a population in which compensatory neural mechanisms to increase volitional tic control ability fail to develop or in which CSTC networks never fully mature). As the study by Pépés et al. (2016) demonstrates, TMS measurements of motor cortex function likely have developmental sensitivity. It will be important to continue to probe the neurophysiology of tics with TMS in cross-sectional or longitudinal cohorts to fully understand the neurodevelopmental pathways by which the disorder progresses, remits, or remains chronic.

Methodological challenges associated with TMS studies of TD neurophysiology should also be carefully considered. The state of the motor cortex likely fluctuates on a brief time scale depending on whether the individual being measured is experiencing tics, actively suppressing tics, or truly in a "no tic resting state." While it is feasible to identify tic vs "no tic" periods using direct observation methods (e.g., Himle et al., 2006), it is far more difficult to ascertain whether "no tic" periods reflect rest or a state of tic inhibition. There is no known method to objectively measure the presence or intensity of premonitory urges. These variables may all influence TMS measurements; for example, SICI may fluctuate and appear more normal "between tics" (Gilbert et al., 2004). Another important methodological challenge has been assessing the effect of medications on TMS measurements. Few studies have specifically addressed this question (Gilbert et al., 2007) despite the impact these medications may have on TMS-related dependent variables. For example, it is possible that the shorter CSP found in the TD group in the Moll et al. (1999) study was influenced by the high rate of neuroleptic utilization in the TD group (62%), as CSP can be modulated by dopaminergic drugs (Priori et al., 1994). Finally, it is important to note that all pediatric studies to date rely on quite small sample sizes ranging from approximately 10–20 TD participants, thus limiting the generalizability of findings and the ability to detect potentially meaningful differences in subgroups of patients.

TMS STUDIES OF OCD NEUROPHYSIOLOGY

There are a handful of studies assessing the neurophysiology of OCD through TMS. An early study by Greenberg et al. (2000) examined motor cortex excitability in 16 adults with OCD and 11 age-matched healthy controls. Paired-pulse TMS to the M1 revealed decreased SICI and decreased active and resting MEP threshold in adults with OCD. These decreases in intracortical inhibition (ICI) and motor threshold were greatest in OCD adults with tics, suggesting either that tics may independently contribute the observed differences in the OCD group or that co-occurring tics and OCD have a distinct neurophysiological profile from either disorder in isolation.

Richter et al. (2012) measured SICI, CSP, and ICF in 34 adults with OCD as compared to 34 healthy controls over left motor cortex. The OCD group showed shortened CSP and increased ICF as compared to healthy controls. The authors note that this was consistent with dysregulation in cortical inhibitory and facilitatory neurotransmission.

There have been even fewer studies in children and adolescents with OCD examining neurophysiological and/or short-term symptom changes after TMS. Pedapati et al. (2015) examined a provocation task in 18 adolescents (ages 12–18 years) with OCD, before and after sham ($n = 6$) or active repetitive TMS (rTMS) to the right dorsolateral prefrontal cortex ($n = 9$; 1800 pulses, 1 Hz, 110% of RMT, 30 minutes). Symptom provocation consisted of personalized images in an aim to evoke OCD-related anxiety, presented during functional magnetic resonance imaging (fMRI) before and after 30 minutes of either sham or active rTMS. There was an increase in blood oxygenation level-dependent (BOLD) response in the right inferior frontal gyrus (IFG) and right putamen in the sham group, but no change in the active group; the authors indicated that, in the OCD group, rTMS failed to activate brain regions involved in processing anxiety-provoking symptoms. There was no impact on subjective anxiety in either group.

TMS TREATMENT STUDIES IN TDs

rTMS has been explored as tic treatment in small trials (Bloch et al., 2016; Mantovani et al., 2007; Münchau et al., 2002; Orth et al., 2005), some of which include children (Kwon et al., 2011; Le et al., 2013; Wu et al., 2014). Interest in rTMS as a treatment option for TDs has emerged based on the need for effective intervention options for severe or treatment-refractory patients. Non-response to current first-line treatments for tics is common: only 30%–70% of those with TDs respond to medication and 50% to behavior therapy (McGuire et al., 2014).

The earliest rTMS trial for TDs was conducted by Münchau et al. (2002). In this single-blinded, placebo-controlled crossover trial, 16 adults with TD received a random sequence of 1 Hz motor cortex, premotor cortex, or sham rTMS in a relatively brief course (two 20-minute sessions over 2 consecutive days, each session consisted of 1200 pulses delivered at 80% AMT). Motor cortex was targeted with inhibitory stimulation based on the rationale that tic improvement would occur by reducing TD-related motor and premotor cortex hyperexcitability. Blinding was reported to be successful, with only one patient correctly identifying the sham condition. However, no significant improvements in tics were observed after any of the rTMS conditions.

Orth et al. (2005) sought to test possible reasons for the null findings in the Münchau et al. (2002) trial, including the high rate of OCD comorbidity and the reliance on a self-report tic rating scale as the primary outcome measure. Their single-blinded, placebo-controlled crossover study included only TD patients without OCD, increased the number of TMS pulses per session to 1800, applied stimulation to premotor cortex of both hemispheres in turn, and assessed tics using the clinician-rated YGTSS and videotape analysis. Despite these changes, rTMS once again showed no significant effect on tic severity. Blinding integrity was not evaluated.

Attention then shifted toward targeting the SMA instead of motor cortex based on the premise that SMA has extensive connections with regions implicated in both cognitive processes and motor control. Neuroimaging research has shown that SMA activity is abnormally elevated in the seconds preceding tics (Bohlhalter et al., 2006) and during periods of higher tic frequency (Hampson et al., 2009; Stern et al., 2000). In addition, SMA functional connectivity abnormalities have been shown to be significantly correlated with tic severity and complexity (Worbe et al., 2012) and premonitory urge severity (Zapparoli et al., 2015).

The first successful clinical use of rTMS over SMA for TDs was reported for two cases (Mantovani et al., 2007). These case studies followed a protocol tested in an OCD-focused study (10 daily sessions, 1 Hz rTMS, 1200 stimuli/day, 100% RMT) (Mantovani et al., 2006). A randomized, sham-controlled double-blind trial of rTMS for adults with severe TDs was recently conducted by this research group (Landeros-Weisenberger et al., 2015). Twenty adults received 15 sessions of 1 Hz or sham rTMS over SMA (1800 pulses/day, 110% RMT). An extended open-label course of an additional 3 weeks was completed by 16 patients. YGTSS reductions did not significantly differ between the active and control groups at the 3-week posttreatment assessment. The additional open-label course resulted in a significant overall reduction in tic severity from baseline for those initially randomized to active rTMS, suggesting that a longer treatment course may be needed.

Collectively, these adult rTMS trials offer very preliminarily support for targeting SMA with low-frequency rTMS for tics but point to the likely need for these protocols to involve a longer course of stimulation at higher intensities in adults with TDs.

TMS TREATMENT STUDIES OF TDs IN CHILDREN

Three studies subsequently tested the benefit of inhibitory rTMS over SMA for TDs in pediatric samples. In a pilot open-label study, Kwon et al. (2011) treated 10 male children (ages 9–14 years) with 1 Hz rTMS over SMA for 10 daily sessions (1200 pulses/day, 100% RMT). rTMS sessions were well-tolerated with minimal side effects, and all youth completed the protocol. A significant reduction in YGTSS-rated tic severity from baseline to day 10 of the study was reported (mean reduction of 7 points, $P = .012$). Significant changes were not observed on measures of ADHD, depression, and anxiety symptoms. The absence of a control group and female patients limits conclusions that can be drawn from this study, but at a minimum, it demonstrated safety, tolerability, and a preliminary signal that SMA may be a useful TMS target for childhood TDs.

A second open-label study of 1 Hz rTMS over SMA was conducted by Le et al. (2013). Twenty-five children (ages 7–16 years) with TD received 20 daily rTMS sessions (1200 pulses/day, 110% RMT). TMS sessions were again well-tolerated; all youth completed the study and the only side effect reported was mild sleepiness in one participant. Significant reduction in YGTSS-rated tic severity was reported at 2 weeks, and reductions remained significant at week 4 (mean reduction of 7 points, $P < .001$). Decreases in depression and anxiety were reported, but these mean scores were not in the clinical range at baseline. Right and left hemisphere RMT significantly increased over the course of treatment (but notably rTMS intensity was based on pretreatment RMT throughout the course of the intervention, which may have diminished stimulation affects over time). At a 6-month follow-up assessment, 17 children (68%) were reported to have maintained tic improvements. Similar to the conclusions that can be drawn from the Kwon et al. (2011) study, the open-label design of this study has limitations, but this study at least replicates the tolerability and safety profile of the procedure.

The most rigorous trial of rTMS for pediatric TDs to date was conducted by Wu et al. (2014) and utilized a randomized, double-blind, sham-controlled trial design. Several other methodological features notably differed from prior trials. First, coil placement was determined using functional neuronavigation, in which SMA was localized on an individual basis using a finger-tapping fMRI task known to activate SMA. This coil placement procedure differs from all previously described rTMS studies for TDs, which instead identified SMA using 10–20 EEG System coordinate measurements. TMS targeting based on individual fMRI is an increasingly preferred method, as it has been shown to enable more precise engagement of specific functional brain networks, account for individual differences in location (which may be particularly important in a developmental population), and increase detectable TMS effects (Luber et al., 2017). Second, continuous theta burst stimulation (cTBS) was used to induce inhibition of SMA; cTBS benefits over 1 Hz rTMS include a much shorter stimulation duration and lower stimulation intensity. Third, outcomes included posttreatment examination of motor cortex areas using fMRI.

Participants in Wu et al. (2014) were 12 youth with TDs (aged 10–22 years) randomized to sham or active stimulation (30 Hz cTBS, 90% RMT, 600 pulses/session) delivered in four sessions per day for 2 consecutive days. A rationale for this schedule was not specified. YGTSS-rated tic severity decreased in both groups and there was no difference between groups. At posttreatment, the active rTMS group showed significant differences in fMRI activation during a finger-tapping task, including decreased activation in SMA and bilateral primary motor cortex posttreatment. Interestingly, decreases were more pronounced under primary motor cortex than directly under the coil in SMA. While the clinical outcomes did not support cTBS protocol efficacy (which could have been impacted by the short duration protocol, lack of opportunity to sleep between sessions, and low sample size), the imaging findings importantly demonstrate that rTMS stimulation can affect distant nodes within the cortical network underlying tic generation.

When considering all of these rTMS treatment trials for TDs together, it becomes difficult to draw firm conclusions about the benefit of rTMS for TDs at this point in time. Put most simply, the existing data suggest that (1) rTMS protocols for TDs appear to have good safety and tolerability profiles, including in pediatric populations, (2) SMA seems to be a more promising stimulation target than primary motor or premotor cortex, and (3) longer duration protocols are likely to yield more clinical benefit. Significant methodological differences between studies make direct comparisons difficult. Open-label studies were more likely to show benefit than randomized, sham-controlled trials, and the two randomized controlled trials (Landeros-Weisenberger et al., 2015; Wu et al., 2014) were likely underpowered to detect significant group differences in clinical outcomes given the very small sample sizes. The range of rTMS parameters used also limits comparisons that can be made between studies, including differences in type of stimulation coil used, pulses delivered per day, number of treatment sessions, and stimulation intensity, duration, and cortical target. Future research is needed to refine these rTMS protocols, including identification of dose–response relationships in rTMS for TDs. The impact of specific rTMS protocols within particular developmental windows (e.g., pre or postpuberty) also remains untested and warrants examination given the known developmental trajectory of TDs. Ongoing research exploring the effect of SMA rTMS (e.g., ClinicalTrials Identifiers NCT02356003; NCT02205918) will hopefully yield more insight into the mechanisms underlying the effect of rTMS on tics and further refinement of treatment protocols.

TMS TREATMENT STUDIES IN OCD

There are a handful of studies examining the use of TMS as a treatment for adults with OCD. The majority of the studies target the dorsolateral prefrontal cortex (DLPFC), based on prior TMS depression studies targeting this area. More recent studies have extended targets to the preSMA and SMA.

DLPFC. As noted above, early studies targeted the DLPFC, based on early depression research targeting this region. Sachdev et al. (2001) randomly assigned 12 participants with resistant OCD to 2 weeks of daily 10 Hz rTMS (30 trains of 5 seconds each, 25 seconds between trains, 110% RMT) to the left or right DLPFC. There was a significant improvement in Yale-Brown Obsessive Compulsive Scale (YBOCS) scores after 2 weeks and at 1 month follow-up, with no difference between right or left rTMS; change in obsessions remained significant after controlling for depression, but total YBOCS score was only at a trend level. A follow-up double-blind study by Sachdev et al. (2007) randomized 18 participants to either active ($n = 10$) 10 Hz rTMS (30 trains of 5 seconds each, 110% motor threshold, with 25 seconds inter-train intervals) or sham ($n = 8$) for 10 daily sessions to the left DLPFC, with a possibility of an open extension of up to 20 sessions. For sham, an inactive coil was placed on the subject's head and an active coil was discharged at the same parameters at least 1 minutes away and out of the patient's line of sight. There was no change in OCD symptomatology after 10 sessions; however, there was a reduction in YBOCS scores after 20 sessions, but not after controlling for depression. Alonso et al. (2001) examined 18 participants who were randomly assigned to 18 sessions (three sessions per week for 6 weeks) active ($n = 10$) or sham ($n = 8$) 1 Hz rTMS (20 minutes) to the right DLPFC. Active rTMS was 110% of the motor threshold, and sham was 20% of the motor threshold. There were no significant changes after treatment. In another 1 Hz study, 33 participants with OCD were randomly assigned to 10 daily sessions of active 1 Hz rTMS (110% of motor threshold) or sham (perpendicular stimulation to active site) to the left DLPFC (Prasko et al., 2006). Thirty participants completed the study. There was no significant difference between active and sham rTMS with regards to global improvement, OCD symptomatology, or anxiety.

Kang et al. (2009) conducted a rater-blinded sham-controlled study of 10 daily sessions of either active ($n = 10$) or sham ($n = 10$) 1 Hz rTMS to the right DLPFC (110% of the motor threshold (MT); 20 minutes) and then the SMA (100% of the MT; 20 minutes) in participants with OCD. There were no significant side effects, but there was also no significant difference between the active and sham groups with regards to OCD symptomatology (YBOCS) and depression Montgomery–Åsberg Depression Rating Scale (MADRS). A larger randomized study was conducted by Sarkhel et al. (2010). Forty-two patients were randomly assigned to 10 sessions (10 Hz, 110% of MT, 4 seconds per train, 20 trains per session) of either right DLPFC active ($n = 21$) or sham ($n = 21$) stimulation. There was no significant effect of treatment and time between the two groups for YBOCS scores. However, there was an improvement in depression and anxiety. Mansur et al. (2011) carried out a double-blind randomized trial of 6 weeks of daily 10 Hz rTMS to the right DLPFC (110% of MT, 1 session/day, 40 trains/session, 5 s/train, 25 seconds inter-train interval). Groups included sham ($n = 14$) or active ($n = 13$). One patient in each group showed a positive response. There was no significant group by time interaction for YBOCS.

Over the past decade, TMS research in OCD has increased. Though groups continued to use the same neural targets as earlier studies, sample sizes in general increased, and investigation of the parameter space (e.g., frequency, intensity) was somewhat more thorough. Badawy et al. (2010) examined rTMS to the left DLPFC in three OCD groups: a sham group ($n = 20$) of never-medicated adults, an active group ($n = 20$) of never-medicated adults, and an active group ($n = 20$) of adults with poor response to SSRIs. After 15 sessions of rTMS, there was a significant increase in the MT in both active groups. Ma et al. (2014) used alpha electroencephalogram to guide rTMS (80% of MT) to the DLPFC bilaterally. Twenty-five patients received active and 21 patients received sham rTMS daily for 10 days. The active group had a significant improvement in YBOCS and HAM-A scores as compared to controls after 2 weeks, and at 1 week follow-up. Haghighi et al. (2015) carried out a randomized crossover study with 21 patients with treatment-resistant OCD. The sham condition consisted of stimulation of the same site as the active treatment, but with the side of the coil resting on the scalp; although participants were not able to see the position of the coil, it is unclear whether the sensation was different for active vs sham.

A total of 20 Hz rTMS was carried out over 20 daily sessions (100% RMT; 25 minutes/session), to first the left and then the right DLPFC, and vice versa. YBOCS values decreased over time in the rTMS-condition but not in the sham condition. A more recent study examined different frequencies of rTMS to the right DLPFC in 45 patients with OCD (Elbeh et al., 2016). Participants were randomly placed into three groups: (1) 1 Hz rTMS at 100% RMT, four trains of 500 pulses each with a 40 seconds inter-train interval; (2) 10 Hz rTMS at 100% RMT in 10 trains of 200 pulses with a 20 second inter-train interval; and (3) sham stimulation comparable to the second group, but with the coil placed perpendicular to the head. Note that the authors did not provide a rationale for the selected pulse sequences. Participants received 2000 pulses per day for 2 weeks (10 days). A total of 1 Hz rTMS showed a greater clinical benefit as compared to 10 Hz or sham, as evidenced by significant time by group interactions for the YBOCS and HAM-A.

Out of these studies targeting DLPFC, there was variable response to TMS, and some effects decreased or were eliminated after controlling for depression.

SMA. As research has continued, though some studies continue to target the DLPFC, some investigators have started examining other targets based on neurocircuitry implicated in OCD and related disorders, particularly tics. For example, Mantovani et al. (2006) conducted 10 daily sessions of 1 Hz rTMS (100% of motor threshold, 1200 pulses/day) to the SMA in 10 adults (8 completers) with OCD and/or TDs. After 2 weeks of treatment, there were significant improvements in OCD symptomatology, depression, anxiety, and tic severity. Three out of five OCD patients without tics had a reduction in YBOCS scores over 40%, and 60% of the total sample had sustained improvements at 3 months. Symptom improvement correlated with an increase of the right RMT, which the authors reported may have indicated that SMA stimulation may have influenced the primary motor cortex. Given that direct stimulation of the primary motor cortex may impact the striatum, the authors noted that improvements may be been secondary to effects on subcortical and cortical regions.

A follow-up randomized sham-controlled double-blind study by Mantovani et al. (2010) again used 1 Hz rTMS to the SMA in 18 medication-resistant adults with OCD. Participants received 4 weeks of active ($n = 9$) or sham ($n = 9$) rTMS (1200 pulses/day, at 1 Hz, 100% of motor threshold). After 4 weeks, there was a 67% improvement in the active group and a 22% improvement in the sham group. Nonresponders to sham and responders to active or sham rTMS were offered 4 additional weeks of open active rTMS, which resulted in an average 49% decrease on the YBOCS. Bilateral RMT and AMT, CSP, SICI, and ICF were assessed at baseline and after the 4 week acute treatment phase (Mantovani et al., 2013). In patients receiving active rTMS, there was an increase in right hemisphere RMT, which correlated with OCD symptom improvement. There was also an increase in right hemisphere SICI, which correlated with OCD symptomatology and global improvement, along with normalization of baseline RMT hemispheric asymmetry.

Gomes et al. (2012) examined 22 patients with OCD who received 2 weeks of daily sham or active (1 Hz, 20-minute trains, 1200 pulses/day, 100% of resting MT) to the SMA. There was a significant difference between active and sham stimulation until week 14, with a 41% response rate in the active group after 14 weeks of stimulation. Though there are only a few studies with a relatively small number of participants, the SMA appears to be a promising target in OCD.

OFC. In addition to motor cortex, researchers have examined other targets, such as the OFC. Ruffini et al. (2009) carried out daily rTMS (1 Hz, 80% of resting MT for 10 minutes) for 3 weeks to the left OFC parallel (considered active, $n=16$) or perpendicular (considered sham, $n=7$) to the scalp. There were reductions in OCD symptomatology for 10 weeks after the end of the rTMS sessions; clinical response was lost after 12 weeks. Fifteen of 16 active patients had a reduction in YBOCS score, with 8/16 having a reduction of $>25\%$, and 4 of 16 having a reduction of $\geq 35\%$.

Nineteen patients with OCD were randomized through a double-blind crossover design to active and sham rTMS (1 Hz, 120% MT, 1200 pulses/session) to the right OFC for 10 days (with a 1 month washout period; Nauczyciel et al., 2014). Both sham and active stimulation groups showed a decline in OCD symptoms (as measured through the YBOCS). Decline in OCD symptoms was greater in the active group as compared to sham, though the authors acknowledged the drawbacks of using a crossover design with regards to the integrity of the blind. In addition, stimulation was related to a bilateral decrease in OFC metabolism on PET. Similar to the SMA studies, although there are only a few studies targeting the OFC, results are promising for use of this area as a target in OCD.

TMS TREATMENT STUDIES OF OCD IN CHILDREN

Though there are numerous studies examining rTMS in adults with OCD, at multiple targets, there is minimal information about the use of rTMS in child or adolescents with OCD. To our knowledge, there are no treatment studies in children or adolescents with OCD at any neural target. Reasons for this are unclear but may reflect the challenges in designing rTMS protocols that adequately account for the neurodevelopmental changes that occur across the course of OCD or simply a less urgent sense of a need to develop TMS as a treatment for pediatric OCD (e.g., in comparison to TDs, there are more treatment options available to severe patients, such as intensive OCD treatment programs).

In summary, the existing data suggest that (1) rTMS protocols for OCD at several neural targets appear to have good safety and tolerability profiles in adults, though there is minimal knowledge about safety in pediatric populations with OCD, (2) supplementary motor area (SMA)/pre-SMA or OFC seem to be a more promising stimulation targets than DLPFC, and (3) longer duration protocols are likely to yield more clinical benefit. As noted within the TD section, significant methodological differences between studies make direct comparisons difficult. As evidenced with the TD studies, the range of rTMS parameters used limits comparisons that can be made between studies.

OVERALL CONCLUSIONS

Though there has been a significant increase in the number of TMS studies over the past 15 years, both in the areas of neurophysiology and treatment, there is still mixed information about the optimal target location and stimulation settings for TDs and OCD. This is especially apparent in the child and adolescent literature, where there were only three studies in TDs and no treatment studies in OCD. Initial indications from the pediatric TD treatment data suggest that low-frequency rTMS and theta burst stimulation are likely to be safe and tolerable in this population. However, it is important to note that these data come from research samples who met strict study inclusion criteria, which limits the generalizability of safety findings.

Much work needs to be done to identify ideal TMS parameters to effectively treat TDs and OCD in youth. Based on the currently available evidence, inhibitory stimulation over SMA seems to be the most promising target for TDs. Patients likely need a longer course of treatment to see a durable effect, although research examining the trajectory of symptom change is needed to identify a more precise number of sessions needed. In the absence of any pediatric data for TMS treatment of OCD, it is premature to draw any firm conclusions about recommended treatment protocols for youth. Based on adult data, it is possible that SMA and OFC may be the most promising targets for pediatric OCD. It is important to note that existing treatment development work has focused on testing single TMS protocols (e.g., single neural target with the same stimulation parameters) within relatively diagnostically homogeneous groups, which precludes examination of dose-response effects or heterogeneous patterns of response across patients.

It will be crucial for future TMS treatment research in TDs and OCD to consider development in protocol design. While it may be tempting to simply downward extend findings, we know that children are not just "little adults" and that the neuropathology underlying these disorders undergoes change across development. For example, neuroimaging research in pediatric OCD has shown unique maturational trajectories of frontal–striatal circuitry: cognitive control circuits are hypoconnected only in young patients near illness onset, while affective control circuits are hyperconnected across development (Fitzgerald et al., 2011). It is therefore plausible that optimal TMS targets and parameters may differ depending on the age of the person being treated.

Given the substantial clinical heterogeneity of TD and OCD samples, it is also plausible that the efficacy of a given TMS protocol may be moderated by factors such as symptomatology, comorbidity profile, and severity. Indeed, neurophysiological biomarkers measured by TMS appear to differ for TDs depending on ADHD status (Gilbert et al., 2004), and rTMS efficacy for OCD differs based on the presence of tics (Mantovani et al., 2006). Conducting research with larger samples that are carefully phenotyped may help better match particular patient subgroups to TMS protocols. Biotyping may also prove beneficial for personalizing TMS delivery. For example, distinct patterns of dysfunctional resting state functional connectivity have been shown to predict individual response to excitatory rTMS over PFC in adult depression (Drysdale et al., 2017). The utility of using neuroimaging to match patients to TMS protocols should be explored in future TD and OCD research.

Although we have focused on TMS here, it is important to acknowledge the work being done to test the treatment efficacy of other noninvasive brain stimulation technologies. A handful of studies have examined the effect of transcranial direct current stimulation (tDCS) protocols on adult OCD (e.g., Bation et al., 2016; D'urso et al., 2016; Todder et al., 2017). In TDs, only one tDCS case study (Mrakic-Sposta et al., 2008) and the protocol for an ongoing randomized trial that includes youth (Eapen et al., 2017) have been published. tDCS involves delivering a weak current (typically between 0.5 and 2 mA) through electrodes placed on the scalp. Compared to TMS, tDCS may have particular advantages for a pediatric population, including a better safety profile and lower cost; as such, it should continue to be examined as an option for these disorders.

Finally, the impact of concurrent treatments on TMS-measured neurophysiology and TMS interventions should be carefully addressed in future research. As demonstrated by the Gilbert et al. (2007) study,

TMS may be a useful tool for measuring the neurophysiological effects of medications or for identifying biomarkers that predict medication response. TMS methods could therefore be used to learn more about the mechanisms underlying the efficacy of commonly prescribed medications for TDs and OCD. From a treatment perspective, it will be essential to understand whether TMS safety and efficacy differs based on an individual's medication status. It will also be interesting to explore whether TMS has a synergistic effect with existing pharmacotherapies or cognitive behavioral interventions.

In conclusion, TMS is a promising technology for pediatric OCD and TDs. Its potential lies both in its applications as a research tool to extend our understanding of the pathophysiology of OCD and TDs in youth and an intervention tool to treat these often debilitating conditions. It is essential for future TMS work to focus specifically on pediatric populations given the neurodevelopmental nature of both disorders.

References

Abramovitch, A., Dar, R., Mittelman, A., Wilhelm, S., 2015. Comorbidity between attention deficit/hyperactivity disorder and obsessive–compulsive disorder across the lifespan: a systematic and critical review. Harv. Rev. Psychiatry 23 (4), 245–262.

Alonso, P., Pujol, J., Cardoner, N., Benlloch, L., Deus, J., Menchón, J.M., et al., 2001. Right prefrontal repetitive transcranial magnetic stimulation in obsessive–compulsive disorder: a double-blind, placebo-controlled study. Am. J. Psychiatry 158 (7), 1143–1145.

American Psychiatric Association, 2013. Diagnostic and Statistical Manual of Mental Disorders (DSM-5®), fifth ed American Psychiatric Association Publishing, Washington, DC.

Badawy, A.A., El Sawy, H., El Hay, M.A., 2010. Efficacy of repetitive transcranial magnetic stimulation in the management of obsessive compulsive disorder. Egypt. J. Neurol. Psychiatry Retrieved from http://ejnpn.org/Articles/582/2010473008.pdf.

Bation, R., Poulet, E., Haesebaert, F., Saoud, M., Brunelin, J., 2016. Transcranial direct current stimulation in treatment-resistant obsessive–compulsive disorder: an open-label pilot study. Prog. Neuropsychopharmacol. Biol. Psychiatry 65, 153–157.

Bäumer, T., Thomalla, G., Kroeger, J., Jonas, M., Gerloff, C., Hummel, F.C., et al., 2010. Interhemispheric motor networks are abnormal in patients with Gilles de la Tourette syndrome. Mov. Disord. 25 (16), 2828–2837.

Baxter Jr., L.R., 1992. Neuroimaging studies of obsessive compulsive disorder. Psychiatr. Clin. North. Am. 15 (4), 871–884.

Bloch, Y., Arad, S., Levkovitz, Y., 2016. Deep TMS add-on treatment for intractable Tourette syndrome: a feasibility study. World J. Biol. Psychiatry 17 (7), 557–561.

Bohlhalter, S., Goldfine, A., Matteson, S., Garraux, G., Hanakawa, T., Kansaku, K., et al., 2006. Neural correlates of tic generation in Tourette syndrome: an event-related functional MRI study. Brain 129 (Pt8), 2029–2037.

Centers for Disease Control and Prevention (CDC), 2009. Prevalence of diagnosed Tourette syndrome in persons aged 6–17 years—United States, 2007. MMWR Morb. Mortal. Wkly. Rep. 58 (21), 581–585.

Conelea, C.A., Woods, D.W., 2008. The influence of contextual factors on tic expression in Tourette's syndrome: a review. J. Psychosom. Res. 65 (5), 487–496.

Conelea, C.A., Freeman, J.B., Garcia, A.M., 2012. Integrating behavioral theory with OCD assessment using the Y-BOCS/CY-BOCS symptom checklist. J Obsessive Compulsive Relat. Disord 1 (2), 112–118.

Conelea, C.A., Walther, M.R., Freeman, J.B., Garcia, A.M., Sapyta, J., Khanna, M., et al., 2014. Tic-related obsessive–compulsive disorder (OCD): phenomenology and treatment outcome in the Pediatric OCD Treatment Study II. J. Am. Acad. Child Adolesc. Psychiatry 53 (12), 1308–1316.

Dell'Osso, B., Marazziti, D., Albert, U., Pallanti, S., Gambini, O., Tundo, A., et al., 2017. Parsing the phenotype of obsessive–compulsive tic disorder (OCTD): a multidisciplinary consensus. Int. J. Psychiatry Clin. Pract. 21 (2), 156–159.

Draper, A., Jude, L., Jackson, G.M., Jackson, S.R., 2015. Motor excitability during movement preparation in Tourette syndrome. J. Neuropsychol. 9 (1), 33–44.

Drysdale, A.T., Grosenick, L., Downar, J., Dunlop, K., Mansouri, F., Meng, Y., et al., 2017. Resting-state connectivity biomarkers define neurophysiological subtypes of depression. Nat. Med. 23 (1), 28–38.

D'urso, G., Brunoni, A.R., Mazzaferro, M.P., Anastasia, A., Bartolomeis, A., Mantovani, A., 2016. Transcranial direct current stimulation for obsessive–compulsive disorder: a randomized, controlled, partial crossover trial. Depress. Anxiety 33 (12), 1132–1140.

Eapen, V., Baker, R., Walter, A., Raghupathy, V., Wehrman, J.J., Sowman, P.F., 2017. The role of transcranial direct current stimulation (tDCS) in Tourette syndrome: a review and preliminary findings. Brain Sci. 7 (12). Available from: https://doi.org/10.3390/brainsci7120161.

Eaton, W.W., Roth, K.B., Bruce, M., Cottler, L., Wu, L., Nestadt, G., et al., 2013. The relationship of mental and behavioral disorders to all-cause mortality in a 27-year follow-up of 4 epidemiologic catchment area samples. Am. J. Epidemiol. 178 (9), 1366–1377.

Elbeh, K.A.M., Elserogy, Y.M.B., Khalifa, H.E., Ahmed, M.A., Hafez, M.H., Khedr, E.M., 2016. Repetitive transcranial magnetic stimulation in the treatment of obsessive–compulsive disorders: double blind randomized clinical trial. Psychiatry Res. 238, 264–269.

Finis, J., Enticott, P.G., Pollok, B., Münchau, A., Schnitzler, A., Fitzgerald, P.B., 2013. Repetitive transcranial magnetic stimulation of the supplementary motor area induces echophenomena. Cortex 49 (7), 1978–1982.

Fitzgerald, K.D., Welsh, R.C., Stern, E.R., Angstadt, M., Hanna, G.L., Abelson, J.L., et al., 2011. Developmental alterations of frontal–striatal–thalamic connectivity in obsessive–compulsive disorder. J. Am. Acad. Child Adolesc. Psychiatry 50 (9), 938.e3–948.e3.

Fontenelle, L.F., Harrison, B.J., Yücel, M., Pujol, J., Fujiwara, H., Pantelis, C., 2009. Is there evidence of brain white-matter abnormalities in obsessive–compulsive disorder?: a narrative review. Top. Magn. Reson. Imaging 20 (5), 291–298.

Freeman, R.D., Fast, D.K., Burd, L., Kerbeshian, J., Robertson, M.M., Sandor, P., 2000. An international perspective on Tourette syndrome: selected findings from 3,500 individuals in 22 countries. Dev. Med. Child Neurol. 42 (7), 436–447.

Gilbert, D.L., Bansal, A.S., Sethuraman, G., Sallee, F.R., Zhang, J., Lipps, T., et al., 2004. Association of cortical disinhibition with tic, ADHD, and OCD severity in Tourette syndrome. Mov. Disord. 19 (4), 416–425.

Gilbert, D.L., Sallee, F.R., Zhang, J., Lipps, T.D., Wassermann, E.M., 2005. Transcranial magnetic stimulation-evoked cortical inhibition: a consistent marker of attention-deficit/hyperactivity disorder scores in Tourette syndrome. Biol. Psychiatry 57 (12), 1597–1600.

Gilbert, D.L., Zhang, J., Lipps, T.D., Natarajan, N., Brandyberry, J., Wang, Z., et al., 2007. Atomoxetine treatment of ADHD in Tourette syndrome: reduction in motor cortex inhibition correlates with clinical improvement. Clin. Neurophysiol. 118 (8), 1835–1841.

Gilbert, A.R., Mataix-Cols, D., Almeida, J.R.C., Lawrence, N., Nutche, J., Diwadkar, V., et al., 2008. Brain structure and symptom dimension relationships in obsessive–compulsive disorder: a voxel-based morphometry study. J. Affect. Disord. 109 (1), 117–126.

Gomes, P.V.O., Brasil-Neto, J.P., Allam, N., Rodrigues de Souza, E., 2012. A randomized, double-blind trial of repetitive transcranial magnetic stimulation in obsessive–compulsive disorder with three-month follow-up. J. Neuropsychiatry Clin. Neurosci. 24 (4), 437–443.

Greenberg, B.D., Ziemann, U., Cora-Locatelli, G., Harmon, D.L., Murphy, J.C., Wasserman, E.M., 2000. Altered cortical excitability in obsessive–compulsive disorder. Neurology 54 (1), 142.

Haghighi, M., Shayganfard, M., Jahangard, L., 2015. Repetitive transcranial magnetic stimulation (rTMS) improves symptoms and reduces clinical illness in patients suffering from OCD—results from a single-blind, randomized clinical trial with sham cross-over condition. J. Psychiatr. Retrieved from http://www.sciencedirect.com/science/article/pii/S0022395615001910.

Hampson, M., Tokoglu, F., King, R.A., Constable, R.T., Leckman, J.F., 2009. Brain areas coactivating with motor cortex during chronic motor tics and intentional movements. Biol. Psychiatry 65 (7), 594–599.

Hanajima, R., Ugawa, Y., 2008. Paired-pulse measures. The Oxford Handbook of Transcranial Stimulation. Oxford University Press Inc., New York, NY, pp. 103–117.

Harrison, B.J., Pujol, J., Cardoner, N., Deus, J., Alonso, P., López-Solà, M., et al., 2013. Brain corticostriatal systems and the major clinical symptom dimensions of obsessive–compulsive disorder. Biol. Psychiatry 73 (4), 321–328.

Hashemiyoon, R., Kuhn, J., Visser-Vandewalle, V., 2017. Putting the pieces together in Gilles de la Tourette syndrome: exploring the link between clinical observations and the biological basis of dysfunction. Brain Topogr. 30 (1), 3–29.

Heise, K.-F., Steven, B., Liuzzi, G., Thomalla, G., Jonas, M., Müller-Vahl, K., et al., 2010. Altered modulation of intracortical excitability during movement preparation in Gilles de la Tourette syndrome. Brain 133 (Pt2), 580–590.

Himle, M.B., Chang, S., Woods, D.W., Pearlman, A., Buzzella, B., Bunaciu, L., et al., 2006. Establishing the feasibility of direct observation in the assessment of tics in children with chronic tic disorders. J. Appl. Behav. Anal. 39 (4), 429–440.

Hirschtritt, M.E., Lee, P.C., Pauls, D.L., Dion, Y., Grados, M.A., Illmann, C., et al., 2015. Lifetime prevalence, age of risk, and genetic relationships of comorbid psychiatric disorders in Tourette syndrome. JAMA Psychiatry 72 (4), 325–333.

Husted, D.S., Shapira, N.A., Goodman, W.K., 2006. The neurocircuitry of obsessive–compulsive disorder and disgust. Prog. Neuropsychopharmacol. Biol. Psychiatry 30 (3), 389–399.

Jackson, S.R., Parkinson, A., Manfredi, V., Millon, G., Hollis, C., Jackson, G.M., 2013. Motor excitability is reduced prior to voluntary movements in children and adolescents with Tourette syndrome. J. Neuropsychol. 7 (1), 29–44.

Jenike, M.A., Breiter, H.C., Baer, L., Kennedy, D.N., Savage, C.R., Olivares, M.J., et al., 1996. Cerebral structural abnormalities in obsessive–compulsive disorder. a quantitative morphometric magnetic resonance imaging study. Arch. Gen. Psychiatry 53 (7), 625–632.

Kang, J.I., Kim, C.-H., Namkoong, K., Lee, C.-I., Kim, S.J., 2009. A randomized controlled study of sequentially applied repetitive transcranial magnetic stimulation in obsessive–compulsive disorder. J. Clin. Psychiatry 70 (12), 1645–1651.

Koch, K., Reess, T.J., Rus, O.G., Zimmer, C., Zaudig, M., 2014. Diffusion tensor imaging (DTI) studies in patients with obsessive–compulsive disorder (OCD): a review. J. Psychiatr. Res. 54, 26–35.

Kwon, H.J., Lim, W.S., Lim, M.H., Lee, S.J., Hyun, J.K., Chae, J.-H., et al., 2011. 1-Hz low frequency repetitive transcranial magnetic stimulation in children with Tourette's syndrome. Neurosci. Lett. 492 (1), 1–4.

Landeros-Weisenberger, A., Mantovani, A., Motlagh, M.G., de Alvarenga, P.G., Katsovich, L., Leckman, J.F., et al., 2015. Randomized sham controlled double-blind trial of repetitive transcranial magnetic stimulation for adults with severe Tourette syndrome. Brain Stimul. 8 (3), 574–581.

Le, K., Liu, L., Sun, M., Hu, L., Xiao, N., 2013. Transcranial magnetic stimulation at 1 Hertz improves clinical symptoms in children with Tourette syndrome for at least 6 months. J. Clin. Neurosci. 20 (2), 257–262.

Leckman, J.F., Riddle, M.A., Hardin, M.T., Ort, S.I., Swartz, K.L., Stevenson, J., et al., 1989. The Yale Global Tic Severity Scale: initial testing of a clinician-rated scale of tic severity. J. Am. Acad. Child Adolesc. Psychiatry 28 (4), 566–573.

Leonard, H.L., Lenane, M.C., Swedo, S.E., Rettew, D.C., Gershon, E.S., Rapoport, J.L., 1992. Tics and Tourette's disorder: a 2- to 7-year follow-up of 54 obsessive–compulsive children. Am. J. Psychiatry 149 (9), 1244–1251.

Luber, B.M., Davis, S., Bernhardt, E., Neacsiu, A., Kwapil, L., Lisanby, S.H., et al., 2017. Using neuroimaging to individualize TMS treatment for depression: toward a new paradigm for imaging-guided intervention. NeuroImage 148, 1–7.

Ma, X., Huang, Y., Liao, L., Jin, Y., 2014. A randomized double-blinded sham-controlled trial of α electroencephalogram-guided transcranial magnetic stimulation for obsessive–compulsive disorder. Chin. Med. J. 127 (4), 601–606.

Mansueto, C.S., Keuler, D.J., 2005. Tic or compulsion?: it's Tourettic OCD. Behav. Modif. 29 (5), 784–799.

Mansur, C.G., Myczkowki, M.L., de Barros Cabral, S., Sartorelli Mdo, C., Bellini, B.B., Dias, A.M., et al., 2011. Placebo effect after prefrontal magnetic stimulation in the treatment of resistant obsessive–compulsive disorder: a randomized controlled trial. Int. J. Neuropsychopharmacol. 14 (10), 1389–1397.

Mantovani, A., Lisanby, S.H., Pieraccini, F., Ulivelli, M., Castrogiovanni, P., Rossi, S., 2006. Repetitive transcranial magnetic stimulation (rTMS) in the treatment of obsessive–compulsive disorder (OCD) and Tourette's syndrome (TS). Int. J. Neuropsychopharmacol. 9 (1), 95–100.

Mantovani, A., Leckman, J.F., Grantz, H., King, R.A., Sporn, A.L., Lisanby, S.H., 2007. Repetitive transcranial magnetic stimulation of the supplementary motor area in the treatment of Tourette syndrome: report of two cases. Clin. Neurophysiol. 118 (10), 2314–2315.

Mantovani, A., Simpson, H.B., Fallon, B.A., Rossi, S., Lisanby, S.H., 2010. Randomized sham-controlled trial of repetitive transcranial magnetic stimulation in treatment-resistant obsessive–compulsive disorder. Int. J. Neuropsychopharmacol. 13 (2), 217–227.

Mantovani, A., Rossi, S., Bassi, B.D., Simpson, H.B., Fallon, B.A., Lisanby, S.H., 2013. Modulation of motor cortex excitability in obsessive–compulsive disorder: an exploratory study on the relations of neurophysiology measures with clinical outcome. Psychiatry Res. 210 (3), 1026–1032.

March, J.S., Franklin, M.E., Leonard, H., Garcia, A., Moore, P., Freeman, J., et al., 2007. Tics moderate treatment outcome with sertraline but not cognitive-behavior therapy in pediatric obsessive–compulsive disorder. Biol. Psychiatry 61 (3), 344–347.

Mataix-Cols, D., van den Heuvel, O.A., 2006. Common and distinct neural correlates of obsessive–compulsive and related disorders. Psychiatr. Clin. North Am. 29 (2), 391–410. viii.

Mataix-Cols, D., Wooderson, S., Lawrence, N., Brammer, M.J., Speckens, A., Phillips, M.L., 2004. Distinct neural correlates of washing, checking, and hoarding symptom dimensions in obsessive–compulsive disorder. Arch. Gen. Psychiatry 61 (6), 564–576.

Mathews, C.A., Grados, M.A., 2011. Familiality of Tourette syndrome, obsessive–compulsive disorder, and attention-deficit/hyperactivity disorder: heritability analysis in a large sib-pair sample. J. Am. Acad. Child Adolesc. Psychiatry 50 (1), 46–54.

McGuire, J.F., Piacentini, J., Brennan, E.A., Lewin, A.B., Murphy, T.K., Small, B.J., et al., 2014. A meta-analysis of behavior therapy for Tourette Syndrome. J. Psychiatr. Res. 50, 106–112.

Mink, J.W., 2006. Neurobiology of basal ganglia and Tourette syndrome: basal ganglia circuits and thalamocortical outputs. Adv. Neurol. 99, 89–98.

Moll, G.H., Wischer, S., Heinrich, H., Tergau, F., Paulus, W., Rothenberger, A., 1999. Deficient motor control in children with tic disorder: evidence from transcranial magnetic stimulation. Neurosci. Lett. 272 (1), 37–40.

Moll, G.H., Heinrich, H., Trott, G.E., Wirth, S., Bock, N., Rothenberger, A., 2001. Children with comorbid attention-deficit-hyperactivity disorder and tic disorder: evidence for additive inhibitory deficits within the motor system. Ann. Neurol. 49 (3), 393–396.

Mrakic-Sposta, S., Marceglia, S., Mameli, F., Dilena, R., Tadini, L., Priori, A., 2008. Transcranial direct current stimulation in two patients with Tourette syndrome. Mov. Disord. 23 (15), 2259–2261.

Münchau, A., Bloem, B.R., Thilo, K.V., Trimble, M.R., Rothwell, J.C., Robertson, M.M., 2002. Repetitive transcranial magnetic stimulation for Tourette syndrome. Neurology 59 (11), 1789–1791.

Nauczyciel, C., Le Jeune, F., Naudet, F., Douabin, S., Esquevin, A., Vérin, M., et al., 2014. Repetitive transcranial magnetic stimulation over the orbitofrontal cortex for obsessive–compulsive disorder: a double-blind, crossover study. Transl. Psychiatry 4, e436.

Orth, M., Rothwell, J.C., 2009. Motor cortex excitability and comorbidity in Gilles de la Tourette syndrome. J. Neurol. Neurosurg. Psychiatry 80 (1), 29–34.

Orth, M., Münchau, A., 2013. Transcranial magnetic stimulation studies of sensorimotor networks in Tourette syndrome. Behav. Neurol. 27 (1), 57–64.

Orth, M., Amann, B., Robertson, M.M., Rothwell, J.C., 2005. Excitability of motor cortex inhibitory circuits in Tourette syndrome before and after single dose nicotine. Brain 128 (Pt6), 1292–1300.

Orth, M., Münchau, A., Rothwell, J.C., 2008. Corticospinal system excitability at rest is associated with tic severity in Tourette syndrome. Biol. Psychiatry 64 (3), 248–251.

Pallanti, S., Grassi, G., Sarrecchia, E.D., Cantisani, A., Pellegrini, M., 2011. Obsessive–compulsive disorder comorbidity: clinical assessment and therapeutic implications. Front. Psychiatry/Front. Res. Found. 2, 70.

Palumbo, D., Kurlan, R., 2007. Complex obsessive compulsive and impulsive symptoms in Tourette's syndrome. Neuropsychiatr. Dis. Treat. 3 (5), 687–693.

Pedapati, E., DiFrancesco, M., Wu, S., Giovanetti, C., Nash, T., Mantovani, A., et al., 2015. Neural correlates associated with symptom provocation in pediatric obsessive compulsive disorder after a single session of sham-controlled repetitive transcranial magnetic stimulation. Psychiatry Res. 233 (3), 466–473.

Pépés, S.E., Draper, A., Jackson, G.M., Jackson, S.R., 2016. Effects of age on motor excitability measures from children and adolescents with Tourette syndrome. Dev. Cogn. Neurosci. 19, 78–86.

Pinto, A., Mancebo, M.C., Eisen, J.L., Pagano, M.E., Rasmussen, S.A., 2006. The Brown Longitudinal Obsessive Compulsive Study: clinical features and symptoms of the sample at intake. J. Clin. Psychiatry 67 (5), 703–711.

Piras, F., Piras, F., Caltagirone, C., Spalletta, G., 2013. Brain circuitries of obsessive compulsive disorder: a systematic review and meta-analysis of diffusion tensor imaging studies. Neurosci. Biobehav. Rev. 37 (10 Pt 2), 2856–2877.

Prasko, J., Paskova, B., Zalesky, R., Novak, T., Kopecek, M., Bares, M., et al., 2006. The effect of repetitive transcranial magnetic stimulation (rTMS) on symptoms in obsessive compulsive disorder. A randomized, double blind, sham controlled study. Neuro Endocrinol. Lett. 27 (3), 327–332.

Priori, A., Berardelli, A., Inghilleri, M., Accornero, N., Manfredi, M., 1994. Motor cortical inhibition and the dopaminergic system: pharmacological changes in the silent period after transcranial brain stimulation in normal subjects, patients with Parkinson's disease and drug-induced parkinsonism. Brain 117 (2), 317–323.

Richter, M.A., de Jesus, D.R., Hoppenbrouwers, S., Daigle, M., Deluce, J., Ravindran, L.N., et al., 2012. Evidence for cortical inhibitory and excitatory dysfunction in obsessive compulsive disorder. Neuropsychopharmacol 37 (5), 1144–1151.

Robinson, D., Wu, H., Munne, R.A., Ashtari, M., Alvir, J.M., Lerner, G., et al., 1995. Reduced caudate nucleus volume in obsessive–compulsive disorder. Arch. Gen. Psychiatry 52 (5), 393–398.

Ruffini, C., Locatelli, M., Lucca, A., Benedetti, F., 2009. Augmentation effect of repetitive transcranial magnetic stimulation over the orbitofrontal cortex in drug-resistant obsessive–compulsive disorder patients: a controlled investigation. Prim. Care Companion J. Clin. Psychiatry 11 (5), 226–230.

Ruscio, A.M., Stein, D.J., Chiu, W.T., Kessler, R.C., 2010. The epidemiology of obsessive–compulsive disorder in the National Comorbidity Survey Replication. Mol. Psychiatry 15 (1), 53–63.

Sachdev, P.S., McBride, R., Loo, C.K., Mitchell, P.B., Malhi, G.S., Croker, V.M., 2001. Right versus left prefrontal transcranial magnetic stimulation for obsessive–compulsive disorder: a preliminary investigation. J. Clin. Psychiatry 62 (12), 981–984.

Sachdev, P.S., Loo, C.K., Mitchell, P.B., McFarquhar, T.F., Malhi, G.S., 2007. Repetitive transcranial magnetic stimulation for the treatment of obsessive compulsive disorder: a double-blind controlled investigation. Psychol. Med. 37 (11), 1645–1649.

Sarkhel, S., Sinha, V.K., Praharaj, S.K., 2010. Adjunctive high-frequency right prefrontal repetitive transcranial magnetic stimulation (rTMS) was not effective in obsessive–compulsive disorder but improved secondary depression. J. Anxiety Disord. 24 (5), 535–539.

Saxena, S., Brody, A.L., Ho, M.L., Alborzian, S., Maidment, K.M., Zohrabi, N., et al., 2002. Differential cerebral metabolic changes with paroxetine treatment of obsessive–compulsive disorder vs major depression. Arch. Gen. Psychiatry 59 (3), 250–261.

Saxena, S., Gorbis, E., O'Neill, J., Baker, S.K., Mandelkern, M.A., Maidment, K.M., et al., 2009. Rapid effects of brief intensive cognitive-behavioral therapy on brain glucose metabolism in obsessive–compulsive disorder. Mol. Psychiatry 14 (2), 197–205.

Schwartz, J.M., Stoessel, P.W., Baxter Jr, L.R., Martin, K.M., Phelps, M.E., 1996. Systematic changes in cerebral glucose metabolic rate after successful behavior modification treatment of obsessive–compulsive disorder. Arch. Gen. Psychiatry 53 (2), 109–113.

Skarphedinsson, G., Compton, S., Thomsen, P.H., Weidle, B., Dahl, K., Nissen, J.B., et al., 2015. Tics moderate sertraline, but not cognitive-behavior therapy response in pediatric obsessive–compulsive disorder patients who do not respond to cognitive-behavior therapy. J. Child Adolesc. Psychopharmacol. 25 (5), 432–439.

Steinberg, T., King, R., Apter, A., 2010. Tourette's syndrome: a review from a developmental perspective. Isr. J. Psychiatry Relat. Sci. 47 (2), 105–109.

Stern, E., Silbersweig, D.A., Chee, K.Y., Holmes, A., Robertson, M.M., Trimble, M., et al., 2000. A functional neuroanatomy of tics in Tourette syndrome. Arch. Gen. Psychiatry 57 (8), 741–748.

Subirà, M., Sato, J.R., Alonso, P., do Rosário, M.C., Segalàs, C., Batistuzzo, M.C., et al., 2015. Brain structural correlates of sensory phenomena in patients with obsessive–compulsive disorder. J. Psychiatry Neurosci. 40 (4), 232–240.

Sukhodolsky, D.G., Woods, D.W., Piacentini, J., Wilhelm, S., Peterson, A.L., Katsovich, L., et al., 2017. Moderators and predictors of response to behavior therapy for tics in Tourette syndrome. Neurology 88 (11), 1029–1036.

Szeszko, P.R., MacMillan, S., McMeniman, M., Lorch, E., Madden, R., Ivey, J., et al., 2004. Amygdala volume reductions in pediatric patients with obsessive–compulsive disorder treated with paroxetine: preliminary findings. Neuropsychopharmacology 29 (4), 826–832.

Todder, D., Gershi, A., Perry, Z., Kaplan, Z., Levine, J., Avirame, K., 2017. Immediate effects of transcranial direct current stimulation on obsession-induced anxiety in refractory obsessive–compulsive disorder: a pilot study. J. ECT. . Available from: https://doi.org/10.1097/YCT.0000000000000473.

Veale, D., Roberts, A., 2014. Obsessive–compulsive disorder. BMJ 348, g2183.

Wang, Z., Maia, T.V., Marsh, R., Colibazzi, T., Gerber, A., Peterson, B.S., 2011. The neural circuits that generate tics in Tourette's syndrome. Am. J. Psychiatry 168 (12), 1326–1337.

Worbe, Y., Malherbe, C., Hartmann, A., Pélégrini-Issac, M., Messé, A., Vidailhet, M., et al., 2012. Functional immaturity of cortico-basal ganglia networks in Gilles de la Tourette syndrome. Brain 135 (Pt6), 1937–1946.

Worbe, Y., Lehericy, S., Hartmann, A., 2015. Neuroimaging of tic genesis: present status and future perspectives. Mov. Disord. 30 (9), 1179–1183.

Wu, S.W., Maloney, T., Gilbert, D.L., Dixon, S.G., Horn, P.S., Huddleston, D.A., et al., 2014. Functional MRI-navigated repetitive transcranial magnetic stimulation over supplementary motor area in chronic tic disorders. Brain Stimul. 7 (2), 212–218.

Yael, D., Vinner, E., Bar-Gad, I., 2015. Pathophysiology of tic disorders. Mov. Disord. 30 (9), 1171–1178.

Zapparoli, L., Porta, M., Paulesu, E., 2015. The anarchic brain in action: the contribution of task-based fMRI studies to the understanding of Gilles de la Tourette syndrome. Curr. Opin. Neurol. 28 (6), 604–611.

Ziemann, U., Paulus, W., Rothenberger, A., 1997. Decreased motor inhibition in Tourette's disorder: evidence from transcranial magnetic stimulation. Am. J. Psychiatry 154 (9), 1277–1284.

CHAPTER

9

tDCS in Pediatric Neuropsychiatric Disorders

Carmelo M. Vicario[1,2,3] *and Michael A. Nitsche*[1,2]

[1]Department Psychology and Neurosciences, Leibniz Research Center for Working Environment and Human Factors, Dortmund, Germany [2]Department of Neurology, University Medical Hospital Bergmannsheil, Bochum, Germany [3]Department of Cognitive Sciences, Educational and Cultural Studies (CSECS), via della concezione 6, Messina, Italy

OUTLINE

Neurotechnological Pediatric Neuropsych
DOI: https://doi.org/10.1016/B978-0-12-812777-3.00009-X

INTRODUCTION

The attempt to identify methodologies for the treatment of pediatric neuropsychiatric disorders that combine both therapeutic effectiveness and tolerability has oriented the interest of large sectors of cognitive and experimental neuroscience toward the use of noninvasive brain stimulation techniques. The interest in these techniques is mainly based on research documenting their effectiveness in modulating neural plasticity (e.g., Kronberg et al., 2017; Ziemann, 2017). The assumption that noninvasive brain stimulation is able to induce synaptic plasticity is supported by pharmacological studies in humans (for an overview see Nitsche et al., 2012) and animal models (Fritsch et al., 2010; Kronberg et al., 2017). For example, Kronberg et al. (2017) reported that cathodal direct current stimulation (DCS) enhanced long-term potentiation (LTP) in apical dendrites, while anodal DCS enhanced LTP in basal dendrites of rat hippocampal slices. Furthermore, a dependency of the aftereffects of transcranial DCS (tDCS) from glutamatergic synapses has been suggested in studies conducted in humans (Nitsche et al., 2003, 2004a). This mechanistic action of noninvasive brain stimulation is a crucial aspect, as pathologically altered neural plasticity is an important component of many neurological and psychiatric diseases, including the failure to achieve a successful (spontaneous vs therapeutically mediated) recovery of brain functions in response to a disease (e.g., Jang, 2013).

The use of noninvasive brain stimulation techniques for the treatment of neurological and psychiatric disorders might be considered useful for at least two reasons:

1. It represents a potential alternative or adjunct to drug treatment. This is particularly relevant for the treatment of pediatric populations, since some classical treatments are highly invasive, or associated with a high probability for side effects, and thus specifically problematic for a developing organism;
2. It might offer a therapeutic option for brain-related diseases for which an effective pharmacologic treatment is currently not available.

The scientific community's interest in noninvasive brain stimulation, including adult (e.g., Kuo et al., 2012, 2014, 2017; Vicario and Rumiati, 2012) and pediatric (Vicario and Nitsche, 2013a; Rivera-Urbina et al., 2017 for review) populations, has resulted in florid research. Yet, although the results of this research line are somewhat striking, as they have revealed promising results, on the other hand, they highlight our limited knowledge in the field, prompting an acknowledged and shared awareness that further and systematic investigation are required to transfer respective interventions into clinical practice.

In this chapter, we will focus on research conducted using tDCS in the field of pediatric neurological and psychiatric disorders. In particular, we aim to provide a general overview about the current state of the art and discuss the main implications and future research directions.

MECHANISMS OF ACTION OF tDCS

tDCS is an established technique in the cognitive and clinical neurosciences that allows safe and noninvasive stimulation of the cerebral cortex. tDCS modulates cortical excitability via small direct electrical currents (usually 1 and 2 mA) passed through the brain (Nitsche and Paulus, 2000; Nitsche et al., 2003, 2008) via two or more electrodes with opposite polarities (i.e., anodal and cathodal) placed on the scalp. At the macroscopic level, anodal (A) stimulation increases cortical excitability, whereas cathodal (C) stimulation decreases it (Stagg and Nitsche, 2011), although specific protocols can result in antagonistic effects (Batsikadze et al., 2013; Monte-Silva et al., 2013). These effects emerge during stimulation, but can last for up to 90 minutesafter a single stimulation session of 13–20 minutes (Nitsche and Paulus, 2001) and can be further extended by repeated stimulation (i.e., cumulative effects) (Monte-Silva et al., 2013).

The primary effects of tDCS, which emerge immediately during stimulation of even a few seconds, can be explained in terms of changes of neural membrane potentials (Stagg and Nitsche, 2011). Short-lasting (application for few seconds) tDCS, which elicits no aftereffects, depolarizes or hyperpolarizes neuronal membranes at a subthreshold level. The direction of the effects depends on the orientation of the electric field in relation to neuronal orientation. This increases or decreases the probability of action potential generation by afferent neuronal activity. Hereby, anodal tDCS is thought to have an overall depolarizing effect, while cathodal tDCS is assumed to have a hyperpolarizing effect at the macroscopic level regarding the neuronal membrane compartments primarily involved in respective effects. In line with these assumptions, blockage of voltage-gated ion channels, which should diminish membrane depolarization, abolishes the respective effects of anodal tDCS (Nitsche et al., 2003). On the other hand, the reduction of excitability caused by cathodal tDCS is not changed by voltage-gated ion channel block, because these channels will be inactivated by the respective stimulation-induced membrane hyperpolarization (Nitsche et al., 2003).

Synaptic mechanisms seem not to be involved in these acute effects elicited by few seconds stimulation, since pharmacologic modulation of the glutamatergic and GABAergic system does not alter these effects (Nitsche et al., 2003, 2004b), and intracortical facilitation (ICF) and short-interval cortical inhibition, which are controlled by GABAergic and glutamatergic interneuronal circuits, are not affected (Nitsche et al., 2005). In summary, during short-lasting tDCS administration (application for a few seconds), both anodal and cathodal tDCS primarily affect resting membrane potentials during stimulation, with no significant effects on synaptic plasticity (Stagg and Nitsche, 2011).

The physiological aftereffects of prolonged (i.e., application for several minutes) anodal and cathodal tDCS, however, are dependent on synaptic modulation. Anodal stimulation increases ICF, and its aftereffects are prevented by *N-methyl-d-aspartate* (NMDA) receptor block, but enhanced by respective receptor agonists (Liebetanz et al., 2002; Nitsche et al., 2003, 2004b, 2005). Moreover, the aftereffects of cathodal tDCS are prevented by NMDA receptor block (Nitsche et al., 2003). It can therefore be assumed that glutamatergic neurons are crucial for the induction of plasticity by tDCS. This is also substantiated by the fact that ICF, which critically depends on glutamatergic receptor activity, is enhanced by anodal and reduced by cathodal tDCS (Nitsche et al., 2005). This tDCS-induced glutamatergic plasticity might, however, be gated by alterations of GABA activity. As shown by a magnetic resonance spectroscopy study, anodal and cathodal tDCS reduced GABA content of the motor cortex (Stagg et al., 2009). This result is in accordance with stimulation polarity-independent enhancement of GABA-dependent I-wave facilitation by tDCS, which was described in an earlier study (Nitsche et al., 2005).

tDCS IN CLINICAL PEDIATRICS

In the following sections, we will outline current research on clinical therapeutic applications of tDCS in neurological and psychiatric populations ranging from childhood to adolescence (i.e., up to 18 years old). In particular, we discuss the available studies on the use of tDCS for the treatment of vascular diseases, epilepsy, ADHD (attention-deficit/hyperactivity disorder), autism spectrum disorder (ASD), schizophrenia, and learning disabilities.

tDCS for the Treatment of Vascular Diseases

tDCS is applied to enhance rehabilitation after brain lesions such as stroke related to vascular diseases so far primarily in adult populations (Flöel, 2014). One of the main pathophysiological substrates after stroke-related brain lesions is hemispheric dysbalance, which refers to a reduction in the activity of the lesioned brain area and enhanced activity of the contra-lesional homologous region (see Grefkes and Fink, 2011 for a recent review). This phenomenon is considered to limit the patient's capacity to regain compromised neural functions. Therapeutic treatments for these conditions are designed to reduce the respective dysbalance. In the following sub-paragraphs, we address two clinical conditions that are related to vascular disease and are specifically relevant for pediatric populations, namely cerebral palsy and dystonia.

Cerebral palsy is characterized by damage of parts of the brain that control for movement, balance, and posture (National Institutes of Health, 2015) and is caused most often by pre- or perinatal vascular disorders. Clinical trials investigating the effectiveness of tDCS for its treatment provide so far inconsistent results. A randomized, sham-controlled trial (Moura et al., 2017) was conducted on 20 children (age range 6–12) with spastic hemiparetic cerebral palsy. The protocol consisted of a 20-minute session of functional training of the paretic upper limb combined with anodal tDCS administered over the primary motor cortex of the hemisphere contralateral to the motor impairment at an intensity of 1 mA. The authors report a reduction in total movement and returning movement duration in both, the paretic and non-paretic limbs, in the group exposed to active tDCS.

In the study by Lazzari et al. (2017), the authors tested 20 children (mean age 7 years and 2 months) by applying anodal (1 mA, experimental group) or sham (control group) tDCS over the primary motor cortex during 10 sessions of a virtual reality mobility training protocol. The results document significant improvements of static and functional balance in the anodal group, at the postintervention and follow-up evaluations in comparison to baseline. Positive effects were also reported by other studies (e.g., Rocha et al., 2016). On the other hand, another recent clinical trial (Kirton et al., 2017) in children (24 participants aged between 6 and 18 years) with cerebral palsy found no significant effects when applying cathodal tDCS (1 mA, 20 minutes per day, 2 weeks) over the contralesional M1. This suggests that an effective treatment protocol might include upregulation of the damaged hemisphere rather than downregulation of the homolog regions of the contralateral neural hemisphere. Moreover, 1 mA cathodal tDCS over the motor cortex in children enhanced excitability in another study (Moliadze et al., 2015a), which might have limited the efficacy of stimulation in that study.

Dystonia, a debilitating movement disorder characterized by involuntary muscle contractions that lead to repetitive movements and/or abnormal postures (Sanger et al., 2010), is another disorder associated with vascular issues in childhood. The current literature provides only one pilot study (Bhanpuri et al., 2015), which reports no significant effects of both, cathodal and anodal tDCS (5-day, sham-controlled, double-blind, crossover study, 2 mA, 9 minutes per day) over the motor cortex (C3 and C4) on these symptoms in a pediatric population (9 patients, age between 10 and 21 years). However, the absence of significant results might be due to the absence of a concomitant motor training, as suggested by Furuya et al. (2014) in adults with focal dystonia. The authors of this study demonstrated that, in absence of a concurrent motor training, bihemispheric stimulation failed to improve fine motor control. This suggests that a protocol combining training and stimulation might be a more effective way to treat dystonia also in pediatric populations.

Overall, the available evidence indicates that research conducted so far, which is characterized by relatively small sample sizes (around 10 patients per stimulation condition), is too preliminary for deriving definitive conclusions about the therapeutic effectiveness of tDCS for the treatment of vascular disorder-related clinical syndrome.

tDCS for the Treatment of Epilepsy

The pathophysiological characteristic of epilepsy is the proneness to develop seizures as effect of an abnormally enhanced cortical excitability (Stafstrom, 2006; Dudek and Sutula, 2007). A reduction of neuronal excitability is the common aim of antiepileptic therapies.

San-Juan et al. (2011) applied cathodal tDCS over the F2 scalp site (2 mA intensity for 60 minutes, 4 sessions) in a 17-year-old patient affected by Rasmussen's encephalitis. The 4 sessions were applied at day 0, 7, 30, and 60. On admission, the patient showed epilepsia partialis continua symptoms characterized by partial motor seizures on the left side of his body (>20 jerks per minute) and right hemiparesis. During tDCS and follow-up, the patient was taking gabapentin 900 mg/day, topiramate 200 mg/day, and valproate 1200 mg/day. Medication was kept constant throughout the tDCS intervention. The patient showed a gradual reduction in the intensity and frequency of seizures following the tDCS sessions. At the follow-up evaluation, 6 months later, the patient was able to walk, with occasional clonic partial seizures of the left arm (approximately one every day) and mild improvement of the partial seizures in the left leg (10–15 jerks per minute).

Yook et al. (2011) applied cathodal tDCS (2 mA, 20 minutes) over the epileptogenic focus (the right temporoparietal area) in an 11-year-old patient diagnosed with congenital bilateral perisylvian syndrome for 2 weeks (5 times a week). The patient suffered from on average 8 seizures per month before the intervention. During the 2-month period after treatment termination, seizure frequency and duration were reduced.

tDCS was repeated for another 2 weeks, 2 months after the first intervention session. For the following 2 months, only one seizure occurred. More recently, Auvichayapat et al. (2013) enrolled 36 children (age range 6–15 years) with focal epilepsy in a crossover sham-controlled study. Participants received a single session of cathodal tDCS (1 mA, 20 minutes) over the epileptogenic region. Active tDCS was associated with significant reductions in epileptic discharge frequency immediately and 24 and 48 hours after tDCS. No change in epileptic discharge frequency was reported for the control group (sham stimulation). Moreover, 4 weeks after treatment, a small (but statistically significant) decrease in seizure frequency was reported in the real stimulation group relative to the sham group. All patients tolerated tDCS well. Finally, a significant reduction of seizure frequency and epileptic activity accomplished by tDCS has recently been reported in children with Lennox–Gastaut syndrome (Auvichayapat et al., 2016). In this case, cathodal tDCS (2 mA, 20 minutes for 5 consecutive days) was applied over the primary motor cortex for 5 consecutive days in 22 children aged between 6 and 15 years. Varga et al. (2011) conducted a study in five pediatric patients (age range 6–11), showing continuous spike-waves during slow-wave sleep (CSWS). One session of cathodal tDCS (1 mA; 20 minutes) over the peak negativity of the epileptogenic pattern revealed an effect (less propagated spikes) on electroencephalographic (EEG) patterns in three of five patients. A possible explanation for this, at least partially negative result, is that the multifocal/diffuse and poorly defined origin of epileptic activity in CSWS makes it difficult to identify the optimal region for stimulation (Brazzo et al., 2012).

Overall, most of the studies mentioned in this paragraph provide preliminary evidence for a possible efficacy of tDCS to treat epilepsy in children. Yet, the heterogeneity of the results, possibly due to different stimulation parameters, and different etiologies of participants, do not allow us to draw definitive conclusions that would justify implementation of tDCS in routine therapy.

Attention-Deficit/Hyperactivity Disorder

ADHD is characterized by inattention, hyperactivity, and executive dysfunction. Neuroimaging studies have shown functional abnormalities in cingulate, frontal and parietal cortical regions, including hypoactivation of these areas (Bush, 2011). Thus, noninvasive brain stimulation procedures to improve ADHD symptoms are oriented toward activity-enhancing protocols.

Only two recently published studies (e.g., Munz et al., 2015; Bandeira et al., 2016) have explored this therapeutic option. The study conducted by Munz et al. (2015) on 14 boys (10–14 years) documents improved inhibitory control skills (monitored via go/no-go reaction time task) after application of slow oscillating (sinusoidal currents oscillating at a frequency of 0.75 Hz) anodal tDCS, that is, oscillating stimulation with no polarity shift, bilaterally over F3-F4 (patient age range between 10 and 14 years), during non Rapid eye movement (REM) sleep (five intervals of each 5 minutes duration). Bandeira et al. (2016) report improved visual attention and inhibitory control skills after five consecutive, once daily, anodal tDCS sessions (2 mA for 28 minutes) over F3 (9 children age range 6–16 years). Both studies suggest that modulation of prefrontal cortex excitability might be effective for the treatment of children with ADHD.

Overall, this research provides exciting preliminary results in support of the therapeutic efficacy of tDCS for the treatment of ADHD symptoms, although further work is required to consolidate this result with regard to other related disorders characterized by high level of comorbidity (e.g., Tourette syndrome, Rizzo et al., 2013).

Autism Spectrum Disorder

ASD is characterized by a marked decrease in social integration and communication. tDCS has been shown to improve executive function and language deficits, which appear to be compromised in this condition (e.g., Kiep and Speck, 2016).

Schneider and Hopp (2011) conducted an open tDCS study in a group of 10 children with autism (aged 6–21 years), to improve language acquisition in patients with minimal verbal language. The anode electrode was applied over F3, while the cathode electrode was applied over the right supraorbital region. The treatment with a current intensity of 2 mA lasted for 30 minutes. For probing language acquisition, the authors adopted a modified version of the bilingual aphasia test, to test only basic canonical subject–verb–object sentences. Patients were first taught the vocabulary that appeared in the syntax test, and then tested to verify successful word learning. This vocabulary test was followed by exposure to scaffolding sentences approximating the syntax to be tested. Finally, the syntax comprehension test was conducted by the patients. The entire procedure was performed both before and after tDCS. The results show a large effect size pre-/post-tDCS, indicating that syntax acquisition was enhanced by the intervention. However, interpretability of these results is limited by the absence of a sham control.

Amatachaya et al. (2014) documented higher scores in the Autism Treatment Evaluation Checklist and Children's Global Assessment Scale after five consecutive sessions of anodal tDCS (1 mA, 20 minutes) applied over F3, compared to sham stimulation (Amatachaya et al., 2014). The age of the examined patient group (20 participants) was between 5 and 8 years. In a further study, the same group (Amatachaya et al., 2015) also reported improvements in social functionality, health and behavioral problem scales of the Autism Treatment Evaluation Checklist after one single sessions of anodal tDCS (1 mA, 20 minutes) over F3 (20 children, age 5–8). Finally, Costanzo et al. (2015) document a reduction of catatonic symptoms after bilateral prefrontal tDCS in an adolescent girl (14 years old) with autism. In this patient, the anode was placed over F3 and the cathode over F4 (1 mA, 20 minutes, 28 sessions).

These studies provide encouraging preliminary evidence for therapeutic benefits of tDCS in children affected by ASD. Like in ADHD, they suggest a relevance of the prefrontal cortex as key cortical region for improving cognitive functions. However, due to the small numbers of studies, and specific design aspects, such as the absence of a placebo condition and/or of a double-blinded protocol (e.g., see the study by Schneider and Hopp, 2011), statements about optimal protocols are premature.

Schizophrenia

Childhood-onset schizophrenia is a rare and severe form of this disorder (Nicolson and Rapoport, 1999) characterized by hallucinations. The only available tDCS-related research has been performed by Mattai et al. (2011), who investigated the tolerability of tDCS in a group of pediatric patients with schizophrenia. From an initial sample of 15 patients, 12 participants (age 10–17) completed the study. They were assigned to one of two groups: bilateral anodal dorsolateral prefrontal cortex (DLPFC) ($n = 8$) stimulation or bilateral cathodal superior temporal gyrus (STG) ($n = 5$) stimulation. Bilateral anodal DLPFC stimulation was adopted to improve cognitive difficulties. The bilateral cathodal STG stimulation was adopted to reduce hallucinations. These electrode arrangements were chosen based on evidence for positive clinical effects of related noninvasive brain stimulation protocols on respective symptoms in adults (e.g., see Cole et al., 2015 for a review). Patients received either 2 mA of active treatment or sham treatment (with the option for future open active treatment) for 20 minutes, for a total of 10 sessions conducted in 2 weeks. No significant improvement was reported regarding mood, arousal, mini-mental state examination performance, and verbal output in response to both stimulation conditions. This might be due to the low number of participants and/or by the use of antipsychotic medication that may have altered cortical excitability, therefore preventing the functional effects of tDCS. Stimulation was well tolerated, and no serious adverse events were reported.

Overall, this research domain remains widely unexplored. For this reason, any statement about the effectiveness of tDCS for the treatment of pediatric schizophrenia is premature.

tDCS in Learning Disabilities

With the term "learning disabilities," we refer to childhood deficits in the ability to execute tasks involving reading, writing, and mathematical skills, in the absence of other intellectual deficits essential for thinking and/or reasoning (See DSM-5 for more details, Association, 2014). So far, the only currently available published research has been conducted in children with dyslexia.

In the first investigation, Costanzo et al. (2016a) tested the effects of tDCS in a group of 19 children and adolescents (age range 10–17 years) with dyslexia. They compared performance in different reading and reading-related tasks at baseline (i.e., without tDCS) and after 20 minutes (1 mA) of exposure to three different tDCS conditions: left anodal/right cathodal tDCS to enhance left lateralization of the parietotemporal region, right anodal/left cathodal tDCS to enhance right lateralization of the parietotemporal region, and sham tDCS. The authors report a significant reduction of text-reading errors after left anodal/right cathodal tDCS and increased errors after left cathodal/right anodal tDCS. The effectiveness of anodal stimulation over the left parietotemporal regions has been corroborated in a subsequent study (Costanzo et al., 2016b) adopting three 20-minute (1 mA) sessions per week for 6 consecutive weeks (18 sessions, 18 participants, age range 10.9–17.1 years). The active group showed reduced errors for low-frequency word reading and non-word reading times. These positive effects were stable even 1 month after the end of treatment. These two studies suggest that tDCS has potential in the rehabilitation of reading skills but, at the same time, shows that stimulation polarity is critical for the direction of performance modulation.

Discussion and Future Directions

In this chapter, we have provided an overview of the effects of tDCS in pediatric neurological and psychiatric disorders (see Table 9.1 for a summary).

Overall, current research provides preliminary evidence that tDCS has therapeutic potential for the treatment of several disorders in children and adolescence. However, we are currently far from having reached a clear picture, as several limitations are currently recognized.

First of all, the literature suffers from a lack of double-blinded sham-controlled studies. Respective limitations of the interpretability of

TABLE 9.1 tDCS for the Treatment of Childhood Brain Disorders

Study	Patients	Stimulation protocol						Outcome
		Polarity/ electrode size	Target electrode position	Return electrode position	Current strength (mA)	Duration (min)/ sessions	Sample size	Effects
VASCULAR DISEASES								
Rocha et al. (2016)	*Cerebral palsy*	A/C/S 35 cm^2	C3 and C4 (anode) M1 unaffected hemisphere (cathode)	Contralateral supraorbital ridge	1 mA	13 min/ anodal 9 min cathodal	21	Significant improvement of motor recovery (assessed by Fugl–Meyer score) in the anodal condition
Lazzari et al. (2017)	*Cerebral palsy*	A/S 25 cm^2	M1 (no details about the hemisphere)	Contralateral supraorbital ridge	1 mA	20 min/ 10 sessions/ once daily, consecutive days	20	Significant improvement of static and functional balance by anodal tDCS
Moura et al. (2017)	*Cerebral palsy*	A/S 25 cm^2	M1 unaffected hemisphere	Contralateral supraorbital ridge	1 mA	20 min/ 1 session	20	Reduction in total duration and returning movement duration
Kirton et al. (2017)	*Cerebral palsy*	C/S 25 cm^2	M1 unaffected hemisphere	Contralateral supraorbital ridge	1 mA	20 min/ 10 sessions/ once daily, consecutive days	24	No significant effects
Bhanpuri et al. (2015)	*Dystonia*	A/C/S 4 x 7 cm	C3 or C4	Contralateral supraorbital ridge	2 mA	9 min/ 5 sessions/ once daily, consecutive days	9	No significant effects

(*Continued*)

TABLE 9.1 (Continued)

		Stimulation protocol						Outcome
Study	Patients	Polarity/ electrode size	Target electrode position	Return electrode position	Current strength (mA)	Duration (min)/ sessions	Sample size	Effects
EPILEPSIA								
San-Juan et al. (2011)	*Rasmussen's encephalitis*	C/S 12 mm length × 0.4 mm diameter	F2	F8	2 mA	60 min/ 4 sessions (days 0, 7, 30, and 60)	1	Cathodal tDCS improved epileptic symptoms, linguistic, and motor functions
Yook et al. (2011)	*Focal epilepsy due to cortical dysplasia*	C 25 cm^2	Right temporoparietal area (between P4 and T4), over the epileptic focus	Contralateral supraorbital ridge	2 mA	20 min/ 5 days per week/ 2 weeks	25	Cathodal tDCS reduced frequency and duration of seizures
Varga et al. (2011)	*Continuous spike waves during slow wave sleep*	C/S 25 cm^2	$N = 1$ T7 $N = 2$ FT7 $N = 3$ T7 $N = 4$ TP8 $N = 5$ T7	Over the area of peak positivity	1 mA	20 min/ 2 sessions/ 1 day	5	Effect on EEG patterns or clinical symptoms only in three patients
Auvichayapat et al. (2013)	Diverse epileptic syndromes	C/S 35 cm^2	Epileptic focus	Contralateral shoulder	1 mA	20 min/ 1 session	36	Significant reductions in epileptic discharge frequency immediately and 24 and 48 h after tDCS
Auvichayapat et al. (2016)	*Lennox–Gastaut syndrome*	C/S 35 cm^2	Left M1	Right shoulder	2 mA	20 m/ 5 sessions/ once daily, consecutive days	22	Significant reduction of seizure frequency and epileptic activity (clinical and EEG measures)

ADHD								
Munz et al. (2015)		A/S 13 mm outer diameter; 8 mm inner diameter: 0.503 cm^2 area	F3–F4	Bilateral over the mastoid	The current strength at the anodal electrodes ranged from 0 to 250 μA	5 × 5 min with an interval of 1 min during REM sleep	14	Improved inhibitory control skills (monitored via go/no-go reaction time task)
Bandeira et al. (2016)		A/S 35 cm^2	F3	Contralateral supraorbital ridge	2 mA	28 min/5 sessions/ once daily/ consecutive days	9	Improved visual attention and inhibitory control skills
SCHIZOPHRENIA								
Mattai et al. (2011)		C/A/S 25 cm^2	Bilateral anodal DLPFC stimulation ($N = 8$), bilateral cathodal STG stimulation ($n = 5$)	Nondominant forearm	2 mA	20 min/ 5 days per week/once daily/ 2 weeks	15 (3 dropouts unrelated to stimulation)	Good tolerability of the intervention
AUTISM								
Schneider and Hopp (2011)		A 25 cm^2	Left DLPFC	Contralateral supraorbital ridge	2 mA	30 min/ 1 session	10	Improved vocabulary score
Amatachaya et al. (2014)		A/S 35 cm^2	F3	Contralateral shoulder	1 mA	20 min/ 5 sessions/ once daily/ consecutive days	20	Higher scores in the Autism Treatment Evaluation Checklist and Children's Global Assessment Scale

(Continued)

TABLE 9.1 (Continued)

		Stimulation protocol						Outcome
Study	**Patients**	**Polarity/ electrode size**	**Target electrode position**	**Return electrode position**	**Current strength (mA)**	**Duration (min)/ sessions**	**Sample size**	**Effects**
Amatachaya et al. (2015)		A/S 35 cm^2	F3	Contralateral shoulder	1 mA	20 min/ 1 session	20	Improvement in social functionality, health and behavioral problem scales of the Autism Treatment Evaluation Checklist
Costanzo et al. (2015)		A/C 25 cm^2	C F4; A F3	Bilateral montage	1 mA	20 min/ 28 sessions/ 5 per week	1	Reduction of catatonic symptoms
LEARNING DISABILITIES								
Costanzo et al. (2016a)		A/C/S 25 cm^2	Parietotemporal regions (between P7/TP7 and P8/ TP8)	Bilateral montage	1 mA	20 min/ 1 session	19	Significant reduction of text-reading errors after left anodal/right cathodal
Costanzo et al. (2016b)		A/C/S 25 cm^2	Parietotemporal regions (between P7/TP7 and P8/ TP8)	Bilateral montage		20 min/ 3 per week/ 6 weeks	18	Reduced low-frequency-word reading errors and non-word reading times after left anodal/right cathodal tDCS

Shown are studies dedicated to treatment of vascular diseases, epilepsy, ADHD, schizophrenia, autism spectrum disorder, and learning disabilities in childhood populations. Study characteristics, details of the stimulation protocols as well as effects of stimulation, including side effects, are shown. Stimulation targets areas are described according to the international 10–20 system. *tDCS*, transcranial direct current stimulation; *A*, anodal transcranial direct current stimulation; *C*, cathodal transcranial direct current stimulation; *S*, sham transcranial direct current stimulation; *ADHD*, attention-deficit/hyperactivity disorder; *DLPFC*, dorsolateral prefrontal cortex; *EEG*, electroencephalographic.

open-label trials should be taken into account, which includes the possibility of placebo effects, especially in the presence of behavioral measures conducted by scientists who are not blinded. This makes it difficult to establish safety and efficacy of the currently implemented protocols. It is important to note that although the examined literature does not report evidence for relevant adverse tDCS effects in children, the number of available studies is still limited. In this regard, it might be crucial keeping in mind that the developing brain is characterized by "sensitive" periods, where the effects of interventions affecting the brain are unusually strong (Knudsen, 2004). This suggests that, during development, the risk to induce maladaptive neural plasticity due to tDCS or other noninvasive brain stimulation techniques might be relatively high (Vicario and Nitsche, 2013a). In consideration of such risk, neuroscientists aiming to apply noninvasive brain stimulation in children should give priority to dose-finding studies, and to longitudinal monitoring of tDCS-induced neural plasticity through neuroimaging and/or electrophysiological techniques. This would allow a valid control for functional and structural changes associated with tDCS. Neuroimaging can be useful for establishing therapeutically meaningful electrode positions and stimulation polarities. This is particularly relevant for treatment of disorders where different patterns of cerebral cortex activity (i.e., hyper- vs hypoactivity) might coexist (e.g., see the case of dyslexia, Vicario and Nitsche, 2013b). Moreover, monitoring of electrophysiological parameters of brain activity represents a crucial aspect for safety and tolerability of the intervention, in addition to subjective reports of side effects, such as performed in a study by Moliadze et al. (2015b), who assessed safety and tolerability of 10 minutes of tDCS via EEG recordings in addition to subjective reports. In this study, the EEG showed no abnormalities after 1 mA tDCS, in addition only a low rate of adverse events, and no major adverse events were documented, suggesting that 10 minutes of tDCS is well tolerated and safe in pediatric populations.

Furthermore, the adoption of tDCS protocols for treatment purposes might benefit from information about the individual neural activity associated with task execution that is relevant for the respective rehabilitative purposes. Such a procedure might help to optimize stimulation protocols at the level of the individual, additional to parameters required for a general optimization of tDCS protocols, which refer to other variables such as duration, frequency, intensity, and cortical target. An example in this regard is provided by the research conducted by Varga et al. (2011) in children affected by epilepsy. The combination of tDCS with neurophysiological monitoring methods appears useful also to establish safety and tolerability of a proposed protocol. An example is offered by a study of Moliadze et al. (2015b), who explored safety aspects of 10 minutes of tDCS via resting state EEG analysis. Moreover, respective physiological parameters obtained before and after intervention will help to rate the efficacy of stimulation beyond behavioral parameters, and relevantly contribute to identify underlying mechanisms.

Apart from the studies, we have referred to here, no published research (including pilot studies) is currently available with regard to therapeutic efficacy of tDCS in childhood neuropsychiatric disorders such as Tourette syndrome, depression, bipolar disorder, anxiety, and other learning disabilities (e.g., dyscalculia and dysgraphia). This is an important gap in our current knowledge that would be worthwhile to cover in future investigations.

Finally, other related transcranial electrical stimulation approaches should be taken into consideration for the treatment of pediatric neuropsychiatric disorders, such as transcranial alternating current stimulation, or transcranial random noise stimulation (tRNS), another approach which might have potential for the treatment for learning disabilities in pediatric populations. This is suggested by a study of Looi et al. (2017), who showed an improvement of performance in children with mathematical learning disabilities by 0.75 mA tRNS applied bilaterally over the DLPFCs of 12 children (age between 8.5 and 10.9 years).

References

Amatachaya, A., Auvichayapat, N., Patjanasoontorn, N., Suphakunpinyo, C., Ngernyam, N., Aree-Uea, B., et al., 2014. Effect of anodal transcranial direct current stimulation on autism: a randomized double-blind crossover trial. Behav. Neurol. 2014, 173073.

Amatachaya, A., Jensen, M.P., Patjanasoontorn, N., Auvichayapat, N., Suphakunpinyo, C., Janjarasjitt, S., et al., 2015. The short-term effects of transcranial direct current stimulation on electroencephalography in children with autism: a randomized crossover controlled trial. Behav. Neurol. 2015, 928631.

Association, A.P, 2014. Diagnostic and Statistical Manual of Mental Disorders, (DSM-5). National Institute of Health. "Cerebral Palsy: Hope Through Research". NINDS 2013, 13–159.

Auvichayapat, N., Rotenberg, A., Gersner, R., Ngodklang, S., Tiamkao, S., Tassaneeyakul, W., et al., 2013. Transcranial direct current stimulation for treatment of refractory childhood focal epilepsy. Brain Stimul. 6, 696–700.

Auvichayapat, N., Sinsupan, K., Tunkamnerdthai, O., Auvichayapat, P., 2016. Transcranial direct current stimulation for treatment of childhood pharmacoresistant Lennox–Gastaut syndrome: a pilot study. Front. Neurol. 7, 66.

Bandeira, I.D., Guimarães, R.S., Jagersbacher, J.G., Barretto, T.L., de Jesus-Silva, J.R., Santos, S.N., et al., 2016. Transcranial direct current stimulation in children and adolescents with attention-deficit/hyperactivity disorder (ADHD): a pilot study. J. Child Neurol. 31, 918–924.

Batsikadze, G., Paulus, W., Kuo, M.F., Nitsche, M.A., 2013. Effect of serotonin on paired associative stimulation-induced plasticity in the human motor cortex. Neuropsychopharmacology. 38, 2260–2267.

Bhanpuri, N.H., Bertucco, M., Young, S.J., Lee, A.A., Sanger, T.D., 2015. Multiday transcranial direct current stimulation causes clinically insignificant changes in childhood dystonia: a pilot study. J. Child Neurol. 30, 1604–1615.

Brazzo, D., Pera, M.C., Fasce, M., Papalia, G., Balottin, U., Veggiotti, P., 2012. Epileptic encephalopathies with status epilepticus during sleep: new techniques for understanding pathophysiology and therapeutic options. Epilepsy Res. Treat. 642725.

Bush, G., 2011. Cingulate, frontal, and parietal cortical dysfunction in attention-deficit/hyperactivity disorder. Biol. Psychiatry 69, 1160–1167.

Cole, J.C., Green Bernacki, C., Helmer, A., Pinninti, N., O'reardon, J.P., 2015. Efficacy of transcranial magnetic stimulation (TMS) in the treatment of schizophrenia: a review of the literature to date. Innov. Clin. Neurosci. 2015 (12), 12–19. Review.

Costanzo, F., Menghini, D., Casula, L., Amendola, A., Mazzone, L., Valeri, G., et al., 2015. Transcranial direct current stimulation treatment in an adolescent with autism and drug-resistant catatonia. Brain Stimul. 8, 1233–1235.

Costanzo, F., Varuzza, C., Rossi, S., Sdoia, S., Varvara, P., Oliveri, M., et al., 2016a. Reading changes in children and adolescents with dyslexia after transcranial direct current stimulation. NeuroReport. 27, 295–300.

Costanzo, F., Varuzza, C., Rossi, S., Sdoia, S., Varvara, P., Oliveri, M., et al., 2016b. Evidence for reading improvement following tDCS treatment in children and adolescents with Dyslexia. Restor. Neurol. Neurosci. 34, 215–226.

Dudek, F.E., Sutula, T.P., 2007. Epileptogenesis in the dentate gyrus: a critical perspective. Prog. Brain Res. 163, 755–773.

Flöel, A., 2014. tDCS-enhanced motor and cognitive function in neurological diseases. NeuroImage. 85, 934–947.

Fritsch, B., Reis, J., Martinowich, K., Schambra, H.M., Ji, Y., Cohen, L.G., Lu, B., 2010. Direct current stimulation promotes BDNF-dependent synaptic plasticity: potential implications for motor learning. Neuron. 66 (2), 198–204.

Furuya, S., Nitsche, M.A., Paulus, W., Altenmüller, E., 2014. Surmounting retraining limits in musicians' dystonia by transcranial stimulation. Ann. Neurol. 75, 700–707.

Grefkes, C., Fink, G.R., 2011. Reorganization of cerebral networks after stroke: new insights from neuroimaging with connectivity approaches. Brain 134, 1264–1276.

Jang, S.H., 2013. Motor function-related maladaptive plasticity in stroke: a review. NeuroRehabilitation 32, 311–316.

Kiep, M., Spek, A.A., 2016. Executive functioning in men and women with an autism spectrum disorder. Autism Res. Available from: https://doi.org/10.1002/aur.1721 [Epub ahead of print].

Kirton, A., Ciechanski, P., Zewdie, E., Andersen, J., Nettel-Aguirre, A., Carlson, H., et al., 2017. Transcranial direct current stimulation for children with perinatal stroke and hemiparesis. Neurology. 88 (3), 259–267.

Knudsen, E.I., 2004. Sensitive periods in the development of the brain and behavior. J. Cogn. Neurosci. 16, 1412–1425.

Kuo, M.F., Nitsche, M.A., 2012. Effects of transcranial electrical stimulation on cognition. Clin. EEG Neurosci. 43, 192.

Kuo, M.F., Paulus, W., Nitsche, M.A., 2014. Therapeutic effects of non-invasive brain stimulation with direct currents (tDCS) in neuropsychiatric diseases. NeuroImage. 85, 948–960.

Kuo, M.F., Chen, P.S., Nitsche, M.A., 2017. The application of tDCS for the treatment of psychiatric diseases. Int. Rev. Psychiatry 29, 146–167.

Kronberg, G., Bridi, M., Abel, T., Bikson, M., Parra, L.C., 2017. Direct current stimulation modulates LTP and LTD: activity dependence and dendritic effects. Brain Stimul. 10, 51–58.

Liebetanz, D., Nitsche, M.A., Tergau, F., Paulus, W., 2002. Pharmacological approach to the mechanisms of transcranial DC-stimulation-induced after-effects of human motor cortex excitability. Brain 125, 2238–2247.

Looi, C.Y., Lim, J., Sella, F., Lolliot, S., Duta, M., Avramenko, A.A., et al., 2017. Transcranial random noise stimulation and cognitive training to improve learning and cognition of the atypically developing brain: a pilot study. Sci. Rep. 7, 4633.

Lazzari, R.D., Politti, F., Belina, S.F., Collange Grecco, L.A., Santos, C.A., Dumont, A.J.L., et al., 2017. Effect of transcranial direct current stimulation combined with virtual reality training on balance in children with cerebral palsy: a randomized, controlled, double-blind, clinical trial. J. Mot. Behav. 49, 329–336.

Mattai, A., Miller, R., Weisinger, B., Greenstein, D., Bakalar, J., Tossell, J., et al., 2011. Tolerability of transcranial direct current stimulation in childhood-onset schizophrenia. Brain Stimul. 4, 275–280.

Moliadze, V., Schmanke, T., Andreas, S., Lyzhko, E., Freitag, C.M., Siniatchkin, M., 2015a. Stimulation intensities of transcranial direct current stimulation have to be adjusted in children and adolescents. Clin. Neurophysiol. 126, 1392–1399.

Moliadze, V., Andreas, S., Lyzhko, E., Schmanke, T., Gurashvili, T., Freitag, C.M., et al., 2015b. Ten minutes of 1 mA transcranial direct current stimulation was well tolerated by children and adolescents: self-reports and resting state EEG analysis. Brain Res. Bull. 119, 25–33.

Monte-Silva, K., Kuo, M.F., Hessenthaler, S., Fresnoza, S., Liebetanz, D., Paulus, W., et al., 2013. Induction of late LTP-like plasticity in the human motor cortex by repeated non-invasive brain stimulation. Brain Stimul. 6, 424–432.

Moura, R.C.F., Santos, C., Collange Grecco, L., Albertini, G., Cimolin, V., Galli, M., et al., 2017. Effects of a single session of transcranial direct current stimulation on upper limb movements in children with cerebral palsy: a randomized, sham-controlled study. Dev. Neurorehabil. 20, 368–375.

Munz, M.T., Prehn-Kristensen, A., Thielking, F., Mölle, M., Göder, R., Baving, L., 2015. Slow oscillating transcranial direct current stimulation during non-rapid eye movement sleep improves behavioral inhibition in attention-deficit/hyperactivity disorder. Front. Cell Neurosci. 9, 307.

National Institute of Health. "Cerebral Palsy: Hope Through Research", NINDS, 2013, 13–159.

Nicolson, R., Rapoport, J.L., 1999. Childhood-onset schizophrenia: rare but worth studying. Biol. Psychiatry 46, 1418–1428.

Nitsche, M.A., Paulus, W., 2000. Excitability changes induced in the human motor cortex by weak transcranial direct current stimulation. J. Physiol. 527, 633–639.

Nitsche, M.A., Paulus, W., 2001. Sustained excitability elevations induced by transcranial DC motor cortex stimulation in humans. Neurology. 57, 1899–1901.

Nitsche, M.A., Fricke, K., Henschke, U., Schlitterlau, A., Liebetanz, D., Lang, N., et al., 2003. Pharmacological modulation of cortical excitability shifts induced by transcranial direct current stimulation in humans. J. Physiol. 553, 293–301.

Nitsche, M.A., Jaussi, W., Liebetanz, D., Lang, N., Tergau, F., Paulus, W., 2004a. Consolidation of human motor cortical neuroplasticity by D-cycloserine. Neuropsychopharmacology. 29, 1573–1578.

Nitsche, M.A., Liebetanz, D., Schlitterlau, A., Henschke, U., Fricke, K., Frommann, K., et al., 2004b. GABAergic modulation of DC stimulation-induced motor cortex excitability shifts in humans. Eur. J. Neurosci. 19, 2720–2726.

Nitsche, M.A., Müller-Dahlhaus, F., Paulus, W., Ziemann, U., 2012. The pharmacology of neuroplasticity induced by non-invasive brain stimulation: building models for the clinical use of CNS active drugs. J Physiol. 590 (19), 4641–4662.

Nitsche, M.A., Seeber, A., Frommann, K., Klein, C.C., Rochford, C., Nitsche, M.S., et al., 2005. Modulating parameters of excitability during and after transcranial direct current stimulation of the human motor cortex. J. Physiol. 568, 291–303.

Nitsche, M.A., Cohen, L.G., Wassermann, E.M., Priori, A., Lang, N., Antal, A., et al., 2008. Transcranial direct current stimulation: state of the art 2008. Brain Stimul. 3, 206–223.

Rivera-Urbina, G.N., Nitsche, M.A., Vicario, C.M., Molero-Chamizo, A., 2017. Applications of transcranial direct current stimulation in children and pediatrics. Rev. Neurosci. 28, 173–184.

Rizzo, R., Gulisano, M., Calì, P.V., Curatolo, P., 2013. Tourette syndrome and comorbid ADHD: current pharmacological treatment options. Eur. J. Paediatr. Neurol. 17, 421–428.
Rocha, S., Silva, E., Foerster, Á., Wiesiolek, C., Chagas, A.P., Machado, G., et al., 2016. The impact of transcranial direct current stimulation (tDCS) combined with modified constraint-induced movement therapy (mCIMT) on upper limb function in chronic stroke: a double-blind randomized controlled trial. Disabil. Rehabil. 38 (7), 653–660.
Sanger, T.D., Chen, D., Fehlings, D.L., Hallett, M., Lang, A.E., Mink, J.W., et al., 2010. Definition and classification of hyperkinetic movements in childhood. Mov. Disord. 11, 1538–1549.
San-Juan, D., Calcáneo, J.D., González-Aragón, M.F., Bermúdez Maldonado, L., Avellán, A.M., Argumosa, E.V., et al., 2011. Transcranial direct current stimulation in adolescent and adult Rasmussen's encephalitis. Epilepsy Behav. 20, 126–131.
Schneider, H.D., Hopp, J.P., 2011. The use of the bilingual aphasia test for assessment and transcranial direct current stimulation to modulate language acquisition in minimally verbal children with autism. Clin. Linguist. Phon. 25, 640–654.
Stafstrom, C.E., 2006. Epilepsy: a review of selected clinical syndromes and advances in basic science. J. Cereb. Blood Flow. Metab. 26 (8), 983–1004.
Stagg, C.J., Best, J.G., Stephenson, M.C., O'Shea, J., Wylezinska, M., Kincses, Z.T., et al., 2009. Polarity-sensitive modulation of cortical neurotransmitters by transcranial stimulation. J. Neurosci. 29 (16), 5202–5206.
Stagg, C.J., Nitsche, M.A., 2011. Physiological basis of transcranial direct current stimulation. Neuroscientist. 17 (1), 37–53.
Varga, E.T., Terney, D., Atkins, M.D., Nikanorova, M., Jeppesen, D.S., Uldall, P., 2011. Transcranial direct current stimulation in refractory continuous spikes and waves during slow sleep: a controlled study. Epilepsy Res. 97, 142–145.
Vicario, C.M., Nitsche, M.A., 2013a. Non-invasive brain stimulation for the treatment of brain diseases in childhood and adolescence: state of the art, current limits and future challenges. Front. Syst. Neurosci. 7, 94.
Vicario, C.M., Nitsche, M.A., 2013b. Transcranial direct current stimulation: a remediation tool for the treatment of childhood congenital dyslexia? Front. Hum. Neurosci. 7, 139.
Vicario, C.M., Rumiati, R.I., 2012. tDCS of the primary motor cortex improves the detection of semantic dissonance. Neurosci. Lett. 518, 133–137.
Ziemann, U., 2017. Thirty years of transcranial magnetic stimulation: where do we stand? Exp. Brain Res. Available from: https://doi.org/10.1007/s00221-016-4865-4 [Epub ahead of print].
Yook, S.W., Park, S.H., Seo, J.H., Kim, S.J., Ko, M.H., 2011. Suppression of seizure by cathodal transcranial direct current stimulation in an epileptic patient—a case report. Ann. Rehabil. Med. 35, 579–582.

Further Reading

Fombonne, E., 2009. Epidemiology of pervasive developmental disorders. Pediatr. Res. 65, 591–598.
Gillick, B., Menk, J., Mueller, B., Meekins, G., Krach, L.E., Feyma, T., et al., 2015. Synergistic effect of combined transcranial direct current stimulation/constraint-induced movement therapy in children and young adults with hemiparesis: study protocol. BMC Pediatr. 15, 178.

CHAPTER

10

Deep Brain Stimulation for Pediatric Neuropsychiatric Disorders

Jennifer L. Quon[1,2], *Lily H. Kim*[1,2], *Caroline A. Quon*[3], *Laura M. Prolo*[1,2], *Gerald A. Grant*[2] *and Casey H. Halpern*[1]

[1]Department of Neurosurgery, Stanford Hospitals and Clinics, Stanford, CA, United States [2]Division of Pediatric Neurosurgery, Lucile Packard Children's Hospital, Palo Alto, CA, United States [3]The PRIME Center, VA Connecticut Healthcare System, West Haven, CT, United States

OUTLINE

Neurotechnological Pediatric Neuropsych
DOI: https://doi.org/10.1016/B978-0-12-812777-3.00010-6

INTRODUCTION

Neurosurgical interventions for psychiatric disorders date back to 1930's when the first report of prefrontal leucotomy was published with the hope of "curing" psychosis. Partly because of the aggressive nature of the resection and partly because of the emotional response elicited by the idea of possibly changing personality through surgery, neuropsychiatric surgery has been controversial since the beginning (Boettcher and Menacho, 2017). Now, we have more treatment options other than radical resective operations and the discussion has become less emotionally charged. On the adult side, deep brain stimulation (DBS), a neuromodulatory technique that targets specific deep brain nuclei to deliver electrical pulses, has largely replaced ablative surgery. But despite many studies reporting symptom relief following DBS implantation for various neuropsychiatric disorders, its success has not yet been fully replicated in pediatric populations. The reason for this may be multifactorial, including spontaneous remission with age, safety concerns, and the higher standards required to demonstrate the utility of experimental therapy in younger patients. But despite the limited evidence to guide clinical decision-making process at present, DBS has been used for different neuropsychiatric illnesses, especially in pediatric patients who have exhausted other treatment options.

INDICATIONS

Conditions treated by DBS can broadly be organized into movement disorders, neuropsychiatric disorders (Garcia-Soriano et al., 2014), and epilepsy (Burke et al., 1985).

Dystonia

The most promising application of DBS in children thus far has been for dystonia. Dystonia is a condition characterized by sustained muscle contractions causing rhythmic movements or abnormal postures. Dystonias are labeled primary if dystonia is the only clinical manifestation of the disease. Primary dystonia can be further broken down into generalized, segmental, or focal (DiFrancesco et al., 2012). A number of different genetic mutations and deletions have been associated with dystonias including torsion dystonia-1 (DYT1) gene, also known as torsin-1A. Secondary dystonias are those that are caused by another disease process (DiFrancesco et al., 2012; van Karnebeek et al., 2015). Current medical management for dystonias includes levodopa, anticholinergics, baclofen, and botulinum toxin injections. Primary dystonias tend to respond better to DBS than secondary dystonias, or at least more reliably, while secondary dystonias respond better to intrathecal baclofen than primary ones (DiFrancesco et al., 2012).

Anatomical Target(s)

For medically refractory dystonia in adults, DBS targeting the globus pallidus internus (GPi) is largely preferred (DiFrancesco et al., 2012). This was identified as a target in adults with Parkinson's disease and generalized dystonia undergoing pallidotomies (Marras et al., 2014). On fMRI, DBS of the GPi decreases cortical activation of areas that have increased activity in dystonia (DiFrancesco et al., 2012). Notably, interest in targeting the subthalamic nucleus (STN) is growing, with evidence suggesting equivalent efficacy to the GPi. Having more than one potential anatomical regions to target is particularly relevant for treating patients who develop secondary dystonias following stroke events; even if stroke alters one of these target regions, the other site will be left amenable to DBS (Ostrem et al., 2016). These same neural targets have resulted in positive outcomes among pediatric dystonia patients. Although GPi has been the target of choice for these patients since early 2000s (Lenders et al., 2006; Marks et al., 2011; Mahoney et al., 2011; Timmermann et al., 2010; Valldeoriola et al., 2010), STN has also been tried on an individual patient basis during the past decade (Ostrem et al., 2016; Starr et al., 2014; Tormenti et al., 2011).

Outcomes

DBS targeting the GPi is most effective in primary generalized dystonias, specifically in patients with the DYT1 gene mutation (Alterman and Tagliati, 2007; Borggraefe et al., 2010; Ghosh et al., 2012; Haridas et al., 2011; Hudson et al., 2017; Krause et al., 2016; Lahtinen, 2017; Owen et al., 2017; Park et al., 2016; Parr et al., 2007; DiFrancesco et al., 2012). In a retrospective study published in 2011, children with primary generalized dystonia and an average age of 13 years old underwent DBS surgery after suboptimal medical treatment response. After a median follow-up time of two years, these patients were observed to have alleviation of their previous motor symptoms despite reduction in oral and intrathecal dystonia-related medications. It is especially worthwhile to highlight that in addition to clinical improvement, these children exhibited better performances in academic and social settings (Haridas et al., 2011). Another study also investigated the short-term and long-term effects of pallidal DBS in adolescent male patients and presented results of motor improvements similar to clinical trials in adults (Krause et al., 2016). DBS in secondary dystonias exhibited modest results (Olaya et al., 2013; Tsering et al., 2017) and may be an option for patients with dystonias due to anoxic brain injury with grossly intact basal ganglia (Alterman and Tagliati, 2007). For most of these studies, the Burke–Fahn–Marsden Dystonia Rating Scale (BFMDRS) was used for quantitative comparison. This scale comprises movement and disability scores, in order to capture both the clinical severity and the ability to perform activities of daily living (Burke et al., 1985). The two components of BFMDRS tended to share similar trends, with improved motor function in these pediatric populations often leading to higher quality of life with better integration to school environment. This impact on patients' daily lives makes DBS an especially attractive treatment option for pediatric dystonia patients, whose disorder can cause a considerable amount of distress during such a sensitive period of development (Air et al., 2011; Alterman and Tagliati, 2007; Marks et al., 2009).

Side Effects

DBS to the bilateral GPi can induce parkinsonian-like symptoms (Kaminska et al., 2017; Lumsden et al., 2012; Miyagi and Koike, 2013). In children, reported postoperative complications are similar to those in adults: hardware complications such as electrode/extension malfunctions (short extension, electrode migration/dislocation), intracranial bleeding, and difficulties maintaining the connection while recharging the neurostimulators (Kaminska et al., 2017). In addition, there have also been a few reported incidences of infection, which have confounded assessments of neurologic improvement (Haridas et al., 2011). Two patients (both 9 year olds with DYT1 mutation) were found to have a recurrence of their dystonic symptoms within a year. Although no complications were noted when the symptoms initially recurred, Patient 1 was found to have lead fractures at 2 and 5 years after the initial implantation, requiring reoperation to replace the DBS leads. Patient 1 subsequently returned to their preoperative level of symptom relief. Patient 2 underwent bilateral pallidal DBS placement after his father successfully underwent the same procedure for dystonia. He reached maximal symptom relief at 3 months, walking with minimal assistance. He subsequently had a gradual worsening of his symptoms and the decision was made to implant leads to the bilateral STN and disconnect his pallidal leads. His dystonia continued to progress such that it was more severe than before his initial surgery, and an emergency reoperation to return to his pallidal DBS was attempted. However, the patient died intraoperatively from multiple organ failure and rhabdomyolysis (Miyagi and Koike, 2013). No direct analysis has been performed to compare complication rates between pediatric versus adult populations.

Tourette Syndrome

Tourette syndrome (TS) is a neuropsychiatric disorder characterized by chronic motor and/or vocal tics, which are movements or sounds that occur intermittently and often unpredictably (Deeb et al., 2016; Hauseux et al., 2017). Tics can be triggered by the patient's level of stress or social awareness and are sometimes suppressible. Usually beginning in early childhood, tics can often progress and then decline in adulthood. For this reason, studies conducted with a primary focus on the pediatric population are limited. TS often occurs in conjunction with other neuropsychiatric conditions such as obsessive–compulsive disorder (OCD) (DiFrancesco et al., 2012). Medical therapy includes alpha-adrenergic agonists (e.g., clonidine or guanfacine), neuroleptics (e.g., haloperidol, risperidone), and tetrabenzine, a catecholamine-blocking agent. Botulinum toxin injection has also been shown to be effective for patients with an isolated motor tic preceded by a strong premonitory sensation.

Anatomical Target(s)

Ablation targets have included the thalamus, cingulate gyrus, as well as prefrontal cortex. DBS in adults and adolescents with TS has focused on the anteromedial and posteroventral GPi, nuclear complex of the thalamus, STN, ventral capsule/ventral striatum, and the nucleus accumbens (NAc) (DiFrancesco et al., 2012). Thalamus remains one of the most popular target regions, which is in alignment with previous work demonstrating structural and metabolic abnormalities in the thalami of patients with TS (Heinz et al., 1998; Hershey et al., 2004; Peterson et al., 1998). In pediatric patient populations, in particular, one study has shown that treatment-naïve male patients with TS have larger left thalamic volume compared to healthy controls, although the implications of this asymmetry are unknown (Lee et al., 2006).

Outcomes

Long-term targeting of DBS to the posteroventral GPi as well as the NAc was observed to decrease the frequency and severity of the motor tics among patients in their early to late teens (Hauseux et al., 2017). Other case reports in pediatric patients have shown conflicting responses to DBS. While one case study documented meaningful symptom alleviation with GPi stimulation (Shahed et al., 2007), another report on a patient within similar age denied any significant clinical improvement even when the same region was targeted (Dueck et al., 2009). As can be inferred from these case reports, currently available studies on DBS use in pediatric populations have focused solely on adolescent patients and excluded younger children, which is understandable as up to half of the patients are expected to see resolution of their symptoms by age 18 (DiFrancesco et al., 2012). But given that TS can be associated with non-tic psychiatric disturbances such as self-harming behaviors and suicidal ideations, trialing DBS implantation in younger patients may be warranted on a case-by-case basis to prevent further morbidity and mortality before they reach adolescence (Cavanna et al., 2011; Storch et al., 2015). This is in light of the fact that DBS implantation can lead to improvement in psychiatric behaviors related to TS as well as motor symptoms in pediatric patients (Shahed et al., 2007; Servello et al., 2008).

Side Effects

One small case series reported postoperative, transient dysarthria in all three pediatric patients undergoing DBS of the posteroventral GPi for TS (Hauseux et al., 2017). In this study, the authors further postulate that targeting the posteroventral GPi to treat motor tics of TS may also exacerbate the symptoms of OCD and impulse control disorder (Hauseux et al., 2017). This is in direct contradiction to previous studies that reported improvement of comorbid psychiatric disorders, further complicating the selection process of surgical candidates in pediatric TS (Servello et al., 2008).

Juvenile Parkinsonism

Juvenile parkinsonism is the onset of parkinsonian symptoms, including bradykinesia, tremor, and focal dystonia, before the age of 21. Unlike its adult counterpart, juvenile parkinsonism is more likely to have a genetic component and is associated with mutations in the parkin and PTEN-induced putative kinase 1 genes. Similar to adults, adolescents with juvenile parkinsonism respond to dopamine replacement with levodopa (DiFrancesco et al., 2012).

Anatomical Target

The DBS targets for adults with Parkinson's disease are either the STN or GPi. DBS has not yet been extensively used to treat a patient with juvenile parkinsonism. This may be partly due to the relative rarity of the condition in pediatric population as the incidence of parkinsonism is known to increase with age (Schrag and Schott, 2006). Although the evidence is sparse, based on the few case studies reported in the pediatric population, the same brain regions as adults—namely, STN and GPi—are targeted in juvenile parkinsonism as well (Genc et al., 2016; Ramirez-Zamora et al., 2017).

Outcomes

In carefully selected adult patients, DBS of the STN or GPi is a safe and efficacious treatment of Parkinson's disease and results in improvements in "off" medication score of the Unified Parkinson's Disease Rating Scale motor part (UPDRS-III), improved ability to perform activities of daily living and the "on" time of good mobility without dyskinesias (Rodriguez-Oroz et al., 2005). Similar results, albeit limited, have been recently reported among pediatric population in the current literature. In 2016, Genc et al. reported a case of a 14-year-old female with juvenile parkinsonism who underwent a DBS placement. In this patient, the device was set to convey stimulation to bilateral STN regions resulting in a significant relief from her previous symptoms, which included tremor, bradykinesia, and gait disturbances (Genc et al., 2016). Another case study in 2017 followed a young woman who has been suffering for several years from juvenile-onset dystonia and parkinsonism for 30 months post-DBS placement. During this time period, the patient's UPDRS-III motor scores improved more than 80%, but despite the marked positive change in dyskinesia and tremor, her spasticity and gait problems persisted (Ramirez-Zamora et al., 2017). It should be noted, however, that although the onset of her parkinsonism was in her adolescence, meeting the criteria for juvenile parkinsonism, the surgery was offered as an adult (the case report does not disclose the patient's exact age), limiting the applicability of its findings to the general pediatric patient population.

Side Effects

Device-related complications in adults include lead fracture, infection, and skin erosion. Non-device-related events include cognitive decline, psychiatric disturbances, depression or mood disturbances, speech difficulties, dysphasia, imbalance, gait disorders, dystonia, dyskinesias, increased PD symptoms, sleep disorders, orthostasis, and double vision. Patients who had adverse events were found to have longer disease duration, as well as more gait and psychiatric disturbances at baseline (Hariz et al., 2008). Given that DBS implantation has been tried in juvenile parkinsonism only recently, the full safety profile cannot be confidently assessed in the absence of long-term follow-up data.

Obsessive Compulsive Disorder

OCD is a neuropsychiatric condition characterized by recurring thoughts (obsessions) and repetitive behaviors (compulsions) that interfere with normal daily activities (Lipsman et al., 2010). The cognitive behavioral model states that the compulsive behaviors are an attempt to reduce the distress caused by the obsessive thoughts. OCD can be diagnosed in childhood even if the patient does not have insight to the unreasonableness of their obsessions, which is a requirement for diagnosis in adulthood (DiFrancesco et al., 2012). OCD is often associated with other neuropsychiatric disorders such as anxiety, depression, and eating disorders. Patients with comorbidities are generally more likely to seek care and may thus be more heavily represented in studies on DBS, although some studies suggest that they tend to show more difficulties in successfully completing treatment (Garcia-Soriano et al., 2014; Levy et al., 2013).

Anatomical Targets

Disruption of the cortico-striato-thalamo-cortico circuit has been implicated in OCD based on imaging studies (DiFrancesco et al., 2012). In stereotactic ablative studies, anterior capsulotomy has been the most effective in treatment followed by cingulotomy, frontal leucotomy, and subcaudate tractotomy. Some centers have also investigated amygdalotomy. DBS for OCD in adults has been approved under a humanitarian device exemption from the FDA in 2009, though there is no study to date that primarily reports DBS use among pediatric patients with OCD (DiFrancesco et al., 2012; Arnold and Kronenberg, 2017; Lipsman et al., 2010).

Outcomes

Patients treated with DBS for TS with comorbid OCD had an improvement in their tic symptoms, but persistent OCD (Hauseux et al., 2017). However, data from several other studies on the efficacy of DBS for OCD from the adult literature are encouraging, suggesting some benefit of stimulation to the ventral capsule/ventral striatum, the STN, and the inferior thalamic peduncle (Holtzheimer and Mayberg, 2011). Even with such reassuring outcomes in adult population, there are multiple factors complicating implementation of DBS use in pediatric OCD. Fortunately, medical treatment in combination with cognitive behavioral therapy is successful in up to two-thirds of patients, foregoing the need for surgical intervention in most cases (Pinto et al., 2006). Nevertheless, an ethical application of DBS in pediatric patients who are severely debilitated by medically refractory OCD would be a meaningful endeavor given that the onset for this psychiatric condition is known to be either in childhood or adolescence for more than half of the cases (Pauls et al., 1995).

Side Effects

Potential complications include adverse effects from the surgery, the hardware, or from stimulation: intracranial hemorrhage, seizures, infection, lead or extension break, autonomic symptoms, hypomania, mood worsening, anxiety, irritability, cognitive abnormalities, and/or increased suicidal ideation (Holtzheimer and Mayberg, 2011). It should be noted, however, these incidents were gathered from studies conducted on adult patients only.

Epilepsy

Epilepsy is defined as the predisposition to develop seizures due to abnormal neuronal firing (Cox et al., 2014; Lipsman et al., 2010; Valentin et al., 2017). Approximately 35% children with epilepsy are refractory to medication, and only those with resectable epileptic foci are candidates for surgical resection (Valentin et al., 2017). Neurostimulation offers a favorable alternative for these patients, with the hopes of reducing the frequency and severity of seizures.

Anatomical Targets

Studies done in adults have targeted the hippocampus and thalamic nuclei (anterior nucleus, centromedian nucleus) (Salanova et al., 2015; Valentin et al., 2017). These regions of the brain, namely centromedian nucleus and hippocampus, have been targeted in pediatric populations as early as early 1990s (Fisher et al., 1992; Velasco et al., 2000, 2007). In one study published in 2005, cerebellum was selected as a DBS target, with one of the included patients being an adolescent (Velasco et al., 2005). Another potential target is the STN, as was presented in the case of a pediatric patient with inoperable dysplasia and refractory epilepsy (Lipsman et al., 2010).

Another neuromodulatory treatment modality in epilepsy is vagus nerve stimulation (VNS). This less invasive alternative also operates through a device that delivers electrical stimulation. But unlike DBS, VNS targets stimulation to the vagus nerve instead of specific deep brain nuclei. In addition, the electrical pulse used in VNS is intermittent and low-frequency rather than constant and high-frequency pulse as in DBS (George et al., 2005).

Outcomes

Targeting the STN in a pediatric patient demonstrated an 86% reduction in the frequency of seizures. Other findings on DBS for epilepsy treatment mostly derive from studies done in adult populations. In adults, DBS to the hippocampus allowed for a reduction in interictal spikes as well as in the frequency of complex–partial and tonic–clonic seizures. DBS for refractory temporal lobe epilepsy has reduced seizures from anywhere between 20% and 100%, with one meta-analysis reporting 59%. Greater seizure reduction after DBS was associated with a unilateral seizure focus and older age of disease onset (Chang and Xu, 2017). The Stimulation of the Anterior Nucleus of the Thalamus for Epilepsy (SANTE) trial investigated the long-term efficacy and safety of DBS to the anterior nucleus of the thalamus for localization-related epilepsy and found a 69% reduction in seizure frequency. Almost half of the patients also reported clinically significant improvement in quality-of-life scores. Although the mean age of study subjects was 36 years old, this trial also included patients who were as young as 18 years old, making the findings somewhat applicable to adolescent patient populations (Salanova et al., 2015). Because the overwhelming majority of the study subjects in the aforementioned "pediatric" trials were mainly adults, children-specific outcome data are currently lacking, other than the observational reports that they tended to tolerate the procedures with some improvement in symptoms.

The success rate of VNS is comparable to that of DBS. In multicenter studies, VNS has shown to be effective in significantly reducing seizure activity, ranging from 40% to 90% in its success rates (Chen et al., 2012; Helmers et al., 2001; Hung, 2012). From a psychiatric perspective, VNS has the added benefit of mood elevation. In fact, the incidental finding that patients with epilepsy demonstrated improvement in mood while on VNS ultimately led to its FDA-approved application in depression (Henderson, 2007). This may be of particular interest when considering the high prevalence of depression reported in pediatric patients with epilepsy, not only when compared to the general population but also patients with other neurologic conditions (Salpekar et al., 2015).

Side Effects

DBS-related device complications are similar to those found in other disease conditions including intracranial hemorrhage, wire disconnection, and infection. Although majority of problems were reversible, more than one-third of patients in the aforementioned SANTE trial experienced some form of device-related adverse event rate at 5 years, including implant site pain, sensory disturbance, and device migration (Salanova et al., 2015). Psychiatric adverse effects were more devastating, even resulting in mortality due to suicide. In fact, a 2017 Cochrane review suggests that DBS to the anterior thalamus, which has been the most commonly chosen target, may lead to higher self-reported rates of depression and subjective memory impairment (Sprengers et al., 2017). However, it is unclear whether DBS treatment is directly responsible for the psychological disturbances as many of the patients with epilepsy are known to have either diagnosed or undiagnosed mood disorders. In addition to these adverse events reported from studies on adult patients, skin erosion appears to be one complication more specific to younger population. In a study that included two children under 10 years old, they were among the three patients who experienced severe skin erosion and subsequently required reoperation. This may be due to the size of device that was originally built for adult use, making younger children more susceptible to subcutaneous trauma from the implantation (Velasco et al., 2000).

SPECIAL CONSIDERATIONS

In children, it is particularly important to consider the natural history of the disease (Air et al., 2011; Austin et al., 2017; Cif and Coubes, 2017; Gardner, 2017; Hudson et al., 2017; Koy and Timmermann, 2017; Koy et al., 2017; Lumsden et al., 2017). For example, in TS, 33%–50% of the patients see a resolution and 75% see a reduction of their tics by 18 years old without treatment (DiFrancesco et al., 2012). Treating medically refractory neuropsychiatric disorders in children using DBS may have additional neuromodulatory benefits as seen in treating other conditions at critical ages (e.g., cochlear implantation at prelingual age) (Hudson et al., 2017). The best timing for DBS in children is unclear though, as even temporary benefit from this treatment may be warranted given the impairment many patients suffer during their development. Another issue relates to the tolerance of stimulation, with speculation that pediatric patients may be more likely to become insensitive to previous stimuli as their brain matures over time (Miyagi and Koike, 2013). Given the various concerns and implications unique to our younger patients, it will be imperative to collect careful long-term outcomes of children treated with DBS to help address questions concerning the best timing for DBS, the risks and benefits in the pediatric population, and refinements in targeting in children.

CONCLUSION

As noted in previous studies (DiFrancesco et al., 2012), the current dearth of information on the topic of DBS for pediatric neuropsychiatric disorders requires that we extrapolate findings from adult data. The handful of studies on DBS use in pediatric population present favorable outcomes. These results, although promising, should be carefully evaluated given that studies on investigational or experimental interventions are known to be especially susceptible to publication bias (Rabins et al., 2009). However, with continued expansion of our knowledge gathered from studies of larger sample sizes, we may be better positioned to recommend DBS as a treatment option for refractory neuropsychiatric conditions in pediatric neurosurgery. The reversibility and titratability of the DBS technique allows us to effectively control debilitating diseases from a young age, while remaining mindful of the plasticity of neural circuits and active neurocognitive development in pediatric patients.

References

Air, E.L., Ostrem, J.L., Sanger, T.D., Starr, P.A., 2011. Deep brain stimulation in children: experience and technical pearls. J. Neurosurg. Pediatr. 8, 566–574.

Alterman, R.L., Tagliati, M., 2007. Deep brain stimulation for torsion dystonia in children. Childs Nerv. Syst. 23, 1033–1040.

Arnold, P., Kronenberg, S., 2017. Biological models and treatments for obsessive–compulsive and related disorders for children and adolescents. In: Abramowitz, J.S., McKay, D., Storch, E.A. (Eds.), The Wiley Handbook of Obsessive Compulsive Disorder. Wiley Blackwell, Hoboken, New Jersey, pp. 1061–1096.

Austin, A., Lin, J.P., Selway, R., Ashkan, K., Owen, T., 2017. What parents think and feel about deep brain stimulation in paediatric secondary dystonia including cerebral palsy: a qualitative study of parental decision-making. Eur. J. Paediatr. Neurol. 21, 185–192.

Boettcher, L.B., Menacho, S.T., 2017. The early argument for prefrontal leucotomy: the collision of frontal lobe theory and psychosurgery at the 1935 International Neurological Congress in London. Neurosurg. Focus 43, E4.

Borggraefe, I., Mehrkens, J.H., Telegravciska, M., Berweck, S., Botzel, K., Heinen, F., 2010. Bilateral pallidal stimulation in children and adolescents with primary generalized dystonia—report of six patients and literature-based analysis of predictive outcomes variables. Brain Dev. 32, 223–228.

Burke, R.E., Fahn, S., Marsden, C.D., Bressman, S.B., Moskowitz, C., Friedman, J., 1985. Validity and reliability of a rating scale for the primary torsion dystonias. Neurology 35, 73–77.

Cavanna, A.E., Eddy, C.M., Mitchell, R., Pall, H., Mitchell, I., Zrinzo, L., et al., 2011. An approach to deep brain stimulation for severe treatment-refractory Tourette syndrome: the UK perspective. Br. J. Neurosurg. 25, 38–44.

Chang, B., Xu, J., 2017. Deep brain stimulation for refractory temporal lobe epilepsy: a systematic review and meta-analysis with an emphasis on alleviation of seizure frequency outcome. Childs Nerv. Syst. 34 (2), 321–327.

Chen, C.Y., Lee, H.T., Chen, C.C., Kwan, S.Y., Chen, S.J., Hsieh, L.P., et al., 2012. Short-term results of vagus nerve stimulation in pediatric patients with refractory epilepsy. Pediatr. Neonatol. 53, 184–187.

Cif, L., Coubes, P., 2017. Historical developments in children's deep brain stimulation. Eur. J. Paediatr. Neurol. 21, 109–117.

Cox, J.H., Seri, S., Cavanna, A.E., 2014. Clinical utility of implantable neurostimulation devices as adjunctive treatment of uncontrolled seizures. Neuropsychiatr. Dis. Treat. 10, 2191–2200.

Deeb, W., Rossi, P.J., Porta, M., Visser-Vandewalle, V., Servello, D., Silburn, P., et al., 2016. The international deep brain stimulation registry and database for Gilles de la Tourette syndrome: how does it work? Front. Neurosci. 10, 170.

DiFrancesco, M.F., Halpern, C.H., Hurtig, H.H., Baltuch, G.H., Heuer, G.G., 2012. Pediatric indications for deep brain stimulation. Childs Nerv. Syst. 28, 1701–1714.

Dueck, A., Wolters, A., Wunsch, K., Bohne-Suraj, S., Mueller, J.U., Haessler, F., et al., 2009. Deep brain stimulation of globus pallidus internus in a 16-year-old boy with severe Tourette syndrome and mental retardation. Neuropediatrics 40, 239–242.

Fisher, R.S., Uematsu, S., Krauss, G.L., Cysyk, B.J., McPherson, R., Lesser, R.P., et al., 1992. Placebo-controlled pilot study of centromedian thalamic stimulation in treatment of intractable seizures. Epilepsia 33, 841–851.

Garcia-Soriano, G., Rufer, M., Delsignore, A., Weidt, S., 2014. Factors associated with non-treatment or delayed treatment seeking in OCD sufferers: a review of the literature. Psychiatry Res. 220, 1–10.

Gardner, J., 2017. Securing a future for responsible neuromodulation in children: the importance of maintaining a broad clinical gaze. Eur. J. Paediatr. Neurol. 21, 49–55.

Genc, G., Apaydin, H., Gunduz, A., Poyraz, C., Oguz, S., Yagci, S., et al., 2016. Successful treatment of Juvenile parkinsonism with bilateral subthalamic deep brain stimulation in a 14-year-old patient with parkin gene mutation. Parkinsonism Relat. Disord. 24, 137–138.

George, M., Nahas, Z., Bohning, D., Kozel, F.A., Anderson, B., Mu, C., et al., 2005. Vagus nerve stimulation and deep brain stimulation. In: Stein M.D., Ph.D., Dan J., Kupfer M.D., David J., Schatzberg M.D., Alan F. (Eds.), The American Psychiatric Publishing Textbook of Mood Disorders. American Psychiatric Association Publishing, Washington, DC, p. 237.

Ghosh, P.S., Machado, A.G., Deogaonkar, M., Ghosh, D., 2012. Deep brain stimulation in children with dystonia: experience from a tertiary care center. Pediatr. Neurosurg. 48, 146–151.

Haridas, A., Tagliati, M., Osborn, I., Isaias, I., Gologorsky, Y., Bressman, S.B., et al., 2011. Pallidal deep brain stimulation for primary dystonia in children. Neurosurgery 68, 738–743. discussion 743.

Hariz, M.I., Rehncrona, S., Quinn, N.P., Speelman, J.D., Wensing, C., 2008. Multicentre Advanced Parkinson's Disease Deep Brain Stimulation Group: Multicenter study on deep brain stimulation in Parkinson's disease: an independent assessment of reported adverse events at 4 years. Mov. Disord. 23, 416–421.

Hauseux, P.A., Cyprien, F., Cif, L., Gonzalez, V., Boulenger, J.P., Coubes, P., et al., 2017. Long-term follow-up of pallidal deep brain stimulation in teenagers with refractory Tourette syndrome and comorbid psychiatric disorders: about three cases. Eur. J. Paediatr. Neurol. 21, 214–217.

Heinz, A., Knable, M.B., Wolf, S.S., Jones, D.W., Gorey, J.G., Hyde, T.M., et al., 1998. Tourette's syndrome: [I-123]beta-CIT SPECT correlates of vocal tic severity. Neurology 51, 1069–1074.

Helmers, S.L., Wheless, J.W., Frost, M., Gates, J., Levisohn, P., Tardo, C., et al., 2001. Vagus nerve stimulation therapy in pediatric patients with refractory epilepsy: retrospective study. J. Child Neurol. 16, 843–848.

Henderson, J.M., 2007. Vagal nerve stimulation versus deep brain stimulation for treatment-resistant depression: show me the data. Clin. Neurosurg. 54, 88–90.

Hershey, T., Black, K.J., Hartlein, J., Braver, T.S., Barch, D.M., Carl, J.L., et al., 2004. Dopaminergic modulation of response inhibition: an fMRI study. Brain Res. Cogn. Brain Res. 20, 438–448.

Holtzheimer, P.E., Mayberg, H.S., 2011. Deep brain stimulation for psychiatric disorders. Annu. Rev. Neurosci. 34, 289–307.

Hudson, V.E., Elniel, A., Ughratdar, I., Zebian, B., Selway, R., Lin, J.P., 2017. A comparative historical and demographic study of the neuromodulation management techniques of deep brain stimulation for dystonia and cochlear implantation for sensorineural deafness in children. Eur. J. Paediatr. Neurol. 21, 122–135.

Hung, K.L., 2012. Vagus nerve stimulation therapy in pediatric epilepsy: current understanding and future directions. Pediatr. Neonatol. 53, 155–156.

Kaminska, M., Perides, S., Lumsden, D.E., Nakou, V., Selway, R., Ashkan, K., et al., 2017. Complications of deep brain stimulation (DBS) for dystonia in children—the challenges and 10 year experience in a large paediatric cohort. Eur. J. Paediatr. Neurol. 21, 168–175.

Koy, A., Timmermann, L., 2017. Deep brain stimulation in cerebral palsy: challenges and opportunities. Eur. J. Paediatr. Neurol. 21, 118–121.

Koy, A., Weinsheimer, M., Pauls, K.A., Kuhn, A.A., Krause, P., Huebl, J., et al., 2017. German registry of paediatric deep brain stimulation in patients with childhood-onset dystonia (GEPESTIM). Eur. J. Paediatr. Neurol. 21, 136–146.

Krause, P., Lauritsch, K., Lipp, A., Horn, A., Weschke, B., Kupsch, A., et al., 2016. Long-term results of deep brain stimulation in a cohort of eight children with isolated dystonia. J. Neurol. 263, 2319–2326.

Lahtinen, M., 2017. A child is not a small adult: complications in deep brain stimulation in children: surgical safety and complications of deep brain stimulation for childhood dystonia. Eur. J. Paediatr. Neurol. 21, 13.

Lee, J.S., Yoo, S.S., Cho, S.Y., Ock, S.M., Lim, M.K., Panych, L.P.:, 2006. Abnormal thalamic volume in treatment-naive boys with Tourette syndrome. Acta Psychiatr. Scand. 113, 64–67.

Lenders, M.W., Vergouwen, M.D., Hageman, G., van der Hoek, J.A., Ippel, E.F., Jansen Steur, E.N., et al., 2006. Two cases of autosomal recessive generalized dystonia in childhood: 5 year follow-up and bilateral globus pallidus stimulation results. Eur. J. Paediatr. Neurol. 10, 5–9.

Levy, H.C., McLean, C.P., Yadin, E., Foa, E.B., 2013. Characteristics of individuals seeking treatment for obsessive–compulsive disorder. Behav. Ther. 44, 408–416.

Lipsman, N., Ellis, M., Lozano, A.M., 2010. Current and future indications for deep brain stimulation in pediatric populations. Neurosurg. Focus 29, E2.

Lumsden, D.E., Kaminska, M., Ashkan, K., Selway, R., Lin, J.P., 2017. Deep brain stimulation for childhood dystonia: Is 'where' as important as in 'whom'? Eur. J. Paediatr. Neurol. 21, 176–184.

Lumsden, D.E., Kaminska, M., Tustin, K., Gimeno, H., Baker, L., Ashkan, K., et al., 2012. Battery life following pallidal deep brain stimulation (DBS) in children and young people with severe primary and secondary dystonia. Childs Nerv. Syst. 28, 1091–1097.

Mahoney, R., Selway, R., Lin, J.P., 2011. Cognitive functioning in children with pantothenate-kinase-associated neurodegeneration undergoing deep brain stimulation. Dev. Med. Child Neurol. 53, 275–279.

Marks, W.A., Honeycutt, J., Acosta Jr, F., Reed, M., Bailey, L., et al., 2011. Dystonia due to cerebral palsy responds to deep brain stimulation of the globus pallidus internus. Mov. Disord. 26, 1748–1751.

Marks, W.A., Honeycutt, J., Acosta, F., Reed, M., 2009. Deep brain stimulation for pediatric movement disorders. Semin. Pediatr. Neurol. 16, 90–98.

Marras, C.E., Rizzi, M., Cantonetti, L., Rebessi, E., De Benedictis, A., Portaluri, F., et al., 2014. Pallidotomy for medically refractory status dystonicus in childhood. Dev. Med. Child Neurol. 56, 649–656.

Miyagi, Y., Koike, Y., 2013. Tolerance of early pallidal stimulation in pediatric generalized dystonia. J. Neurosurg. Pediatr. 12, 476–482.

Olaya, J.E., Christian, E., Ferman, D., Luc, Q., Krieger, M.D., Sanger, T.D., et al., 2013. Deep brain stimulation in children and young adults with secondary dystonia: the Children's Hospital Los Angeles experience. Neurosurg. Focus 35, E7.

Ostrem, J.L., San Luciano, L., Dodenhoff, K.A., Ziman, N., Markun, L.C., Racine, C., et al., 2016. Subthalamic nucleus deep brain stimulation in isolated dystonia: a 3-year follow-up study. Neurology 88, 25–35.

Owen, T., Adegboye, D., Gimeno, H., Selway, R., Lin, J.P., 2017. Stable cognitive functioning with improved perceptual reasoning in children with dyskinetic cerebral palsy and other secondary dystonias after deep brain stimulation. Eur. J. Paediatr. Neurol. 21, 193–201.

Park, H.R., Lee, J.M., Ehm, G., Yang, H.J., Song, I.H., Lim, Y.H., et al., 2016. Long-term clinical outcome of internal globus pallidus deep brain stimulation for dystonia. PLoS ONE 11, e0146644.

Parr, J.R., Green, A.L., Joint, C., Andrew, M., Gregory, R.P., Scott, R.B., et al., 2007. Deep brain stimulation in childhood: an effective treatment for early onset idiopathic generalised dystonia. Arch. Dis. Child 92, 708–711.

Pauls, D.L., Alsobrook II, J.P., Goodman, W., Rasmussen, S., Leckman, J.F., 1995. A family study of obsessive–compulsive disorder. Am. J. Psychiatry 152, 76–84.

Peterson, B.S., Skudlarski, P., Anderson, A.W., Zhang, H., Gatenby, J.C., Lacadie, C.M., et al., 1998. A functional magnetic resonance imaging study of tic suppression in Tourette syndrome. Arch. Gen. Psychiatry 55, 326–333.

Pinto, A., Mancebo, M.C., Eisen, J.L., Pagano, M.E., Rasmussen, S.A., 2006. The Brown Longitudinal Obsessive Compulsive Study: clinical features and symptoms of the sample at intake. J. Clin. Psychiatry 67, 703–711.

Rabins, P., Appleby, B.S., Brandt, J., DeLong, M.R., Dunn, L.B., Gabriels, L., et al., 2009. Scientific and ethical issues related to deep brain stimulation for disorders of mood, behavior, and thought. Arch. Gen. Psychiatry 66, 931–937.

Ramirez-Zamora, A., Gee, L., Youn, Y., Shin, D.S., Pilitsis, J.G., 2017. Pallidal deep brain stimulation for the treatment of levodopa-responsive juvenile dystonia and parkinsonism secondary to SPG11 mutation. JAMA Neurol. 74, 127–128.

Rodriguez-Oroz, M.C., Obeso, J.A., Lang, A.E., Houeto, J.L., Pollak, P., Rehncrona, S., et al., 2005. Bilateral deep brain stimulation in parkinson's disease: a multicentre study with 4 years follow-up. Brain 128, 2240–2249.

Salanova, V., Witt, T., Worth, R., Henry, T.R., Gross, R.E., Nazzaro, J.M., et al., 2015. Long-term efficacy and safety of thalamic stimulation for drug-resistant partial epilepsy. Neurology 84, 1017–1025.

Salpekar, J.A., Mishra, G., Hauptman, A.J., 2015. Key issues in addressing the comorbidity of depression and pediatric epilepsy. Epilepsy Behav. 46, 12–18.

Schrag, A., Schott, J.M., 2006. Epidemiological, clinical, and genetic characteristics of early-onset parkinsonism. Lancet Neurol. 5, 355–363.

Servello, D., Porta, M., Sassi, M., Brambilla, A., Robertson, M.M., 2008. Deep brain stimulation in 18 patients with severe Gilles de la Tourette syndrome refractory to treatment: the surgery and stimulation. J. Neurol. Neurosurg. Psychiatry 79, 136–142.

Shahed, J., Poysky, J., Kenney, C., Simpson, R., Jankovic, J., 2007. GPi deep brain stimulation for Tourette syndrome improves tics and psychiatric comorbidities. Neurology 68, 159–160.

Sprengers, M., Vonck, K., Carrette, E., Marson, A.G., Boon, P., 2017. Deep brain and cortical stimulation for epilepsy. Cochrane Database Syst. Rev. 7, CD008497.

Starr, P.A., Markun, L.C., Larson, P.S., Volz, M.M., Martin, A.J., Ostrem, J.L., 2014. Interventional MRI-guided deep brain stimulation in pediatric dystonia: first experience with the ClearPoint system. J. Neurosurg. Pediatr. 14, 400–408.

Storch, E.A., Hanks, C.E., Mink, J.W., McGuire, J.F., Adams, H.R., Augustine, E.F., et al., 2015. Suicidal thoughts and behaviors in children and adolescents with chronic tic disorders. Depress. Anxiety 32, 744–753.

Timmermann, L., Pauls, K.A., Wieland, K., Jech, R., Kurlemann, G., Sharma, N., et al., 2010. Dystonia in neurodegeneration with brain iron accumulation: outcome of bilateral pallidal stimulation. Brain 133, 701–712.

Tormenti, M.J., Tomycz, N.D., Coffman, K.A., Kondziolka, D., Crammond, D.J., Tyler-Kabara, E.C., 2011. Bilateral subthalamic nucleus deep brain stimulation for dopa-responsive dystonia in a 6-year-old child. J. Neurosurg. Pediatr. 7, 650–653.

Tsering, D., Tochen, L., Lavenstein, B., Reddy, S.K., Granader, Y., Keating, R.F., et al., 2017. Considerations in deep brain stimulation (DBS) for pediatric secondary dystonia. Childs Nerv. Syst. 33, 631–637.

Valentin, A., Selway, R.P., Amarouche, M., Mundil, N., Ughratdar, I., Ayoubian, L., et al., 2017. Intracranial stimulation for children with epilepsy. Eur. J. Paediatr. Neurol. 21, 223–231.

Valldeoriola, F., Regidor, I., Minguez-Castellanos, A., Lezcano, E., Garcia-Ruiz, P., Rojo, A., et al., 2010. Efficacy and safety of pallidal stimulation in primary dystonia: results of the Spanish multicentric study. J. Neurol. Neurosurg. Psychiatry 81, 65–69.

van Karnebeek, C., Horvath, G., Murphy, T., Purtzki, J., Bowden, K., Sirrs, S., et al., 2015. Deep brain stimulation and dantrolene for secondary dystonia in x-linked adrenoleukodystrophy. JIMD Rep. 15, 113–116.

Velasco, F., Velasco, M., Jimenez, F., Velasco, A.L., Brito, F., Rise, M., et al., 2000. Predictors in the treatment of difficult-to-control seizures by electrical stimulation of the centromedian thalamic nucleus. Neurosurgery 47, 295–304. discussion 304–305.

Velasco, F., Carrillo-Ruiz, J.D., Brito, F., Velasco, M., Velasco, A.L., Marquez, I., et al., 2005. Double-blind, randomized controlled pilot study of bilateral cerebellar stimulation for treatment of intractable motor seizures. Epilepsia 46, 1071–1081.

Velasco, A.L., Velasco, F., Velasco, M., Trejo, D., Castro, G., Carrillo-Ruiz, J.D., 2007. Electrical stimulation of the hippocampal epileptic foci for seizure control: a double-blind, long-term follow-up study. Epilepsia 48, 1895–1903.

CHAPTER

11

Using a Novel Approach to Assess Dynamic Cortical Connectivity Changes Following Neurofeedback Training in Children on the Autism Spectrum

Hristos S. Courellis[1,3], *Asimina S. Courelli*[1], *Elisabeth V.C. Friedrich*[2] and *Jaime A. Pineda*[2,4]

[1]Bioengineering Department, University of California, San Diego, CA, United States [2]Cognitive Science Department, University of California, San Diego, CA, United States [3]Swartz Center of Computational Neuroscience, University of California, San Diego, CA, United States [4]Group in Neurosciences, University of California, San Diego, CA, United States

OUTLINE

Neurotechnological Pediatric Neuropsych

DOI: https://doi.org/10.1016/B978-0-12-812777-3.00011-8

INTRODUCTION

Electroencephalography (EEG)-based neurofeedback (NF) has emerged over the last decade as a potential platform for the noninvasive treatment of pediatric populations diagnosed with neuropsychiatric disorders (Marzbani et al., 2016). Despite a number of methodological issues, there is a growing recognition that the benefits of this methodology outweigh the concerns (Orndorff-Plunkett et al., 2017). NF entails the use of operant conditioning in an active, EEG-driven brain–computer interface. It allows patients to control some aspect of a computerized task by volitionally modulating electrocortical activity in response to the task (Hwang et al., 2009). Such control is achieved via real-time feedback provided to the patient so that preferred electrophysiological states are positively reinforced through visual and auditory reward cues, in the context of task performance (Gruzelier, 2014; Pineda et al., 2008). When NF is employed as a treatment modality, its effectiveness hinges on the presence of abnormal electrophysiological activity associated with the condition, which can be influenced consciously by the individual undergoing the training. Over the course of a treatment regimen, patients learn to normalize this abnormal activity and gradually (via operant conditioning) switch the brain to an electrophysiological state where the activity is more similar to that of a neurotypical individual.

NF has been used with individuals, including children with neurological problems, diagnosed with attention-deficit/hyperactivity disorder (ADHD), epilepsy, and autism spectrum disorder (ASD) as a method for both treatment and rehabilitation (Marzbani et al., 2016). For children with ADHD, the imbalance between theta (θ) and beta (β) rhythms over the right frontal and prefrontal regions of the cortex allows for NF to be conducted with high spectral and spatial specificity, driving feedback only for the frequency bands of interest and utilizing small and targeted electrode montages (Bink et al., 2015; Arns et al., 2009; Leins et al., 2007). Children with epilepsy tend to exhibit significant negative deflection in their slow cortical potential (SCP) prior to the onset of an epileptic seizure. The SCP, detectible from a single electrode placed at the Cz position, can be trained through NF so that fewer

SCP deflections, and therefore fewer seizures, occur (Legarda et al., 2011; Walker and Kozlowski, 2005; Strehl et al., 2014). An emerging foundation has also been constructed for the application of NF as a treatment modality for children with ASD that are designated as high-functioning autistic (HFA) (Pineda et al., 2008; Fishman et al., 2014). An HFA diagnosis is required so that the child is able to comprehend the parameters of the training, able to complete relatively lengthy recording and training sessions, effectively interact with the administrator of the training, and understand what is required of them at each stage of that training (Friedrich et al., 2015; Pineda et al., 2014).

ASD was once considered to be of psychogenic origin but is now widely recognized to have a biological source with genetic, neurobiological, and perhaps environmental underpinnings. Although a wide range of behavioral, pharmacological, and alternative medicine strategies have been reported to ameliorate symptoms for some individuals, there is presently no cure for the condition, and the inherent heterogeneity of endophenotypical presentation makes clinical management challenging. One reported characteristic exhibited by children with HFA ASD is diminished functionality of the mirror neuron system (MNS), a collection of neurons present in the somatosensory, premotor, and motor areas of the human cortex (Fabbri-Destro and Rizzolatti, 2008; Oberman et al., 2007; Hadjikhani, 2007; Pineda, 2008). When not actively processing visuomotor and socially relevant information, motor neurons in these regions are presumed to be in an "idling" state firing synchronously at a frequency of 8–13 Hz. This creates a strong current dipole whose associated electric field potentials originate in sensorimotor circuits, travel to the surface of the scalp, and oscillate at the mu (μ) band frequency (Yao and Dewald, 2005; Oberman et al., 2008). This synchronous firing of the MNS circuit is readily detected from the surface of the scalp using noninvasive neuroimaging techniques such as EEG and magnetoencephalography (MEG). In contrast to neurotypical children, the MNS in an ASD child does not generally exhibit EEG-detectable desynchronization in the μ band (also called μ suppression or event-related desynchronization) during information processing, especially when the child observes another agent perform a similar, meaningful action (Oberman et al., 2007). Interestingly, children with ASD exhibit normal levels of MNS desynchronization or suppression of the μ-rhythm when observing actions conducted by people with whom the child is familiar (Oberman et al., 2007, 2008). This suggests that the MNS circuit is not totally broken and will work normally under certain circumstances. With that in mind, it has now been amply demonstrated that given enough time and training, children with ASD can learn to regulate the power of their μ-rhythm and volitionally alter the dynamic features of the MNS, producing more normal μ suppression during

EEG-monitored observation of meaningful movements of strangers (Pineda et al., 2008; Friedrich et al., 2015). Thus, NF that targets the μ-rhythm has demonstrated a good degree of efficacy when it is used as a treatment modality for HFA children (Pineda et al., 2008; Friedrich et al., 2014, 2015; Oberman et al., 2005, 2007, 2008).

In this chapter, we provide an overview of spectral power NF centered on the μ-rhythm and its administration as a treatment protocol to children diagnosed with HFA. More importantly, we discuss a novel analysis methodology for assessing the effects of NF on cortical connectivity, wherein changes in spectro-temporal causal interactions between cortical regions are examined before and after the administration of NF. Changes in the connectivity of specific subnetworks are evaluated during an emotion imitation task (EIT) composed of faces of different adult faces with a happy, fearful, or an angry expression. Furthermore, changes in behavioral characteristics exhibited by the HFA children after completing the NF are captured by cognitive assessment instruments such as the Social Responsiveness Scale (SRS) and the Autism Treatment Evaluation Checklist (ATEC). We hypothesized that different functional networks associated with different EEG rhythms would correlate with the processing of positive and negative emotional faces. In addition, we predicted that high-level, complex brain dynamics significantly altered by the application of spectral power NF would correlate with behavioral improvement.

Analysis Methodologies

During the past several decades, there has been an agreed upon common pipeline developed for EEG signal processing. The first part involves preprocessing of the data. This includes acquisition of the signal, removal of artifacts, signal averaging, thresholding of the output, enhancement of the resulting signal, etc. The second part in most analysis pipelines is the feature extraction process. A feature is an individual measurable property or characteristic of a phenomenon being observed. Choosing informative, discriminating, and independent features is a crucial step for effective algorithms in pattern recognition, classification, and regression. Once the key features are extracted, the final stage is signal classification, which can be solved by linear analysis, nonlinear analysis, adaptive algorithms, clustering and fuzzy techniques, and neural networks (Courellis et al., 2016, 2017; Bell and Sejnowski, 1995; Ojeda et al., 2014).

Relevant to this chapter are the variety of methods developed and used to extract key features from EEG signals. Some of the more common ones involve time-frequency distributions, fast Fourier transform (FFT), eigenvector methods, wavelet transform (WT), and autoregressive method. All these techniques, for frequency and time-frequency domain analysis, have major advantages and disadvantages. For example, FFT is a good tool for stationary signal processing (e.g., of sine waves) and has enhanced speed over other methods in real-time applications. However, it is weak in analyzing nonstationary signals such as EEG, while also suffering from large noise sensitivity. The WT method is better suited for analysis of sudden and transient signal changes (as occurs in EEG) but requires selection of a proper mother wavelet. Another approach, autoregression analysis, suffers from poor speed and is not always applicable in real-time analysis. Given these various advantages and disadvantages, the researchers have argued that it is crucial to make clear the signal to be analyzed in the application of the method, whenever the performance of analyzing method is discussed. Furthermore, the optimum method for any single application might be different (Courellis et al., 2016, 2017; Bell and Sejnowski, 1995; Ojeda et al., 2014).

Spectral Power Neurofeedback

We recently utilized a novel analysis methodology, which allows us to quantify cortical connectivity by examining the spectro-temporal interactions between regions, in a study to assess the effectiveness of μ-rhythm spectral power NF in a group of participants composed of 10 children diagnosed with HFA between the ages of 6 and 17 years. All of the participating children were independently evaluated by health-care professionals and diagnosed with HFA within a year of beginning the treatment. This diagnosis was verified by the administration of the Autism Diagnostic and Observation Scale, the Autism Diagnostic and Interview Scale-Revised, and the Wechsler Abbreviated Scale of Intelligence. Each participant completed 16 hours of NF over the course of 2 months in 1-hour sessions at least twice a week for 8 weeks. All participants underwent pre- and post-assessments using 32-channel cap EEG to quantify changes in brain activity and cortical connectivity among brain regions of interest (ROIs). NF training consisted of subjects rewarded for modulating μ-rhythm power (Friedrich et al., 2015). Basically, EEG was recorded from one electrode over right sensorimotor cortex (C4); sampled at 256 Hz; and filtered for mu (μ: 8–12 Hz), theta (θ: 4–8 Hz), and high-beta (β: 12–20 Hz) frequency bands. To increase mu power, children were asked to relax, while to decrease mu power, they were asked to imagine socially interacting with friends. The beta and theta frequency bands, typically associated with blinks and overall muscle movements, were inhibitors of positive feedback on the NF when the amplitudes in these frequency bands exceeded a specified threshold. Electromyography (EMG) was recorded from the superficial flexor muscles of the left forearm, sampled at 2048 Hz, and filtered between 100 and 200 Hz. If a movement exceeded a specific threshold ($\sim$10 mV), the positive feedback was inhibited in order to avoid children receiving positive feedback while moving or learning to modulate mu activity over the right sensorimotor cortex by moving the left hand.

To keep the children attentive and engaged over the course of the NF training, three different "nonsocial gaming stages" were employed, separated by three "social interaction gaming" stages, all of which required threshold-based control of the μ, β, and θ rhythms in order to progress from one stage to the next. These stages included interactive portions, where the subject undergoing NF influenced the progress of the game by achieving the desired rhythm thresholds. These thresholds were computed for every training session during a 3-minute baseline period of electrical activity over the region of training (C4)[1] prior to the commencement of the training session (Friedrich et al., 2015). During the three "social interaction gaming" stages, the child's avatar would stand before a non-playable character (NPC) and be required to meet or exceed the μ, β, and θ power thresholds in order to receive the positive feedback, which included increasing their progress bar and having their avatar imitate the facial expression of the NPC. The game would proceed to the next stage if the child held the aforementioned positive interaction with the NPC for at least 30 seconds. During the three "nonsocial gaming" stages, the child would play mini-games involving various objectives, such as apple and coin collection in the first two nonsocial gaming stages. Maintaining threshold power during the "nonsocial gaming" stage would allow the child to collect more tokens while playing each game and motivate the child with visual and auditory feedback, as they successfully collected virtual apples or tokens. The score achieved in each game was tracked across sessions to provide children with an incentive to perform better and surpass previous high scores. The third "nonsocial gaming" stage involved "combat" with a dragon that launched fireballs toward the child's avatar. If the child was able to maintain the desired μ, β, and θ threshold for 4 seconds, the fireball was deflected and hit the dragon instead, providing positive reinforcement through the avatar's happy facial expression and the animation of the dragon when struck by the fireball.

Assessments of Effects

An evaluation methodology for the effects of NF on HFA children similar to that used in the present study has been used by our group in the past (Friedrich et al., 2014, 2015). This methodology employs the

[1]Training involved primarily the right hemisphere (C4) since this hemisphere shows greater μ rhythm dysfunction in children with autism.

administration of extensive pre- and post-NF tests. One of the first is the recording of resting state EEG while the children remain still and relaxed with their eyes open for approximately 6 minutes while looking at a blank monitor screen, and with their eyes closed for an additional 6 minutes. Children were only asked to relax and had no problems following the instructions. The remaining three tasks involve the μ-suppression index (MSI), Reading the Mind in the Eyes test (RMET), and EIT and were administered in a randomized order for each pre/post-NF evaluation session. The MSI test (Pineda et al., 2008; Oberman et al., 2005) was used to quantify event-related μ-rhythm desynchronization exhibited by the children in response to observing short movement–related videos and conducting motor actions themselves. Baseline videos were presented to normalize power across subjects and evaluation sessions. The RMET was used to evaluate social and emotional acuity of the children by presenting only the eye region of many different faces, asking the children to identify the emotion of the person or their gender based on the eyes alone, and counting the correct responses. The intent of the task is to evaluate the ability of HFA children to "tune in to the mental state" of the individual whose eyes are presented, thus interrogating "theory of mind" capabilities in the children, that is, their ability to ascribe a mental state to another person (Friedrich et al., 2015). It has been demonstrated that HFA individuals perform significantly worse on this test than typically developed individuals and is associated with their inability to relate to others or understand the intentions of others in a social setting (Baron-Cohen et al., 2001). Thus, the RMET may be considered as a performance metric for evaluating effects of the administered NF with respect to social and emotional generalization. The EIT was used to assess child responsiveness specifically in the context of emotional mirroring, yet another area in which HFA children display deficits relative to typically developing children (Beall et al., 2008). This task consists of passive viewing of full faces exhibiting positive (happiness) or negative (anger/fear) emotions, while the electrophysiology and facial EMG activity of the zygomaticus major and corrugator supercilii muscles (see Fig. 11.1) are monitored. The EIT task required that participants observe a sequence of 68 short videos—each up to 2 seconds in length, with a period of 4–5 seconds between videos displaying a black screen, presented to them in a randomized order—composed of faces of different adult men and women with a happy, fearful, or an angry expression for the duration of each video, as shown in Fig. 11.1. The 68 short videos were divided into two groups: (1) 34 videos with happy facial expressions constituting the positive emotion (PE) group and (2) 34 videos with angry or fearful facial expressions constituting the negative emotion (NE) group. The EIT differs from the RMET in that it specifically targets mirroring activity,

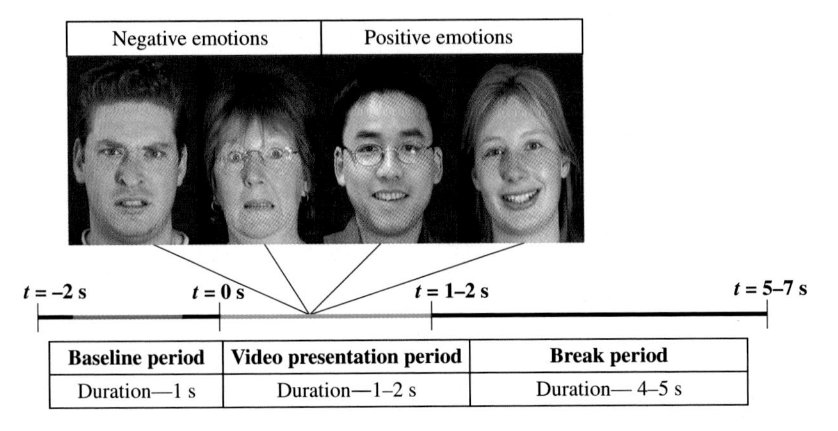

FIGURE 11.1 Epoch structure for the emotion imitation task. Includes a fixed length baseline period of 1 s, beginning 1.5 s before and ending 0.5 s before the onset of every video clip, during which a blank screen was displayed, the duration of the video clip (1 and 2 s) and a break period lasting 4 and 5 s, also with a blank screen displayed. Clips [examples of negative (left) and positive (right) emotions above] were presented at the beginning of the video presentation period and were accompanied by a tone. Connectivity analysis is conducted only during the 1- and 2-s window while the video was presented.

since muscular activity in the face mirroring the presented emotion appears in typically developing children to a greater degree than in HFA children. Successful μ-rhythm NF elicits increased cortical processing as indicated by a decrease in μ-power (or increased mu suppression) and increased muscular activation while viewing the videos, associated with an increase in emotional mirroring.

Analysis of the Electrophysiology

The effects of μ-rhythm NF on the cortical dynamics of the HFA children who received the training were quantified using cortical network connectivity analysis based on spectro-temporal causal connectivity between different cortical regions, which were computed from pre- and post-NF EEG recordings during the various tasks. For the purposes of reducing the amount of detail, only the analysis of the EIT results will be discussed in this chapter (for additional related results, see Friedrich et al., 2015). Spectro-temporal causal cortical connectivity was quantified across all the patients for each condition (PE and NE) in the EIT, by adapting the computational pipeline described previously (Courellis et al., 2017). In short, that pipeline consisted of the raw data collected from the EIT processed using automated rejection and reconstruction of noisy electrodes (Courellis et al., 2016), extraction of the first 500 ms of EEG after the presentation of each of the 68 videos as epochs of interest, and manual rejection of epochs with a large amount of artifactual data. The data were then projected into component space using Infomax-Independent Component Analysis (Bell and Sejnowski, 1995), and components exhibiting non-cortical characteristics (e.g., muscle artifacts and eye blinks) removed from the data before projecting the cleaned components back to the channel space. The cleaned channel space data were projected through spatial source localization onto a cortical boundary element model (BEM), constructed using the MoBILab toolbox (Ojeda et al., 2014) that served as a common fixed source space for each child. This BEM was segmented into a number of anatomical locations based on the Talairach Brain Atlas (Lancaster et al., 2000), and a small subset of these anatomical locations, henceforth referred to as ROIs, were selected based on the presence of a cortical dipole within a given anatomical identity.[2]

Representative current source density from each of the ROIs was extracted using a spatially regularized eigenspace beamformer (Sekihara et al., 2002), and causal information flow within cortical networks in the PE and NE groups was identified by treating ROIs as network nodes and quantifying causal interactions between those nodes in the form of spectro-temporal maps (Delorme et al., 2011) by fitting multivariate autoregressive models to the ROI data, and employing the short-time direct directed transfer function (SdDTF) causality metric (Korzeniewska et al., 2008). Changes in connectivity, separately for the PE and NE networks, were determined by subtracting the pretest connectivity maps from the posttest connectivity maps and averaged across children to generate a group-average spectro-temporal profile for the change in connectivity associated with NF.

Analysis of the Behavior

Quantitative behavioral scales, for example, the SRS (Constantino et al., 2003) and the ATEC, were completed by the children's caretakers before and after the NF treatment in order to assess whether or not any observable changes had occurred in the behavior of the children after the completion of the training. The correlation between changes in connectivity and behavior was evaluated by conducting Principal Component Analysis (PCA) to extract high-power connectivity eigenvectors, henceforth referred to as connectivity modes, which represent spectro-temporal connectivity dynamics present across all subjects. The degree to which each connectivity mode is present in a given child's data is determined by taking the vector projection of each child's

[2]Although a larger number of channels would provide greater resolution, it is possible to obtain source separation with only 32 EEG channels.

connectivity profile onto that mode, thus generating projection coefficients that were correlated with post-NF to pre-NF score differences of the behavioral questionnaires in order to determine which connectivity modes, and associated cortical ROIs, act as the best correlates of behavioral change.

RELEVANT FINDINGS

The BEM anatomical locations, with the presence of a cortical dipole, gave rise to two small ROI networks, one network of four ROIs for the positive (happy) emotion condition (PE network), and a separate network of four ROIs for the negative (angry/fearful) emotion condition (NE network). The selected ROIs presented an encapsulation of the cortical networks involved in processing the visual information and highlighted the effects of NF on those pathways. The PE network included the following ROIs: left anterior cingulate cortex (L-ACC), left precentral gyrus (L-Mot), left middle occipital gyrus (L-Vis), and right superior frontal gyrus (R-SFG). In contrast, the NE network included the following ROIs: R-SFG, right middle temporal gyrus (R-Temp), left supplementary motor area (L-SMA), and left middle occipital gyrus (L-Vis). Although the resulting PE network was different from that of the NE network, they included ROIs with overlapping functional involvement. The geometry of the ROIs for the PE and NE networks is shown in Figs. 11.2 and 11.3, where the colored regions on the surface of the BEMs indicate the area covered by the vertices associated with each region according to the information provided by the MNI-aligned anatomical atlas.

The statistically significant changes in connectivity (post-NF − pre-NF) averaged across all 10 patients are shown in Fig 11.4 ($P < .05$). Fig. 11.4A shows the changes in the connectivity matrix for the PE network and Fig. 11.4B shows the changes in the connectivity matrix for the NE network. Each rectangular figure in the connectivity matrix indicates the relative amount of causal influence an ROI from the top row (indicated as "FROM" and often referred to as a source) has on an ROI to the left column (indicated as "TO" and often referred to as a sink). This influence is measured by the spectro-temporal profile of the SdDTF connecting each source to each sink, giving rise to the asymmetry present in the matrix. The connectivity computed on the diagonal was set by default to 0 since investigation of the causal influence of an ROI onto itself does not constitute part of this study and can effectively be ignored.

As predicted, different functional networks were associated with different facial emotions. In the PE network (Fig. 11.4A), the largest increases in

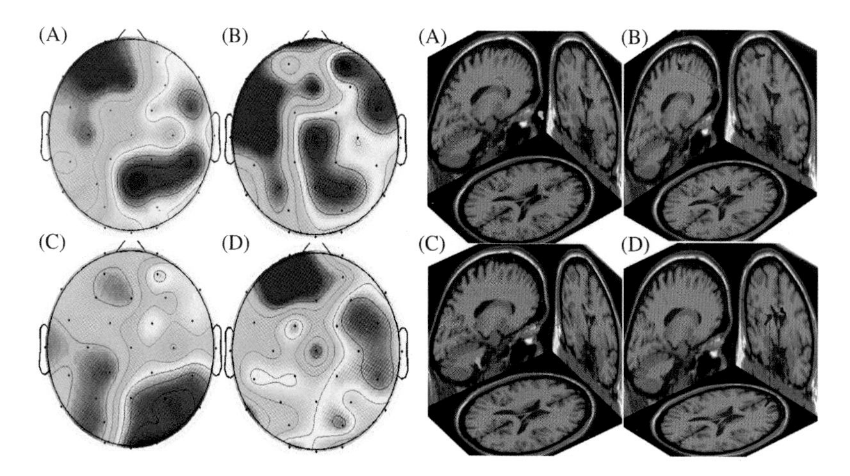

FIGURE 11.2 Positive condition (happy): sample independent components, represented by the potential scalp maps, and equivalent current dipoles. Scalp maps show relative contributions of electrodes to each source, where red is a positive signal contribution from the underlying channel and blue is a negative signal contribution. Anatomical locations and associated Talairach coordinates: (A) left anterior cingulate cortex (−1, 15, 27), (B) left precentral gyrus (−27, 3, 52), (C) left middle occipital gyrus (−6, −85, 23), (D) right superior frontal gyrus (9, 66, 28).

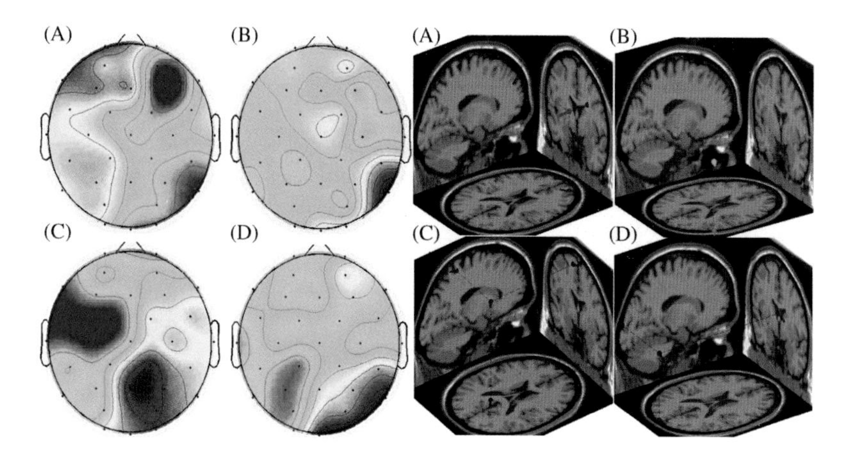

FIGURE 11.3 Negative condition (anger/fear): sample independent components, represented by the potential scalp maps, and the equivalent current dipoles that lie within the ROIs selected for source localization on the BEM. Anatomical locations and associated Talairach coordinates: (A) right superior frontal gyrus (24, 60, 26), (B) right middle temporal gyrus (64, −55, 16), (C) left supplementary motor area (−10, −40, 63), (D) left middle occipital gyrus (−1, −97, 18).

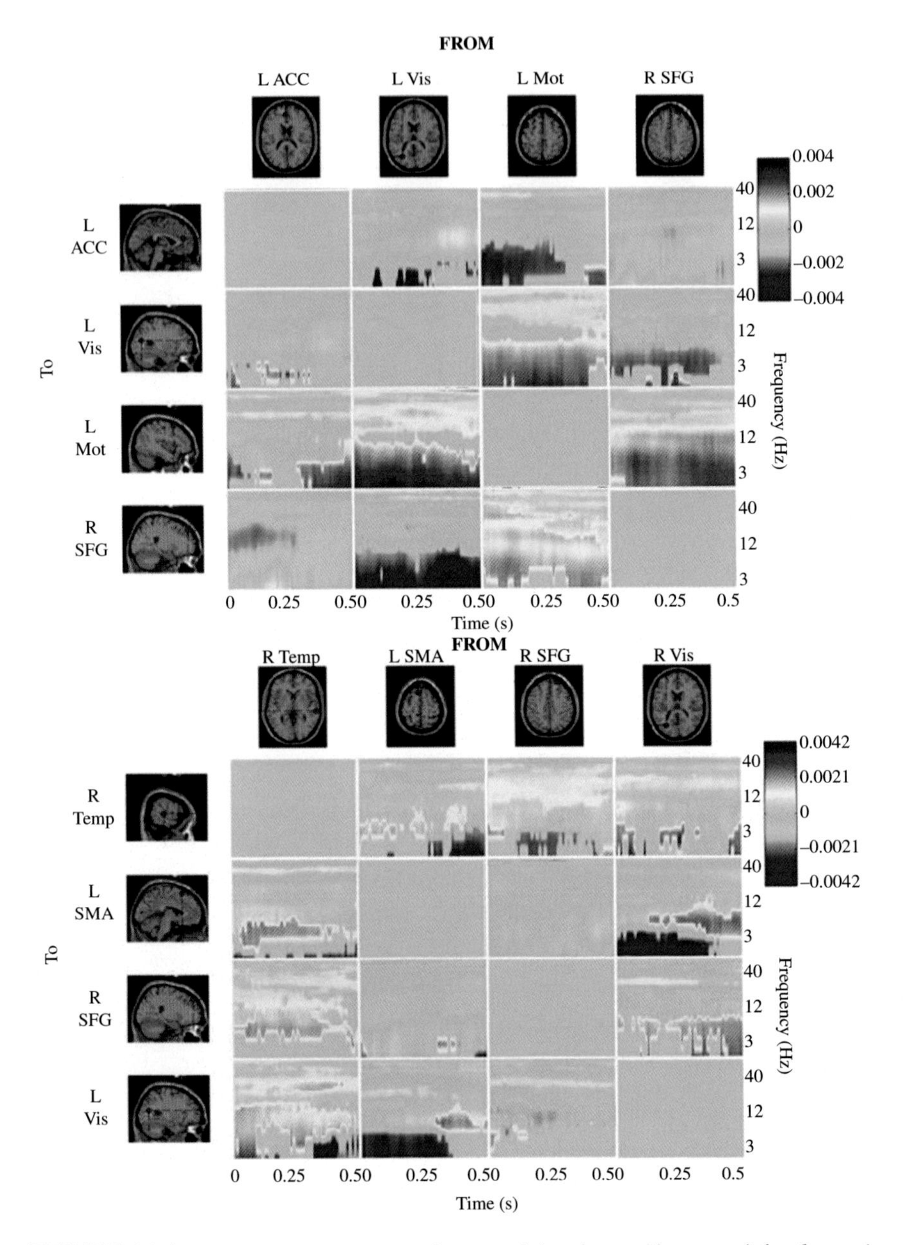

FIGURE 11.4 Average spectro-temporal connectivity changes (post–pre) for the positive condition (A) and the negative condition (B). Positive spectral power indicates an increase in connectivity following NFT training, and negative spectral power indicates a decrease in connectivity. All changes in connectivity shown above are significant with $P < .05$.

connectivity occurred in the precentral (L-Mot)−occipital gyrus (L-Vis) connection and in the R-SFG−precentral gyrus (L-Mot) connection. The *increase* in causal connectivity at the (L-Mot) $\leftrightarrow$ (L-Vis) connection occurred across all frequencies, bidirectionally, while the increase in causal connectivity at the (R-SFG) $\rightarrow$ (L-Mot) was more unidirectional, that is, greater from R-SFG to L-Mot than it was in the opposite direction and dominated by lower frequency bands. The greatest *decrease* in connectivity occurred in the L-Vis to R-SFG connection, mainly at the δ, θ, and low-α bands, unidirectionally. The R-SFG to L-Vis connectivity change was much smaller in magnitude (about half) at the high-θ/α band. Another notable change in connectivity occurred between L-Mot and L-ACC where, bidirectionally, early epoch θ connectivity decreased followed by an increase later in the epoch. The magnitude of these changes varied depending on direction, as L-Mot to L-ACC exhibited a much larger early-epoch connectivity decrease, while the reverse direction (L-ACC to L-Mot) exhibited a larger late-epoch connectivity increase.

In contrast, in the NE network, the average connectivity changes displayed markedly different dynamics, with heavier recruitment of the Temporal Lobe (Fig. 11.4B). Much of the *increase* in this network's connectivity arose from R-Temp, which featured increases in α/β and low-γ (35−40 Hz) connectivity to the L-SMA, R-SFG, and L-Vis (Fig. 11.4B, first column). Narrow-band low-frequency *decreases* in connectivity were also present in the connection from R-Temp to L-SMA (intermittent δ) and L-Vis (late-epoch δ/θ) regions. This decrease appeared bidirectional as it was reciprocated by a low-frequency late-epoch decrease in connectivity from L-SMA to R-Temp but was not observed from L-Vis to R-Temp, as it exhibited only low-frequency intermittent increases in connectivity. L-Vis also exhibited increased connectivity to R-SFG at the low-α band early in the epoch that widens to encompass δ and θ frequencies as the epoch progresses. Variable change in connectivity was also noted in the bidirectional L-Vis $\leftrightarrow$ L-SMA connection, both of which exhibit a sharp decrease in δ/θ connectivity early in the epoch followed by an increase in α-band connectivity later in the epoch.

Decomposition of the SdDTF connectivity data from both the PE and NE networks using PCA yielded a set of orthogonal connectivity modes. When individual subject connectivity data were projected back onto these connectivity modes, five high-power modes (three from the NE network and two from the PE network) were found to correlate strongly with the behavioral changes exhibited by the test subjects (see Fig. 11.5). It should be noted that behavioral improvements are associated with changes in score for the SRS and ATEC questionnaires. Thus, connectivity modes whose projections correlate negatively with changes in questionnaire scores are suppressed when improvement is observed, whereas connectivity modes whose projections correlate positively with

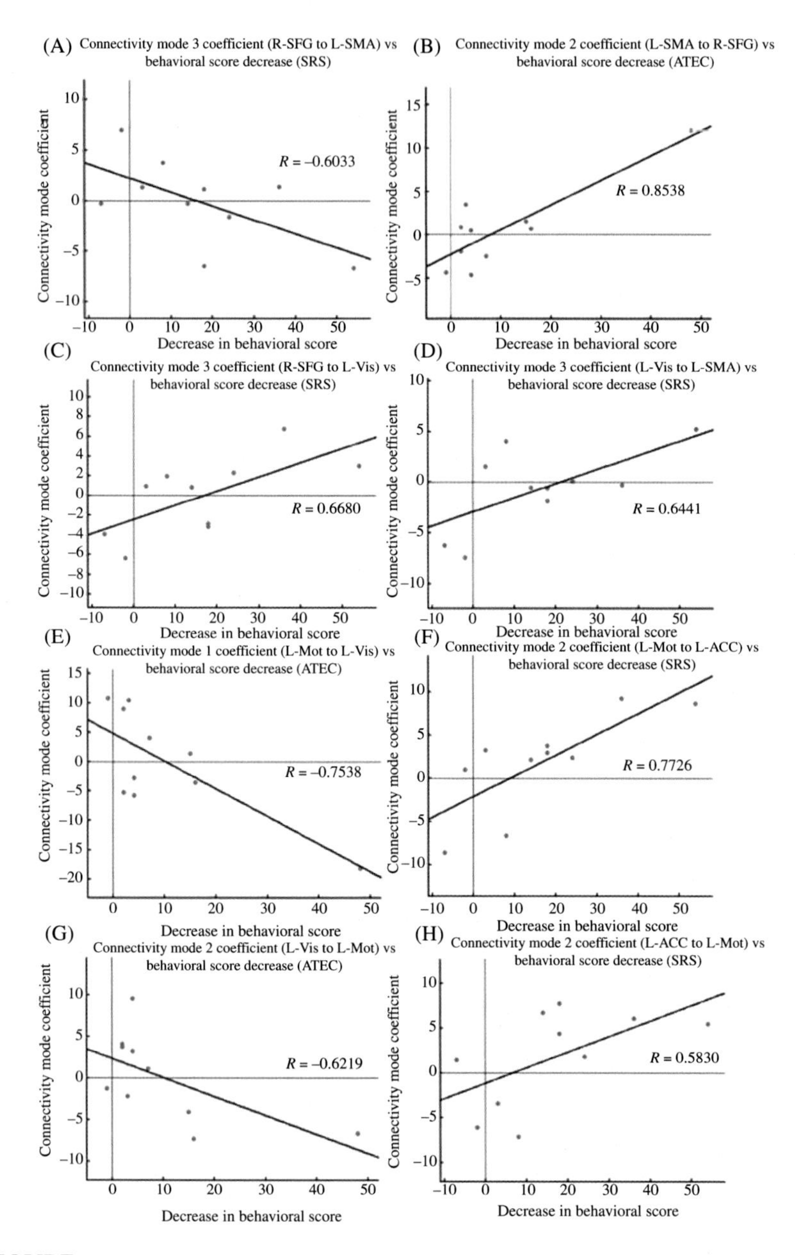

FIGURE 11.5 Connectivity mode projection correlations with both SRS and ATEC behavioral metrics. The modes selected represent negative condition (A–D) and positive condition (E–H) cortical dynamics that exhibit the largest magnitude correlations with the behavioral data. Note: negative projection values correspond to an inversion of the connectivity mode as an explanatory connectivity motif (e.g., for a mode with an increase in α connectivity, a negative projection corresponds to a decrease instead).

changes in questionnaire scores are dominant when improvement is observed after training.

For the NE condition, one connectivity mode exhibited negative correlation with changes in behavioral scores and two exhibited positive correlations. The third R-SFG to L-SMA connectivity mode (Fig. 11.5A) displayed a negative correlation, $R = -0.60$, with the SRS behavioral changes, intermittent low-frequency connectivity increase in the δ and θ bands, and consistent high-frequency (low-γ) connectivity decrease throughout the epoch. The third R-SFG to L-Vis connectivity mode (Fig. 11.5C) exhibited positive correlation with SRS measured behavioral changes ($R = 0.67$) and exhibited narrow-band intermittent connectivity increases in the δ and θ frequency bands, and a strong increase connectivity in mid-β band from the middle of the epoch, rising in strength until the end of the epoch. The third L-Vis to L-SMA mode (Fig. 11.5B) also exhibited a positive correlation with changes in SRS measured behavioral data ($R = 0.64$) and showed increases in connectivity in the δ, low-β, and low-γ frequency bands, followed by a sharp decrease in the θ and low-α bands late in the epoch.

For the PE condition, two connectivity modes associated with reciprocal connectivity between L-Mot and L-ACC exhibited positive correlation with changes in SRS behavioral scores. The second mode (Fig. 11.5D) of the L-Mot to L-ACC connection correlated positively ($R = 0.77$) with SRS changes and showed low-frequency intermittent connectivity increases in the θ and α band. The second mode of L-ACC to L-Mot (Fig. 11.5E) also correlated positively ($R = 0.58$) with changes in SRS and exhibited intermittent, high-power, low-frequency connectivity motif that dominated the high-θ and low-α bands, along with a relatively smaller magnitude connectivity increase in the low-β band during the middle of the epoch.

The subnetworks that exhibit high correlation with the behavioral data are summarized in Fig. 11.6 where solid arrows indicate positive correlation while dashed arrows indicate negative correlation for the PE (Fig. 11.6A) and NE (Fig. 11.6B) conditions. The summary figure highlights the bidirectionally connected L-ACC/L-Mot subnetwork in the PE condition, and the R-SFG/L-Vis/L-SMA subnetwork in the NE condition.

DISCUSSION

All EEG-based NF approaches are predicated on changing the dynamics of brain activity to produce beneficial and long-lasting changes in function and structure (Marzbani et al., 2016; Orndorff-Plunkett et al., 2017). Distinct protocols have been developed by

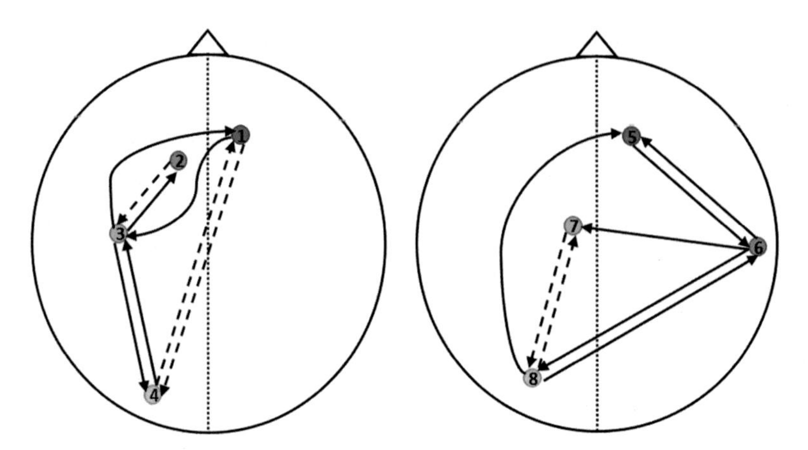

FIGURE 11.6 (A) Positive emotion connectivity change summary. Solid lines indicate increased overall connectivity, and dotted lines indicate decreased overall connectivity. The network nodes are (1) right superior frontal gyrus, (2) left anterior cingulate cortex, (3) left precentral gyrus, (4) left middle occipital gyrus. (B) Negative emotion connectivity change summary. Solid lines indicate increased overall connectivity, and dotted lines indicate decreased overall connectivity. The network nodes are (5) right superior frontal gyrus, (6) right medial temporal gyrus, (7) left supplementary motor area, (8) left middle occipital gyrus.

clinicians and researchers to target different brain signals and their concomitant physiological processes (for reviews, see Marzbani et al., 2016; Orndorff-Plunkett et al., 2017). These protocols are applied to a variety of psychiatric problems, including ADHD, bipolar disorder, schizotypal personality disorder, dissociative identity disorder, major depressive disorder, post-traumatic stress disorder, addiction, autism, and others. Generally speaking, NF protocols can be placed into categories based on the signal and/or the focus of training, including (1) basic frequency training, (2) deep-state or hypnogogic training, (3) infra-low frequency and SCP training, (4) synchrony/cross-frequency coupling, (5) network-based training, and (6) normative and symptom-based selection. This scope of potential therapeutic protocols has been recently expanded to include control of hemodynamic (e.g., blood oxygen level dependent signal, hemoencephalographic signal) as well as magnetic fields. Unlike EEG-based NF, which relies upon electrical activity recorded from the scalp, functional magnetic resonance imaging, functional near-infrared spectroscopy, and MEG NF provide more precise localization and modulation of relevant brain structures (Marzbani et al., 2016; Orndorff-Plunkett et al., 2017). Individuals undergoing training learn to volitionally control these new signals rapidly at a pace similar to the EEG approach. However, their main disadvantage is cost. Power spectrum NF, such as used in our study, offers a noninvasive, well-tolerated, non-pharmacologic treatment modality that can be rapidly disseminated at low cost.

Results from the cortical network connectivity analysis, a novel method of capturing changes in spectro-temporal causal connectivity between different cortical regions, showed that four ROIs were associated with the PE network. These included L-ACC, L-Vis, L-Mot, and R-SFG, while the four ROIs associated with the NE network included R-Temp, L-SMA, R-SFG, and L-Vis. The L-Vis ROI accounted for the visual component of the task. Previous studies have shown that several of these areas, such as the primary motor cortex, SMA, and the precentral gyrus, contain mirror neurons (Fabbri-Destro and Rizzolatti, 2008; Oberman et al., 2008; Mukamel et al., 2010). The middle temporal gyrus is similarly relevant from a functional standpoint due to its involvement in emotional and facial processing; thus, participating in network connectivity known to be abnormal in children with ASD (Sabatinelli et al., 2011; Pierce et al., 2001; Cheng et al., 2015). This region has been demonstrated to be functionally connected to motor and premotor regions that are active during mirroring of emotions (Budell et al., 2015) and, thus, would be expected to exhibit commensurate changes in spectro-temporal connectivity after NF is conducted on the upstream MNS. The SFG is highly relevant to the task due to its involvement in both default mode processing and the theory of mind network (ToMN), the latter being implicated in empathy deficits exhibited by children with ASD (Plitt et al., 2015; Kana et al., 2015; Li et al., 2013). The ACC has twofold relevance to the EIT task due to its abnormal activation in children with ASD as well as its involvement in decision-making and discrimination of the positive and negative emotional stimuli presented during the pre/post-NF EEG evaluations (Salmi et al., 2013; Zhou et al., 2016). The decreased functional connectivity of SFG and ACC with each other and with the motor cortices specifically in HFA individuals (Salmi et al., 2013; Minshew and Keller, 2010) and their involvement in the mirroring task confirm the relevance of the emergence of the corresponding ROIs as key PE and NE network nodes following conducting MNS NF.

Changes in network dynamics differed significantly between the PE and NE conditions in the EIT task following NF. The most prominent of these differences was the reciprocal connectivity between the left middle occipital gyrus and the motor-related region involved in mirroring. In the PE condition, the generalized increase in information flow between the precentral gyrus and L-Vis suggests that after training, visual information more heavily drives processes that occur in the precentral gyrus during empathic facial processing, indicating an increase in utilization of visual information by the MNS for the entirety of the epoch. In the NE condition, this connectivity increase is only reflected later in the epoch in the α band, which is associated with inactivity in both the SMA (μ rhythm) and in the middle temporal gyrus (occipital α), with a significant decrease in lower frequency connectivity. These reciprocal connection dynamics suggest that significant increases in visuomotor connections may reflect ASD improvement when processing and mirroring NEs. In fact, communication within this portion of the cortical network may be more suppressed when presented with NEs as a result of the training, indicated by the increase in α connectivity, suggesting that occipital inactivation causes MNS inactivation during the end of the epoch, whereas in the PE condition, broadband connectivity was maintained between the visual and motor areas until the end of the epoch.

The second prominent reciprocal connection in the PE cortical network, between the SFG and the precentral gyrus (Fig. 11.3A), indicates increased broadband communication between portions of the MNS (L-Mot) and the ToMN (R-SFG), which is consistent with prior research (Fishman et al., 2014; Kana et al., 2015). This increased reciprocal connection suggests that the subjects may be appraising emotional information received from the PE videos to a greater degree after NF, implying that the subjects are able to attribute the emotional state to the observed individual in the video to a greater degree. The increased task-positive flow of information between sections of the MNS and ToMN suggests that NF can rectify one of the major functional problems present in children with ASD, the inability to effectively process and appraise the emotions of others, leading to the social interactions deficits presented by HFA individuals (Fishman et al., 2014; Kana et al., 2015). Interestingly, the NE condition does not reflect this change in connectivity directly. The increase in information flow from L-Vis to R-SFG in the NE condition, coupled with the increase in information flow from the R-SFG to R-Temp and increase from R-Temp to L-SMA, provides support for the idea that through an indirect pathway, information regarding the presented emotion arrives in the MNS (L-SMA) after being processed in the ToMN (R-SFG) network via a narrow frequency band (θ and low-α). The medial temporal gyrus generally showed an increase in information outflow to all other nodes of the cortical network, a finding consistent with existing literature since the R-Temp specifically has been implicated in the processing of faces that exhibit emotions when evaluating typically-developing (TD) individuals (Cheng et al., 2015). The increase in information outflow from R-Temp demonstrates that following NF, the emotional-face processing dynamics of the children with ASD align more closely with those of typically developed individuals (Sabatinelli et al., 2011), providing more support for the beneficial effects of the training. The middle occipital gyrus exhibited a number of instances of θ connectivity decreases in both the PE and NE networks, a motif that can be explained by the involvement of the occipital θ rhythm in efficacy of monitoring during repetitive tasks (Beatty et al., 1974).

The identification of connectivity modes with high behavioral correlation reveals the presence of condition-dependent cortical networks, whose causal dynamic interactions are associated with behavioral improvement following NF. For the PE condition, the cortical network involves bidirectional communication between L-ACC and L-Mot, implicating both the MNS-relevant motor region, the precentral gyrus, and the decision-making and emotional processing region, the ACC, of the cortical network (Zhou et al., 2016). The reciprocal L-Mot/L-ACC connections exhibited positive correlation (Table 11.1—rows D and E), indicating that a higher magnitude, positive involvement of the second connectivity mode for each connection (Fig. 11.5D and E) is associated with larger behavioral improvements. Inspection of the nodes reveals that the increase in low-α connectivity constitutes a key connectivity motif in both directions for the connection between the L-ACC and L-Mot regions. Strong bidirectional connectivity in the α band implies that subject modulations of the α power in the L-Mot region of the MNS cause similar α power modulations in the ACC and vice versa. Therefore, α suppression in the MNS associated with mirroring (Oberman et al., 2008) causes subjects who exhibit greater improvement to also exhibit a greater alpha desynchronization in the ACC when attempting to mirror an emotion, implying that more emotional processing–related information is being recruited by the MNS during the mirroring.

The behavioral correlation network associated with the NE condition (Fig. 11.6B) involves the visual node (L-Vis), the MNS-related node (L-SMA), and the ToMN-related node (R-SFG), the L-SMA acting as a sink node, and the R-SFG acting as a common source node for both

TABLE 11.1 Summary of Correlation Between Connectivity Mode Projection Coefficients and Changes in Behavior as Quantified by the SRS Questionnaire

Connectivity mode	Emotion condition	ROI source node	ROI sink node	Correlation
A	Negative	R-SFG	L-SMA	−0.60
B	Negative	L-Vis	L-SMA	0.64
C	Negative	R-SFG	L-Vis	0.67
D	Positive	L-Mot	L-ACC	0.77
E	Positive	L-ACC	L-Mot	0.58

The five columns from left to right indicate the connectivity mode shown in Fig. 11.5, the emotion condition for which the correlation exists, the cortical ROI that acts as the source node for the causal interaction, the cortical ROI that acts as the target, or sink, node for the causal interaction, and the correlation exhibited by the listed connection's connectivity mode coefficients with the change in SRS score for each subject. *L-SMA*, Left supplementary motor area; *R-SFG*, Right superior frontal gyrus; *L-ACC, Left anterior cingulate cortex.*

L-SMA and L-Vis. This network exhibits different behavioral correlations from that of the PE condition since the indirect connection from R-SFG to L-SMA through L-Vis features positively correlated connectivity mode coefficients (Table 11.1 rows B and C), and the direct R-SFG to L-SMA connection exhibits the only connectivity mode that is negatively correlated with improvements in behavior (Table 11.1, row A).

This network exhibits two separate pathways of information flow, which originate in the R-SFG and terminate in the L-SMA. The first pathway, the direct R-SFG to L-SMA connection, features elevated δ/θ frequency connectivity and diminished narrow-band high-β frequency connectivity in the corresponding connectivity mode (Fig. 11.5A). The negative correlation between the connectivity mode and the behavioral data indicates that improvements in behavior are associated with opposite changes in the connectivity mode, that is, when information flow between these regions increases in higher frequencies and decreases in lower frequencies. This result could be reflective of the NF in the sense that, throughout the training, activity in the β band was to be suppressed during the task-positive condition, suggesting a possible direct manifestation of NF parameters in subnetwork communication associated with behavioral improvement. It may be viewed as a sufficient condition on connectivity between these two ROIs for behavioral improvements to be observed. It may not provide a necessary condition because the tasks the subjects were undertaking during training were not emotional mirroring tasks, and single-location downregulation of frequency content during training may not necessarily lead to similar changes in the frequency content of the communication between the two ROIs, especially when the task used for evaluation is different from the task used for training. The second information pathway of this network indicates that, for information flowing into the network from visual stimuli (Fig. 11.5B), θ connectivity is initially increased and subsequently strongly suppressed. This θ suppression occurs on a similar timescale to the mid-β connectivity from R-SFG to L-Vis (Fig. 11.5C). A possible explanation for this observation is the presence of cross-frequency band causal influence, where emotional mirroring information processed by the SFG drives β oscillations in L-Vis, and subsequently influencing the α information flow from L-Vis back into the frontal regions of the subnetwork. Although it is not straightforward to provide conclusive evidence for cross-frequency interactions using the computational methods we employed, it appears that connectivity modes featuring long-range increases in connectivity both in the NE and PE subnetwork are associated with behavioral improvements. This observation suggests that the under-connectivity hypothesis of ASD, which posits that long-range connectivity decreases in children with ASD (Schipul et al., 2011; Just et al., 2012), although see Fishman et al.

(2014) for a different perspective, is being rectified over the course of the NF, and increases in long-range connectivity are coupled with beneficial behavioral changes in the HFA children undergoing training (Boutros et al., 2015).

LIMITATIONS OF THE STUDY

The promising results produced by spectral power NF as a treatment modality for ASD must be judged with respect to a number of limitations associated with this study. The use of a 32-channel EEG headset and a spatially regularized source localization technique limited the functional spatial resolution available for resolving current density on the cortical surface. Large anatomical ROIs were selected to compensate for this resolution limit, and the use of Laplacian regularization prevented current density from spilling through adjacent sulci to proximal gyri, thus maintaining the functional divisions generally found between gyri and across the interhemispheric fissure. The choice of large ROIs limited the detail to which connectivity results may be interpreted, since a broad number of functionally distinct subregions of a gyrus were grouped together.

Another limitation was the relatively small number of participants involved in the study. This issue arose mainly from the extensive, high-commitment treatment schedule being employed, requiring that a child and their caretaker be available twice a week over the course of 2 months to complete all 16 training sessions. These strict scheduling parameters and the experimental nature of the training, which could not guarantee improvement in the condition of the child, limited the number of potential participants who were willing and able to partake in the training.

A third limitation was the lack of an age- and gender-matched control group. A control group proved even more difficult to incentivize since this subject group did not receive any immediate benefit from the training, and finding age- and gender-matched neurotypical children who were willing to participate in the full training was prohibitively difficult. This difficulty, coupled with the fact that limited existing literature has quantified spectro-temporal connectivity dynamics to the extent shown in this paper, prompted the consideration of changes between pre- and post-NF cortical connectivity dynamics within the NF-treated group rather than comparison of cortical connectivity dynamics between the NF-treated group and a neurotypical reference group. It might be viable to consider a wait-list control group. Other types of control groups, such as sham/placebo or alternate type of NF training, would suffer the same problems of participation. More monetary incentive may work, combined with a reduction in training duration. Nonetheless, continuing to quantify cortical dynamics in studies such as the one presented here, using similar analytical methods will help create a knowledge base against which neuropathological data, such as the kind acquired in this study, may be compared in order to draw stronger conclusions about the neurological state of the evaluated subject group.

References

Arns, M., Ridder, S., Strehl, U., Breteler, M., Coenen, A., 2009. Efficacy of neurofeedback treatment in ADHD: the effects on inattention, impulsivity, and hyperactivity: a meta-analysis. Clin. EEG Neurosci. 40 (3), 180–189.

Baron-Cohen, S., Wheelwright, S., Hill, J., Raste, Y., Plumb, I., 2001. The "Reading the mind in the eyes" test revised version: a study with normal adults, and adults with Asperger syndrome or high-functioning autism. J. Child Psychol. Psychiatry 42 (2), 241–255.

Beall, P.M., Moody, E.J., McIntosh, D.N., Hepburn, S.L., Reed, C.L., 2008. Rapid facial reactions to emotional facial expressions in typically developing children and children with autism spectrum disorder. J. Exp. Child Psychol. 101 (3), 206–223.

Beatty, J., Greenberg, A., Deibler, W.P., O'Hanlon, J.F., 1974. Operant control of occipital theta rhythm affects performance in a radar monitoring task. Science 183 (4127), 871–873.

Bell, A.J., Sejnowski, T.J., 1995. An information-maximisation approach to blind separation and blind deconvolution. Neural Comput. 7 (6), 1004–1034.

Bink, M., van Nieuwenhuizen, C., Popma, A., Bongers, I.L., van Boxtel, G.J., 2015. Behavioral effects of neurofeedback in adolescents with ADHD: a randomized controlled trial. Eur. Child Adolesc. Psychiatry 24 (9), 1035–1048.

Boutros, N.N., Lajiness-O'Neill, R., Zillgitt, A., Richard, A.E., Bowyer, S.M., 2015. EEG changes associated with autistic spectrum disorders. Neuropsychiatr. Electrophysiol. 1 (1).

Budell, L., Kunz, M., Jackson, P.L., Rainville, P., 2015. Mirroring pain in the brain: emotional expression versus motor imitation. PLoS ONE 10 (2), e0107526.

Cheng, W., Rolls, E.T., Gu, H., Zhang, J., Feng, J., 2015. Autism: reduced connectivity between cortical areas involved in face expression, theory of mind, and the sense of self. Brain 138 (5), 1382–1393.

Constantino, J.N., et al., 2003. Validation of a brief quantitative measure of autistic traits: comparison of the social responsiveness scale with the autism diagnostic interview-revised. J. Autism Dev. Disord. 33 (4), 427–433. 2003.

Courellis, H., Iversen, J.R., Poizner, H., Cauwenberghs, G., 2016. EEG channel interpolation using ellipsoid geodesic length. Presented at The IEEE Biomedical Circuits and Systems, Shanghai, China.

Courellis, H., Mullen, T., Poizner, H., Cauwenberghs, G., Iversen, J.R., 2017. EEG-based quantification of cortical current density and dynamic causal connectivity generalized across subjects performing BCI-monitored cognitive tasks. Front. Neurosci. 11 (180).

Delorme, A., et al., 2011. EEGLAB, SIFT, NFT, BCILAB, and ERICA: new tools for advanced EEG processing. Comput. Intell. Neurosci. 2011, 130714.

Fabbri-Destro, M., Rizzolatti, G., 2008. Mirror neurons and mirror systems in monkeys and humans. Physiology (Bethesda) 23, 171–179.

Fishman, I., Keown, C.L., Lincoln, A.J., Pineda, J.A., Muller, R.A., 2014. Atypical cross talk between mentalizing and mirror neuron networks in autism spectrum disorder. JAMA Psychiatry 71 (7), 751–760.
Friedrich, E.V., Suttie, N., Sivanathan, A., Lim, T., Louchart, S., Pineda, J.A., 2014. Brain–computer interface game applications for combined neurofeedback and biofeedback treatment for children on the autism spectrum. Front. Neuroeng. 7, 21.
Friedrich, E.V., et al., 2015. An effective neurofeedback intervention to improve social interactions in children with autism spectrum disorder. J. Autism Dev. Disord. 45 (12), 4084–4100.
Gruzelier, J.H., 2014. EEG-neurofeedback for optimising performance. I: a review of cognitive and affective outcome in healthy participants. Neurosci. Biobehav. Rev. 44, 124–141.
Hadjikhani, N., 2007. Mirror neuron system and autism. In: Carlisle, P.C. (Ed.), Progress in Autism Research. Nova Science Publishers, Inc, pp. 151–166.
Hwang, H.J., Kwon, K., Im, C.H., 2009. Neurofeedback-based motor imagery training for brain-computer interface (BCI). J. Neurosci. Methods 179 (1), 150–156.
Just, M.A., Keller, T.A., Malave, V.L., Kana, R.K., Varma, S., 2012. Autism as a neural systems disorder: a theory of frontal-posterior underconnectivity. Neurosci. Biobehav. Rev. 36 (4), 1292–1313.
Kana, R.K., et al., 2015. Aberrant functioning of the theory-of-mind network in children and adolescents with autism. Mol. Autism 6, 59.
Korzeniewska, A., Crainiceanu, C.M., Kus, R., Franaszczuk, P.J., Crone, N.E., 2008. Dynamics of event-related causality in brain electrical activity. Hum. Brain Mapp. 29 (1), 1170–1192.
Lancaster, J.L., et al., 2000. Automated Talairach Atlas labels for functional brain mapping. Hum. Brain Mapp. 10, 120–131.
Legarda, S.B., McMahon, D., Othmer, S., Othmer, S., 2011. Clinical neurofeedback: case studies, proposed mechanism, and implications for pediatric neurology practice. J. Child Neurol. 26 (8), 1045–1051.
Leins, U., Goth, G., Hinterberger, T., Klinger, C., Rumpf, N., Strehl, U., 2007. Neurofeedback for children with ADHD: a comparison of SCP and Theta/Beta protocols. Appl. Psychophysiol. Biofeedback 32 (2), 73–88.
Li, W., et al., 2013. Subregions of the human superior frontal gyrus and their connections. NeuroImage 78, 46–58.
Marzbani, H., Marateb, H.R., Mansourian, M., 2016. Neurofeedback: a comprehensive review on system design, methodology and clinical applications. Basic Clin. Neurosci. 7 (2), 143–158.
Minshew, N.J., Keller, T.A., 2010. The nature of brain dysfunction in autism: functional brain imaging studies. Curr. Opin. Neurol. 23 (2), 124–130.
Mukamel, R., Ekstrom, A.D., Kaplan, J., Iacoboni, M., Fried, I., 2010. Single-neuron responses in humans during execution and observation of actions. Curr. Biol. 20 (8), 750–756.
Oberman, L.M., Hubbard, E.M., McCleery, J.P., Altschuler, E.L., Ramachandran, V.S., Pineda, J.A., 2005. EEG evidence for mirror neuron dysfunction in autism spectrum disorders. Brain Res. Cogn. Brain Res. 24 (2), 190–198.
Oberman, L.M., McCleery, J.P., Ramachandran, V.S., Pineda, J.A., 2007. EEG evidence for mirror neuron activity during the observation of human and robot actions: toward an analysis of the human qualities of interactive robots. Neurocomputing 70 (13–15), 2194–2203.
Oberman, L.M., Ramachandran, V.S., Pineda, J.A., 2008. Modulation of mu suppression in children with autism spectrum disorders in response to familiar or unfamiliar stimuli: the mirror neuron hypothesis. Neuropsychologia 46 (5), 1558–1565.

Ojeda, A., Bigdely-Shamlo, N., Makeig, S., 2014. MoBILAB: an open source toolbox for analysis and visualization of mobile brain/body imaging data. Front. Hum. Neurosci. 8, 121.

Orndorff-Plunkett, F., Singh, F., Aragon, O.R., Pineda, J.A., 2017. Assessing the effectiveness of neurofeedback training in the context of clinical and social neuroscience. Brain Sci. 7 (8).

Pierce, K., Muller, R.A., Ambrose, J., Allen, G., Courchesne, E., 2001. Face processing occurs outside of the fusiform 'face area' in autism: evidence from functional MRI. Brain 124, 2059–2073.

Pineda, J.A., 2008. Sensorimotor cortex as a critical component of an 'extended' mirror neuron system: does it solve the development, correspondence, and control problems in mirroring? Behav. Brain Funct. 4, 47.

Pineda, J.A., et al., 2008. Positive behavioral and electrophysiological changes following neurofeedback training in children with autism. Res. Autism Spectr. Disord. 2 (3), 557–581.

Pineda, J.A., Friedrich, E.V., LaMarca, K., 2014. Neurorehabilitation of social dysfunctions: a model-based neurofeedback approach for low and high-functioning autism. Front. Neuroeng. 7, 29.

Plitt, M., Barnes, K.A., Martin, A., 2015. Functional connectivity classification of autism identifies highly predictive brain features but falls short of biomarker standards. Neuroimage Clin. 7, 359–366.

Sabatinelli, D., et al., 2011. Emotional perception: meta-analyses of face and natural scene processing. NeuroImage 54 (3), 2524–2533.

Salmi, J., et al., 2013. The brains of high functioning autistic individuals do not synchronize with those of others. Neuroimage Clin. 3, 489–497.

Schipul, S.E., Keller, T.A., Just, M.A., 2011. Inter-regional brain communication and its disturbance in autism. Front. Syst. Neurosci. 5, 10.

Sekihara, K., Nagarajan, S.S., Poeppel, D., Marantz, A., Miyashita, Y., 2002. Application of an MEG eigenspace beamformer to reconstructing spatio-temporal activities of neural sources. Hum. Brain Mapp. 15 (4), 199–215.

Strehl, U., Birkle, S.M., Worz, S., Kotchoubey, B., 2014. Sustained reduction of seizures in patients with intractable epilepsy after self-regulation training of slow cortical potentials—10 years after. Front. Hum. Neurosci. 8, 604.

Walker, J.E., Kozlowski, G.P., 2005. Neurofeedback treatment of epilepsy. Child Adolesc. Psychiatr. Clin. N. Am. 14 (1), 163–176. viii.

Yao, J., Dewald, J.P., 2005. Evaluation of different cortical source localization methods using simulated and experimental EEG data. NeuroImage 25 (2), 369–382.

Zhou, Y., Shi, L., Cui, X., Wang, S., Luo, X., 2016. Functional connectivity of the caudal anterior cingulate cortex is decreased in autism. PLoS ONE 11 (3).

CHAPTER

12

Neurofeedback for Pediatric Emotional Dysregulation

Jason Kahn[1,2,3], *Michaela Gusman*[4] *and Suzanne Wintner*[3,5]

[1]Department of Psychiatry, Boston Children's Hospital, Boston, MA, United States [2]Department of Psychiatry, Harvard Medical School, Boston, MA, United States [3]Neuromotion Labs, Boston, MA, United States [4]Boston Children's Hospital, Boston, MA, United States [5]Simmons College School of Social Work, Boston, MA, United States

OUTLINE

Neurotechnological Pediatric Neuropsych
DOI: https://doi.org/10.1016/B978-0-12-812777-3.00012-X

WHAT IS EMOTIONAL REGULATION?

We have all seen emotional regulation (ER) in children. We watch it as a baby learns to self-soothe, when a child pushes through a disappointment without a tantrum, or when an adolescent thinks better of a particularly risky decision. We see it as adults, as we navigate through social and professional lives, balancing reactions as not everything goes our way. ER can be seen as the absence, or inhibition, of extreme or maladaptive behaviors in the face of potentially overwhelming emotional experience.

ER requires some shared definitions, starting with a brief understanding of what we mean by emotion. Emotions are a very basic part of the human experience; however, defining emotions is a complex undertaking. ER, then, involves the monitoring, modulation, or maintenance of emotional states. This provides a framework to discuss the neurological, biological, and behavioral dimensions of regulation.

HEALTHY EMOTIONAL REGULATION

ER is commonly understood as the modulation of effective sensation and corresponding urges to take action, toward achieving greater physical or psychological comfort, or in the service of other goals (Denham, 2007). Regulation requires action across multiple systems and includes neurological processes, cognition, perception/interpretation, and behavior management (Zeman et al., 2006). Effective regulation relies on some combination of cognitive and behavioral strategies. One common framework for understanding ER is in terms of antecedent-focused versus response-focused regulation strategies, the former consisting of modification or avoidance of triggering events and altered attention or appraisal, and the latter embodied by modulation of emotional experience or behavioral responses (Gross, 1998). Another potentially useful distinction in ER mechanisms recognizes explicit, conscious forms of regulation versus implicit, largely unconscious or subconscious regulation (Etkin et al., 2015).

Researchers have found that some ER strategies are more adaptive over the long-term than others. Top-down strategies are those that rely on cognitive control of emotions. Cognitive reappraisal of antecedent events, problem-solving, reflection, distraction, social sharing, and acceptance of emotions have been recognized as adaptive ER strategies, associated with fewer anxiety and depression symptoms and more positive affect, while strategies such as avoidance, suppression, and rumination are considered maladaptive and associated with more depression and anxiety symptoms and negative affect (Brans et al., 2013; Schäfer et al., 2017). These findings should be considered within a developmental context. Though reappraisal and problem-solving are generally more adaptive strategies, they would not be considered developmentally accessible until late childhood or early adolescence (Gullone et al., 2010; Rawana et al., 2014).

Bottom-up, or automatic, approaches to regulation stand in contrast to top-down approaches. The limbic system can, without higher level cortical processes, reject responses in well-defined clinical tasks (Gospic et al., 2011). Ochsner and Gross (2005) note that physiological and neurological differences predict which individuals will have less reactivity to negative images. This indicates that certain individuals have more regulation resources (Vohs and Heatherton, 2000) available to them. Perhaps most tantalizing, if ER becomes conceptualized as a resource, a possible implication is that development and growth is possible. ER ability is not fixed in time, but something that can change, and ideally grow, over a lifetime.

EMOTIONAL DYSREGULATION

Difficulties understanding emotions, modulating affect, and controlling emotionally driven behavioral impulses contribute to emotional dysregulation (Bjureberg et al., 2016). Viewed through the lens of Gross's (1998) emotion regulation framework, emotional dysregulation can be understood as impairment in one or more regulation processes, including situation selection and problem-solving, antecedent appraisal, management of physiological or somatic sensations, and containment of behavioral responses. Information processing abilities, which facilitate an individual's ability to interpret and resolve social and emotional problems, are highly influenced by age and health. In neurotypical children, social information processing (SIP) abilities develop noticeably from the preschool years through middle childhood (Crick and Dodge, 1994). The development of SIP competencies can be delayed or impeded by intellectual disabilities, developmental, and neurological disorders such as autism spectrum disorder (ASD) and attention deficit hyperactivity disorder (ADHD), and childhood socioeconomic disadvantage (Andrade et al., 2012; Larkin et al., 2013; Schultz and Shaw, 2003). Heart rate variation governed by the autonomic nervous system (ANS) has been implicated in the management of physiological or somatic sensations (Scarpa, 2015). A combination of factors, such as childhood adversity, biological sensitivity, and sensory processing disorders, contribute to hyperarousal and other difficulties in physiological arousal management (Aron and Aron, 1997; Dunn et al., 2002).

Chronic emotional dysregulation, "pervasive dysfunction in the ER system" (Miller et al., 2007), affects individuals across the lifespan. Many children who struggle frequently with emotional dysregulation display externalizing behaviors such as noncompliance, aggression, anger, and poor regulation of impulses (Campbell et al., 2000). These symptoms cut across many different pediatric mental health diagnoses, including ADHD, anxiety, ASD, disruptive mood dysregulation disorder and other mood disorders, behavioral disorders, and schizophrenia (Brotman et al., 2006; Hinshaw, 1992; Bianchi et al., 2017). Young children with impulsive externalizing behavior challenges are at risk for ongoing school problems, mental health challenges, and difficulties in other domains (Campbell et al., 2000; Brotman et al., 2006). In school settings, children with dysregulation symptoms and related behaviors are more likely to struggle with classroom adjustment and to come into conflict with peers (Miller et al., 2004). In adolescence, struggles with dysregulation are associated with more negative self-perceptions about academic competencies and more negative views of school (Oram et al., 2017).

Early intervention shows clear value in externalizing disorders (Stormont, 2002), while inaction has a clear negative feedback loop as stress builds for a child who is increasingly removed from class, falling behind academically, isolated from peers, experiencing negative family interactions (Eisenberg et al., 2000), and at risk for negative health outcomes (Correll and Carlson, 2006; Correll et al., 2009; Nocavo, 2002).

Children who struggle with ER are also at greater risk for internalizing problems, such as anxiety and depression (Kim-Spoon et al., 2013). Childhood and adolescent anxiety are associated with increased risk of adult anxiety, depression, and substance use disorders (Beesdo et al., 2009). Children who develop severe depression before adolescence are more likely to have later experiences with suicidality, conduct disorder, substance use, and bipolar disorder (Weissman et al., 1999). Dysregulation and internalizing symptoms can also negatively influence children's school experiences. In adolescence, struggles with dysregulation are associated with more negative self-perceptions about academic competencies and more negative views of school (Oram et al., 2017). A variety of interventions have shown success in the treatment of internalizing disorders, across school and mental health settings (Herman et al., 2011; Elkins et al., 2011; Oswald and Mazefsky, 2006).

EMOTIONAL REGULATION NEUROFEEDBACK—A PROMISING BUT EARLY TECHNOLOGY

While psychotherapeutic options for treating ER are available, they are burdened by high attrition rates (Pina et al., 2003; Grant et al., 2012) and uneven treatment integrity in the real world (Herschell et al., 2010). Additionally, less than half of children and adolescents in need of these types of mental health treatments are likely to access them (Olfson et al., 2015), and when they do, skills and insights gained do not always transfer to external situations (Berking et al., 2011). Neurofeedback, on the other hand, offers several potential advantages. Under certain designs, it can adapt to idiosyncrasies in individuals, potentially increasing response rates and effect sizes. The treatment can be given algorithmically, increasing fidelity and again potentially improving response rates. And by directly training neural pathways, conceivably the same cognitive resources built during neurofeedback would be available in external, demanding settings beyond the clinician's walls (Arns et al., 2009; Johnstone, 2008; Niv, 2013; Teplan, 2002). Here, we explore the underlying neurology of ER and look at several of the ER neurofeedback paradigms that have emerged.

WHAT IS NEUROFEEDBACK?

Neurofeedback (NF) is the process by which an individual is presented with a real-time view of his or her own neurological patterns, learns strategies to manipulate brain waves, and aims to self-regulate different brain functions (Teplan, 2002; Johnstone, 2008).

The underlying theory of NF posits that individuals can control their brain activity once exposed to real-time feedback of that activity. In NF interventions, technology monitors some type of brain activity, be it electrical signals in the brain, blood flow, or other markers, and translates these physical fluctuations into a representation. This representation is typically graphically displayed on a digital screen (i.e., computer monitor or tablet). The user then exerts cognitive control over brain activity in an effort to manipulate the representation (e.g., Lubar et al., 1995a).

When the aim of the intervention is to provide symptomatic or functional gains for a clinical need, NF becomes therapy, sometimes also referred to as neurotherapy or neurobiofeedback. In this process, a certified NF therapist instructs a participant to observe and manipulate his or her own brain behavior in a way that aims to build or change brain biology and function to improve the patient's overall health (Cramer et al., 2011; Lubar, 1995a). When used correctly, NF can be a valuable tool in treating neurocognitive or psychological disorders, like ADHD or depression, or in enhancing mental strategies, like inhibitory control or alertness. (Lubar, 1991; Niv, 2013). However, evidence of NF's clinical efficacy remains nascent, with disorder-specific evidence remaining decidedly mixed (Cortese et al., 2016, Sitaram et al., 2017).

For the purposes of this chapter, we are excluding biofeedback from NF. Biofeedback is the process of observing and exerting cognitive control over physical biomarkers, such as heart rate, or respiration. Much like in NF, the frequency, amplitude, or variability of these signals is made visible, and a user manipulates the signals. Also like NF, biofeedback has clinical utility (Schoenberg and David, 2014). Rooted in a longer literature history, symptomatic and functional gains with biofeedback have been observed in ADHD (Lubar, 1991), anxiety (Wenck et al., 1996), depressive disorders (Knox et al., 2011), and others (Thomas, 2016). Because these processes are rooted in the brain, cognitive control over these markers could, under some interpretations, be considered NF. While we will adopt the definition that NF exclusively looks at manipulation of signals that are located within the brain, these clinical gains provide reason to believe that NF could be as effective, if not more, than biofeedback.

A NEUROLOGICAL ACCOUNT OF EMOTIONAL REGULATION

As NF is centered upon manipulation of neurophysiological signals within the brain, a prerequisite for therapeutic NF for ER is an understanding of the neurobiology of ER.

Humans are fundamentally social creatures (Vygotsky, 1980; Rogoff et al., 1993). ER provides an adaptive tool, which allows us to rein in base impulses and participate in society around us (Porges, 2001) as well as quiet down our internal monologue and act to reach richer and more complex goals (Satpute and Lieberman, 2006). ER networks within the brain have been well-studied. On a very high level, sensory systems pass stimuli information to the amygdala. From there, the signal passes through the anterior cingulate cortex (ACC) (Bush et al., 2000) and the prefrontal cortex (PFC). The PFC is broadly thought to be a higher order region of the brain responsible for some of the more uniquely human behaviors we display (Davidson et al., 2000). Within the PFC, various regions are responsible for modulating the original emotional signal (Banks et al., 2007; Gospic et al., 2011), determining goals for target behavior (Bissonette et al., 2013; Goel and Dolan, 2003), and appraising and reappraising strategies for achieving those goals (Dennis and Chen, 2007; Phelps et al., 2004).

The Major Pathway for Emotional Regulation: Amygdala—PFC Connectivity

The amygdala is a small region located deep within the brain. Universally present in the mammalian brain, the amygdala plays a deep-seated role in behavior and points to neurological resources dating back to reptilian ancestry coexisting with more evolutionarily modern adaptations (Porges, 1995). The amygdala directly receives sensory input from the thalamus and sensory cortices (Janak and Tye, 2015), before passing the signal onward to both emotional processing and inhibitory control systems (Kohn et al., 2014). Fig. 12.1 illustrates the major pathways in ER.

The amygdala's clear role was first revealed through lesioning studies (Davidson, 2000; Janak and Tye, 2015). Damage to the amygdala results in changes to fear (Davidson, 2000; Ochsner et al., 2012; Kohn et al., 2014) and reward conditioning (Zotev et al., 2011). The region of the brain also plays a major role in the fight or flight reflex (Applehans and Luecken, 2006; Metcalfe and Mischel, 1999), pointing to the deep relationship between emotions and the ANS (Ferrer and Helm, 2013). As the physiological changes in heart rate and breathing (Berntson et al., 1993) are called into action, we experience emotions such as fear or anxiety (Panksepp, 2004). These physiological changes then give way to behavior that can be maladaptive in social situations (Shweder et al., 2008). Overstimulation of "fight" impulses can lead to outward expressions of anger or aggression, a.k.a. externalizing behaviors, while the "flight" response can lead to anxiety and withdrawal—internalizing behaviors.

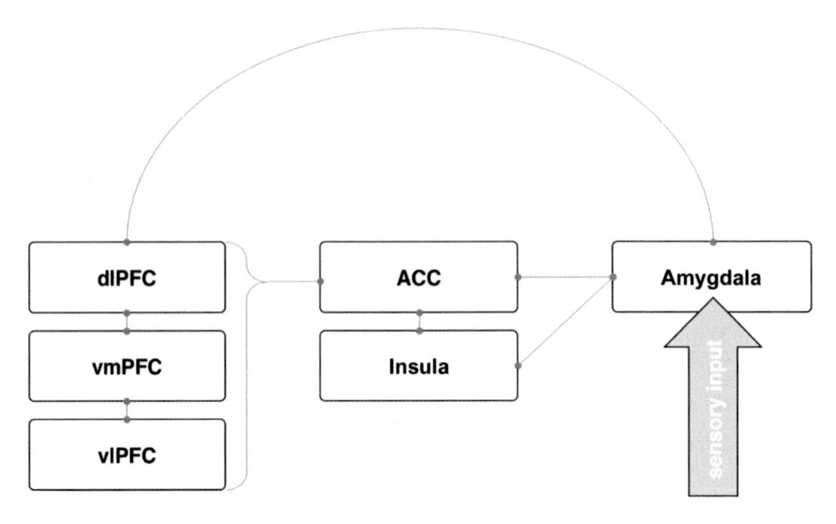

FIGURE 12.1 An illustration of neural pathways used in emotional regulation.

We know by looking at the world around us that anxiety and fear do not dominate every social or novel situation we find ourselves in. The brain has specific circuitry to modulate these impulses and allow new strategies to emerge. Functionally, these are generally thought of as executive functions. These include inhibitory control, working memory, and cognitive flexibility (Diamond, 2013; Garon et al., 2008). Collectively, they allow for us to pause, examine possibilities, and determine alternate strategies when faced with challenges (Gross, 2002). Unsurprisingly, strong executive functioning skills are associated with strong ER skills (Diamond et al., 2007).

Neurologically, current understanding is that executive functions arise specifically in the dorsolateral PFC (dlPFC) (Banich, 2009). This region in the front of the brain has been shown to connect to the ER system (Delgado et al., 2008; Phillips et al., 2008), and the strength of these connections predicts ER skills (Gee et al., 2013; Kim et al., 2011a). This indicates a path for intervention—strengthening connection between the amygdala and the dlPFC. As we consider ER, the cortico-limbic coordination is as important as the functioning of the regions themselves (Banks et al., 2007). This has implications across a wide variety of mental health disorders, including anxiety (Kim et al., 2011b; Monk et al., 2008; Roy et al., 2013), impulsive aggression (Coccaro et al., 2007; Davidson, 2000; Winstanley, 2004), depression (Johnstone et al., 2007), and personality disorders (Yang and Raine, 2009).

The PFC develops and changes significantly across infancy, childhood, and adolescence, providing a biological foundation to changes in executive functions, impulsivity, and risk-taking behaviors that that are routinely observed across development (Fuster, 2002). A notable early example is the development of the A-not-B error. The A-not-B task originates with Piaget (1954) and asks an infant to find an object hidden in one of two locations. The A-not-B error is made when the infant continues to search in the previous hidden location, even after seeing the object get hidden in the new location. Infants reliably stop making the A-not-B error at about 8 months of age (Smith et al., 1999).

The biological development underlies the behavioral change. Dendritic neurons in the dlPFC grow rapidly (Diamond, 2002; Koenderink et al., 1994), introducing neural circuitry that allows for the inhibition of the original behavior (reaching for the "A" target), and development of an alternative strategy (reach for the "B" target) to reach a goal (find the desired object). Lesion studies of rhesus monkeys (Diamond, 2002) have shown that the A-not-B error continues to be made when development of the dlPFC has been delayed or interrupted. While growth of the dlPFC continues through childhood, myelination does not complete until the very end of adolescence, often in the early to mid-twenties (Blakemore and Choudhury, 2006). A hypothesis around adolescent risk-taking behavior suggests that they are based in incomplete neurological rein on impulsive actions (Olson et al., 2009).

Other Connections That Activate During Emotional Regulation

Reappraisal is a well-studied framework for ER. As would be expected, reappraisal strategies elicit upregulation of the PFC and downregulation of amygdala (Ochsner et al., 2002; Ochsner and Gross, 2005). The amygdala connects to the insula, which is also downregulated during reappraisal. The ACC plays roles in both cognitive and affective processing, mediates connections between the amygdala and PFC, and is associated with attention, goal selection (through the ventral medial PFC), and language processing (through the ventral lateral PFC) (Banks et al., 2007; Goldin et al., 2008). Goal selection and language processing are key cognitive tools for enabling reappraisal.

Reappraisal is just one strategy, and one that runs into limitations. Behaviorally, many individuals do not have access to regulation strategies under moments of stress (Arnsten, 2009). Neurologically, the PFC functions become "unavailable" during times of stress (Kim et al., 2011b). Neurochemically, the PFC is sensitive to the balance of norepinephrine and dopamine, both being balanced in such a way that too much or too little leads to impaired functioning. During times of stress, the amygdala releases catecholamines, beginning a chemical cascade resulting in impaired PFC activation (Arnsten, 2011). If the initial response to stressors is overwhelming or prefrontal-limbic connectivity development is compromised, an individual can have significant difficulty accessing reappraisal, or any top-down regulation strategy.

TECHNOLOGIES ENABLING EMOTIONAL REGULATION NF

The ability to look into the workings of the brain and provide real-time feedback has grown remarkably over the past 50 years (Hammond, 2007). Each of the technologies provides some amount of benefit and trade-off, and potential to look deeply in the workings of the brain. Here, we look at the various technologies that have enabled this change, with the goal of providing the reader and understanding of how we peer into the brain.

Electroencephalography

Electroencephalography (EEG) requires placing conductive electrodes on the scalp, which in turn detect the tiny electrical impulses generated as neurons fire (Gruzelier, 2014). These impulses fire at certain rates and amplitudes, and the signal can be spatially localized along the surface of the scalp as increasing numbers of electrodes are placed. Though the spatial resolution can be relatively poor, the temporal precision of EEG is quite high, capable of registering changes in electric potential on the order of milliseconds (Teplan, 2002).

EEG NF produces a waveform of electrical activity in the brain that is typically broken into power bands, shown in Table 12.1.

A number of controlled studies have validated EEG as a dynamic method for NF, demonstrating the potential to modulate waves at faster or slower rates and with different training strategies (e.g., Egner and Gruzelier, 2004; Hanslmayer et al., 2006; Keizer et al., 2010a,b; Strehl et al., 2006; Gruzelier, 2009).

TABLE 12.1 Summary of EEG Bands Based on Gruzelier(2014), Teplan (2002), and Egner and Gruzelier (2004)

Band name	Frequencies (Hz)	Associated behaviors
Delta	0.5–4	Rises during sleep
Theta	4–8	Appears during meditation, sleep, relaxation
Alpha	8–13	Increases with cognitive activity and when closing eyes
SMR	12–15	Increases with semantic processing and sustained attention
Beta 1 (low beta)	15–18	Associated with concentration
Beta 2 (high beta)	22–30	Appears during complex thought and excitatory states

SMR, Sensorimotor rhythm.

In NF, researchers typically use ratios of band values to define intervention goals. These ratios are computed in real-time, and the resulting value is visualized by the participant. EEG NF for ADHD, for example, often targets the elevated theta/beta ratio frequently seen in individuals with ADHD (Arns et al., 2009; Masterpasqua and Healey, 2003). The upregulation of sensorimotor rhythm (SMR) and beta waves combined with the downregulation of theta has been shown to strengthen inhibitory function (e.g., Sterman, 1996), attentional processes and motor system excitability (e.g., Studer et al., 2014), and learning and memory (e.g., Lévesque et al., 2006).

As previously discussed ADHD and ER have a high degree of overlap, making the work done in NF for ADHD relevant. ADHD has clinical symptoms that are distinct, particularly in the selection of attention that leads to symptoms that are distinct from ER; patients with combined type are more likely to demonstrate the symptoms associated with emotional dysregulation (Clarke et al., 2011; Wheeler Maedgen and Carlson, 2000). Patients with ADHD often show less frontal volume (Krain and Castellanos, 2006). While hypotheses have arisen that these cortical areas are understimulated (Schatz and Rostain, 2006), the physiological differences are accompanied by differences in electrical activity, specifically increased resting theta activity and decreased beta activity compared to healthy controls (Arns et al., 2013). Pharmacological treatment changes EEG activity along with symptomatic improvement (Lubar et al., 1995b), forming the theoretical basis for directly intervening with NF and intentionally modulating brain activity. Moving farther afield of ER, a newer ADHD EEG approach involves NF of slow cortical potentials (SCP; Strehl et al., 2006). SCP are the depolarization of large cortical cell assemblies and reflect internal or external events. Quantitative EEG (Arns et al., 2012) has made progress in assessing how different presentations of ADHD will respond to different EEG approaches, leading to the promise of far more precision in the application of EEG approaches to ADHD treatment. While these gains in ADHD treatment have not yet translated to ER interventions, there is reason to be hopeful.

Blood Oxygen Level–Dependent (BOLD) Imaging: Real-Time Functional Magnetic Resonance Imaging (RTfMRI) and Near-Infrared Spectroscopy (NIRS)

The working of the brain consumes energy, which is transported by oxygen (HBO_2; Attwell and Laughlin, 2001). Real-time functional magnetic resonance imaging (RTfMRI; Weiskopf, 2012) and near-infrared spectroscopy (NIRS) are tools that can enable a real-time window into the transport and movement of oxygen throughout the brain, providing feedback that shows specific regions of the brain being used. While the blood oxygen signal that both approaches image is the same, the underlying technologies are different. A magnetic resonance imaging (MRI) machine uses a large magnet that interacts with the iron in blood and can image the resultant magnetic field. NIRS illuminates the blood with infrared light, which is differentially reflected and refracted based upon the amount of oxygen within the blood. MRI offers an ability to look at the entire brain structure, while NIRS is limited to looking at blood oxygen at the surface.

Unlike EEG, blood oxygen level–dependent (BOLD) imaging offers an ability to provide far greater spatial resolution (even, in the case of NIRS, when that spatial resolution is limited to the surface). The tradeoff comes in cost and speed. MRI relies on large, expensive magnets that are typically confined to specialized providers. The clinical utility of RTfMRI for NF will rely on the device becoming more accessible (Gorgolewski et al., 2017). NIRS is still an emerging technology. The underlying technology of NIRS is simpler, and smaller than MRI, giving the technology the potential to become far more widely available as an intervention. However, right now NIRS-based interventions do not have the same widespread access that EEG-based interventions (Safaie et al., 2013), and there is less data of clinical interventions using NIRS compared to EEG.

BOLD imaging can localize effects within a region of interest (ROI), accounting for idiosyncratic differences within each participant. Typically, before engaging in BOLD NF, the practitioner will isolate the NF procedure to hone in on specific ROI through eliciting neural activity through stimulus before beginning the feedback. The location of the activity then becomes the focus for the NF (Ochsner et al., 2012).

EEG relies on electrical signals, which can be measured on the order of milliseconds after they are produced. BOLD imaging does not have as high temporal resolution as EEG. Both BOLD approaches rely on the physical transport of blood, which is a mechanical system within the body. By nature, a mechanical system will operate slower than an electrochemical system. MRI adds even more latency. The mechanical and processing limitations of typical real-time MRI leads to refresh rates on the order of 1 Hz (Zotev et al., 2011), though the theoretical maximum latency is on the order of 100 Hz (Uecker et al., 2010). The resultant signal is thousands of times slower than EEG, as the time from neurological signal to representation to the patient is seconds delayed. Previous work on real-time representations from the education field have shown the effects to be significantly degraded when the representation is delayed from action, even if by only seconds (Brasell, 1987). This provides a serious potential limitation to RTfMRI approaches to NF.

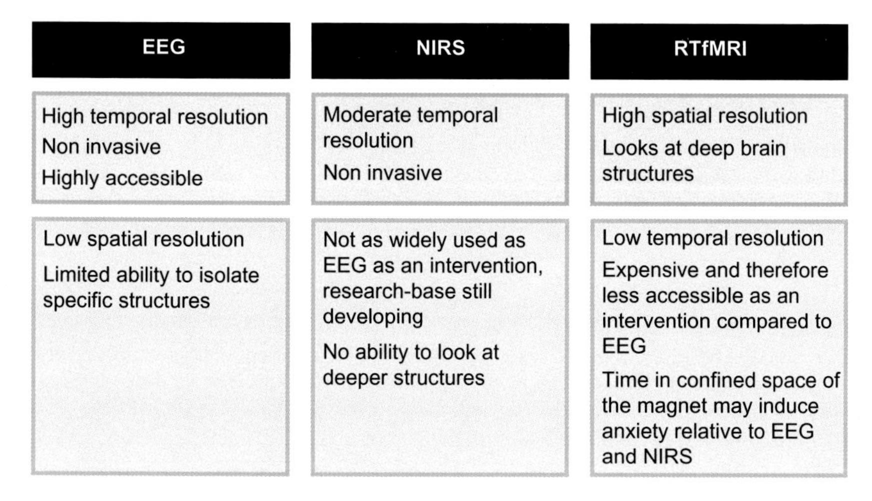

FIGURE 12.2 Summary of major technologies used in NF for emotional regulation.

On a final note, obtaining an MRI signal in itself can be a relatively invasive and anxiety inducing event. Subjects must lie down in a small cylinder, which houses the magnet. The magnet itself produces loud noises in its operation. Most subjects quickly get over any anxiety that comes with the procedure, though these considerations should still be acknowledged (Murphy and Brunberg, 1997) (Fig. 12.2).

NF-BASED THERAPEUTIC APPROACHES FOR THE TREATMENT OF EMOTIONAL DYSREGULATION

In this section, we present a current review of recent literature of approaches to treating symptoms of emotional dysregulation. The articles cited here have been published in the past 10 years and include methodological efforts to control for placebo effects or assess transfer post NF. The goal of this chapter is to provide insight on NF in pediatric populations. However, because the field is still emerging, we present the literature with a wider lens, including the work with adult populations. The relatively small number of studies the small number of studies, particularly randomized controlled trials (RCT), show the field's nascence, and the research presented here in some ways reminiscent of early cartography, with researchers actively drawing early maps of how NF can change the brain and ultimately behaviors.

APPROACHES FOCUSING ON SYMPTOMS OF EMOTIONAL REGULATION

<table>
<tr><th>Author(s) and year</th><th>Region of interest</th><th>Technology</th><th>Study design</th></tr>
<tr><td>Hudak et al. (2017)</td><td>dlPFC activation</td><td>NIRS</td><td>RCT

$N = 20$</td></tr>
<tr><td colspan="4">Intervention design: NIRS NF where participants controlled overhead lighting in a VR classroom. The NF condition, overhead lighting in the VR classroom was controlled with average amplitude of HBO_2 within the bilateral dlPFC over a 5 second window. In the control condition, lighting in the VR classroom was controlled by muscle function electromyography (EMG) collected at the shoulder

Participants: 20 healthy adult (mean age = 23.4 years) participants randomized into intervention or control conditions

Key results: Behaviorally, there were group effects for the stop signal task, a reaction time measure, a trend for group effects in the go/no-go task, a measure of inhibitory control, and no effects in the n-back task, a measure of working memory. Biologically, the experimental group saw higher concentrations of HBO_2 after training during the go/no-go task, but no group effects in other tasks</td></tr>
<tr><td>Zotev et al. (2011)</td><td>Amygdala activation</td><td>RTfMRI</td><td>RCT

$N = 28$</td></tr>
<tr><td colspan="4">Intervention design: RTfMRI NF where participants recalled happy memories and received basic graphical feedback on activation in the ROI. In the intervention condition, NF reflected activation of the left amygdala. A sham-controlled condition gave NF from a region of the brain unassociated with ER (the intraparietal sulcus was used)

Participants: Healthy male adults (median 28 years) randomized into intervention and control conditions

Key results: NF of upregulation of the left amygdala leads to increased control over the left amygdala during training runs. Effects of this intentional control are further reflected in increased functional connectivity with the right medial frontal polar cortex, bilateral dorsomedial prefrontal cortex, left anterior cingulate cortex, and bilateral superior frontal gyrus. This study did not directly address behavioral or clinical impact, showing that purposeful upregulation of networks associated with emotional regulation</td></tr>
<tr><td>Hamilton et al. (2011)</td><td>ACC activation</td><td>RTfMRI</td><td>Open-label

$N = 17$</td></tr>
<tr><td colspan="4">Intervention design: RTfMRI biofeedback of the ACC. Participants asked to downregulate the ACC via increasing positive mood and received graphical feedback. Participants received either NF or sham NF with "yoked" data</td></tr>
</table>

(*Continued*)

(Continued)

Author(s) and year	Region of interest	Technology	Study design
Participants: 17 adult females between 18 and 50 years (mean age 29.8). The first eight participants received NF while the remainder received sham NF. The entire experimental group was conducted before the control group			
Key results: NF led to participants being able to downregulate ACC; however, this did not transfer when NF was removed			
Paret et al. (2016)	Amygdala	RTfMRI	RCT $N = 32$
Intervention design: RTfMRI NF of amygdala functioning. Participants were asked to downregulate their amygdala in response to aversive picture over one session. The control group received NF from a different region			
Participants: 22 healthy adult females (mean age 24.56 years) split into two groups			
Key results: Group effects were observed for connection between the amygdala and vmPFC; however, these effects did not persist into a transfer run			
Zotev et al. (2013)	rACC, left amygdala	RTfMRI	$N = 28$
Intervention design: RTfMRI NF of the amygdala over one single-day session consisting of eight runs. Participants were given graphical feedback on BOLD activity in the left amygdala and directed to recall happy memories. Control participants received sham NF from a region not associated with emotional regulation			
Participants: 28 healthy male volunteers (median age: 28 years) randomized into intervention on sham-controlled groups			
Key results: The methods of the study allowed the researchers to show group effects on not just the ROI but also networks modulating the left amygdala. The rACC modulated the amygdala and PFC structures			
Sulzer et al. (2013)	SN/VTA	RTfMRI	$N = 32$
Intervention design: RTfMRI NF of the SN/VTA as a dopaminergic region of the brain. An intervention group was asked to upregulate the region (feedback proportional to BOLD in SN/VTA) while a second group was asked to do the opposite (NF *inversely* proportional to BOLD in SN/VTA). Participants were told they could use any cognitive strategy they liked (this excludes strategies like breathing), but seeded with ideas to produce happy memories			
Subjects: 22 healthy male subjects, between 24 and 35 years (mean not reported)			
Key results: Group effects were observed for the proportional feedback group being able to upregulate compared to the inverse group. However, when controlling for baseline, these effects did not transfer. Qualitative feedback on strategies used revealed substantial differences: the proportional group relied on romantic or sexual imagery while the inverse group relied on memories of time with community (family, friends, sports, travel) or personal success			

(*Continued*)

(Continued)

Author(s) and year	Region of interest	Technology	Study design
Johnston et al. (2010)	Amygdala, insula	RTfMRI	$N = 13$
Intervention: RTfMRI NF based on an ROI determined during a calibration task. Participants were shown images, and an individualized ROI was determined based upon the regions with the highest BOLD response. The participations regulated graphical feedback in the individualized ROI while being shown pictures in a single session			
Subjects: Thirteen healthy adults in a single group (mean age not reported, range 21–52)			
Key results: Participants were not given strategies, and as a result reported developing trying multiple strategies. Participants improved at the NF task over repeated runs. They showed an ability to regulate and improved positive and negative affect after NF			
Zotev et al. (2014)	Amygdala	RTfMRI and EEG	$N = 6$
Intervention design: Combined RTfMRI and EEG NF, with the MRI NF using an ROI of the left amygdala and the EEG NF giving feedback on the high-beta band. Participants asked to upregulate both the MRI and EEG "knobs" being simultaneously given graphical feedback on the two, which were normalized to the same scale. Participants were also given a target; they were asked to be recall happy memories, count, or rest			
Participants: Six healthy adults, mean age 24 years old			
Key results: The technologies were successfully combined, and the authors found simultaneous upregulation of the amygdala and high-beta symmetry that led to a possible conclusion that the technologies are not redundant. NF led to successful regulation but were not maintained in a transfer run			
Brühl et al. (2014)	Amygdala	RTfMRI	$N = 6$
Intervention design: A pilot study during which participants were asked to downregulate the amygdala upon viewing negative faces by using RTfMRI color-based NF. Four training sessions of NF included one "regulate" condition and one "view" condition			
Participants: Six healthy adults (mean age 26 years old) completed the experimental condition			
Key results: Participants reported using cognitive and attentional strategies to regulate. Downregulation improved significantly over the four training sessions. No functional or transfer behaviors measured			
Kadosh et al. (2016)	Bilateral insula, amydala, left insula, mid-cingulate cortex, supplementary motor area, inferior parietal lobe	RTfMRI	$N = 17$ (7–16 year)
Intervention design: RTfMRI NF in children and adolescents of right insula region. Participants completed an upregulation task during which strategies for regulation during NF were provided, and a rest condition that included NF without instructions for regulation strategies. Examined bottom-up and top-down Granger information flow between the different brain regions in the emotion regulation network between tasks			

(Continued)

(Continued)

Author(s) and year	Region of interest	Technology	Study design
Participants: A total of 17 healthy children and adolescents (ages 7–16 years, mean 11.6) received the experimental condition			
Key results: No functional differences were observed, but NF training was successful in facilitating insula activation and increasing bottom-up driven Granger information flow in the emotion regulation network. The study also found that increased bottom-up information flow between the amygdala and insula was associated with success in NF during the upregulation trials			
Caria et al. (2010)	Anterior insula	RTfMRI	$N = 27$
Intervention design: RTfMRI NF of upregulation of the anterior insula, with four training sessions in a single day. Participants were guided to use emotional imagery to move a thermometer that changed with BOLD activity, though they were free to use other strategies that would also move the thermometer. Control conditions included nonspecific BOLD response over a large brain area and emotional imagery without NF			
Participants: 27 participants split into three groups, nine in each group. Age and demographic information was not given			
Key results: NF led to participants being able to increase activity in anterior insula during transfer. Participants who showed greater reaction to negative imagery were also able to show better ability to regulate the anterior insula. No functional outcomes were measured			
Marxen et al. (2016)	Amygdala	RTfMRI	$N = 32$
Intervention design: RTfMRI NF training of the bilateral amygdala to determine if participants can learn to regulate the functional magnetic resonance imaging (fMRI) signal without instruction. Participants completed 4 days of NF training that included two sessions of continuous feedback and one session of End-of-Block feedback, and a fourth day of no feedback and continuous feedback			
Participants: A total of 32 healthy adults (15 female, 18–40 years old, mean 24.7) were recruited to complete all sessions			
Key results: Subjects were able to manipulate the BOLD signal from the bilateral amygdala without instruction, and there was a transfer effect during the last day with no feedback. There was a stronger learning effect over training sessions with End-of-Block feedback than sessions with continuous feedback. However, the study did not include a pre-intervention baseline and therefore could demonstrate an effect of increased self-regulation. No functional or behavioral outcomes were observed			

dlPFC, Dorsolateral PFC; *NIRS*, Near-infrared spectroscopy; *VR*, Virtual reality; *ER*, Emotional regulation; *ACC*, Anterior cingulate cortex; *RTfMRI*, Real-time functional magnetic resonance imaging; *rACC*, Rostral anterior cingulate cortex; BOLD, Blood oxygen level–dependent; ROI, Region of interest; *PFC*, Prefrontal cortex; *SN/VTA*, Substantia nigra/ventral tegmental area complex; *EEG*, Electroencephalography.

These studies consistently rely on prior work that establishes ER pathways and then work to establish NF-induced purposeful control of these pathways. As a group, these studies do not then tell us the behavioral implications of modulation of these circuits. They are very much a first step, validating core technologies, and establishing the potential for NF to induce change. These studies do not typically try to establish that this control leads to behavioral changes

ER NF technologies have emerged over the past 10 years. Building on the models that have enabled ADHD NF, ER NF has primarily focused on specific neural structures related to the ER. Zotev et al. (2010, 2013), Paret et al. (2016), Johnston et al. (2010), and Brühl et al. (2014) have focused on amygdala activation. The Zotev studies focused on amygdala downregulation, where Paret focused on upregulation of the amygdala. In both studies, participants were asked to regulate and give feedback while observing the same set pictures calibrated for affect. The results follow a typical pattern for NF studies. During the NF training, the participants quickly learn to up- or downregulate (respectively) the ROI. The NF is then withdrawn but demonstrated with fMRI the participants continue the regulation even without the feedback as new pictures are presented. This change projects into the rest of the ER network, raising hope that continued NF can lead to change.

A remaining puzzle is how up- versus downregulation impacts both regulation networks and informs future treatment direction. Several studies (Zotev et al., 2011, 2013, 2014; Johnston et al., 2010; Marxen et al., 2016) asked participants to upregulate amygdala, while others (Brühl et al., 2014; Hamilton et al., 2011; Paret et al., 2016) asked participants to downregulate the amygdala. Yet, the outcomes of these studies seem remarkably similar, as if intentional control in either direction provides some amount of effect.

Many researchers have focused NF on regions associated with reappraisal. Hamilton et al. (2011) looked at downregulation of ACC activation in a picture task, given the ACC's central role in ER networks. Participants were able to regulate ACC activation during NF, though this ability to regulate did not persist in the task once the NF was withdrawn. Sulzer et al. (2013) similarly looked at the mid brain, similarly found ability to regulate BOLD response during NF and likewise did not see transfer once the NF was withdrawn. Several studies focused on NF on insular activation (Kadosh et al., 2016; Caria et al., 2010; Johnston et al., 2010), finding that individuals can train both up- and downregulation of the insula with training. Kadosh et al. (2016) trained children and adolescents to upregulate the insula, providing evidence that the same plasticity observed in adults is seen in children. The same study offered multiple training sessions, through the NF tasks varied by session.

NF Leads to Changes in How the Brain Approaches Cognitive Tasks

In total, the work on ER NF to date paints a clear picture that the approach can target specific regions of the brain and produce changes in brain function that last after the NF is withdrawn. This is an exciting first step, demonstrating that how our brains can change and that effortful cognitive control can incite the brain to modulate activity in different regions and networks.

A Reliance on RTfMRI

A clear pattern in the research to date is the use of BOLD imaging for ER NF, particularly RTfMRI. Because the neurology of ER depends upon deep-seated structures and can be idiosyncratically developed even within specific structures of the brain, this approach makes sense. RTfMRI allows direct imaging of the amygdala and ACC, which would not otherwise be possible with EEG. However, the approach is not without limitations. Beyond the accessibility concerns, there are questions about the implications of amygdala and ACC regulation. Paret et al. (2016) and Zotev et al. (2013) getting similar results while differentially asking for up- versus downregulation leads to ongoing puzzles about what exactly these types of up or downregulations mean for behavior. More work is needed, especially with NF applied over multiple sessions. This will provide more insight into how up or downregulation can affect cognitive performance, symptoms, and behaviors with any clinical utility.

Single-Session NF

Perhaps because of the cost and difficulty associated with MRI-based techniques, the NF done to date focuses on single-session NF. This has established that NF typically leads to biological changes when the NF is withdrawn, at least at the end of the session, demonstrating that the structures of the brain are plastic. It also opens the door to further questions, including what happens when more sessions are added? How long do the effects persist? And, what clinical or therapeutic benefits can we expect from both single and ultimately longer NF treatments?

Questions About Networks

Much of the research highlighted shows that networks themselves act differently after NF—the changes are not confined to ROIs. However, none of the NF research thus far focuses on networks. It will be exciting as RTfMRI technology advances to provide real-time NF on functional connectivity and observe how this NF leads to changes in the brain and eventually in behaviors. While the technology for real-time functional connectivity is still on the horizon (Monti et al., 2017), its emergence will open up new directions.

Focus on Adults

The work to date has focused nearly exclusively on ER in young adults, while the symptoms of dysregulation emerge in childhood (Bradley et al., 2011; Gratz and Roemer, 2004). This is not surprising given the early state of ER NF, with both the major technology (RTfMRI) and underlying conceptualizations of brain plasticity still emerging. Future research will need to extend the findings thus far to pediatric population.

NF FOR IMPULSIVE BEHAVIORS ASSOCIATED WITH SPECIFIC MENTAL HEALTH DISORDERS

One current direction of mental health is to look away from diagnostic categories in favor of looking at specific symptoms and then exploring the underlying psychology and biology of those symptoms (Insel et al., 2010). Emotional dysregulation is present in a number of disorders and, as discussed previously, seems to have the same neurobiological makeup across these disorders. However, work still continues on specific disorders. It is helpful to note other disorders where NF has emerged as a potential therapeutic effect.

Author and year	Primary disorder	Technology	Study design
Fuchs et al. (2003)	ADHD	EEG (SMR/low beta ratio)	Open-label NF versus stimulants
Intervention design: EEG NF for was differentially applied based on ADHD subtype. Children received three training sessions per week over a 12-week period. SMR NF was used for children of the hyperactive-impulsive type, while beta1 NF was used for children with inattentive type			

(*Continued*)

(Continued)

Author and year	Primary disorder	Technology	Study design
Participants: Pediatric participants diagnosed with ADHD. 22 children opted into NF treatment, and 12 children opted into methylphenidate treatment. The mean age was 9.7 years and all but one of the children were male			
Key findings: Both NF and methylphenidate led to symptomatic improvements. No group effects for impulsivity were found between NF and methylphenidate. No group effects were seen for behavior. Results were limited by nonrandomized group assignment and lack of blinding			
Schmidt and Martin. (2016)	Binge rating	EEG (suppress beta)	Waitlist controlled RCT
Intervention design: Participants randomized into either EEG NF, mental imagery, or waitlist control groups. In 10-session NF, participants were asked to reduce high-beta activity shown through a bar graph			
Participants: 75 adult women, mean age 44 years			
Key findings: Group effects were observed showing differences between NF and waitlist for improvement in symptoms, and mental imagery and waitlist for improvement in symptoms. A larger effect was observed for the NF group			
Bluschke et al. (2016)	ADHD	EEG (theta/beta)	Waitlist controlled RCT
Intervention design: Biweekly NF sessions over a period of 8 weeks (16 sessions total). Participants were asked to downregulate theta while upregulating beta and given graphical feedback. Changes in inhibitory performance were measured via a go/no-go task. Participants were randomized into either an intervention or waitlist control group			
Participants: A total of 19 patients (18 male, mean age 11.2 years) participated in NF, while 17 patients (all male, mean age 11.3 years) were recruited to be part of the waiting list control group			
Key findings: NF improved inhibitory control but not perceptual gating, attentional selection, and resource allocation processes. Impulsive behavior was reduced in NF participants. The study is notable for its measuring of both impulsive behavioral changes (via go/no-go task) and the use of event related potential (ERPs) during the task to understand what processes were being trained in the NF (inhibitory performance versus perceptual gating, attention, and resource allocation)			
Linden et al. (2012)	Depression	RTfMRI ventral prefrontal cortex (vPFC)	Open label
Intervention: Participants were asked to upregulate the PFC via RTfMRI NF over four sessions in the period of 1 month. A localizer method was used to isolate the ROI as one that reacted with affective pictures. A control group looked at affective pictures while using regulation strategies but did not receive NF			

(Continued)

(Continued)

Author and year	Primary disorder	Technology	Study design
Participants: A waitlist-controlled open-label intervention where eight subjects (mean age 48 years) with MDD were recruited to participate in NF. Subsequently, a control group was identified **Key findings:** Group effects were seen on symptomatic measures for depression and positive and negative affect. In pre–post analysis, the NF group showed increased ability to upregulate the ROI. The study is limited, however, by lack of a strong control group, which did not receive sham NF			
Scheinost et al. (2013)	Contamination anxiety	RTfMRI orbitofrontal cortex (OFC)	Sham RCT
Intervention design: A sham-controlled RCT with RTfMRI of the OFC. Participants completed four scanning sessions over a 3-week period. Participants were asked to upregulate the OFC via graphical feedback while viewing emotional images **Participants:** 20 adult participants (mean age and age range not reported) with high contamination–related anxiety were randomized into active or sham groups **Key findings:** Group effects were observed for connectivity within the brain. Group effects were seen for symptomatic measures of contamination anxiety measured several days after NF. OFC activity in the absence of NF, pre- and post-session was measured, with a trend suggesting changed activity post-NF			
Nicholson et al. (2017)	Post-traumatic stress disorder (PTSD)	RTfMRI (downregulate amygdala)	Open label
Intervention design: PTSD patients completed three sessions of RTfMRI NF to downregulate the amygdala and to determine if NF could shift activity to prefrontal emotion regions (dlPFC and ventro-lateral prefrontal cortex (vlPFC)). During the "regulate" condition, participants were instructed to downregulate the ROI while viewing a personalized trauma word. The "view" condition required participants to view the trauma word without attempting to regulate the amygdala, and the "neutral" condition presented a neutral word while the participants were instructed to view it without attempting to regulate. This training run was presented three consecutive times, followed by a transfer run without the visual feedback. Participants then rated their ability to regulate their "emotion center" **Participants:** 10 participants (mean age 49.6) completed three sessions of NF. **Key findings:** Participants successfully downregulated the amygdala during the "regulate" condition when presented with a personalized trauma word, and the transfer run demonstrated a transfer effect. Downregulating the amygdala during NF and the transfer run was associated with dlPFC and vlPFC activation. PFC, rACC, and insula activation negatively correlated with PTSD dissociative symptoms during the transfer run. No functional or behavioral outcomes were measured			

ADHD, Attention deficit hyperactivity disorder; *EEG*, Electroencephalography; *SMR*, Sensorimotor rhythm; *RTfMRI*, Real-time functional magnetic resonance imaging; *PFC*, Prefrontal cortex; *MDD*, Major depressive disorder; *ROI*, Region of interest; *dlPFC*, Dorsolateral PFC; *rACC*, Rostral anterior cingulate cortex.

Attention Deficit Hyperactivity Disorder

Impulsivity is a significant symptom for many children with ADHD (Coghill et al., 2014), which typically presents with symptoms related to children's inability to reign in disruptive behaviors. NF is a growing modality of treatment for ADHD (Gevensleben et al., 2014; Holtmann et al., 2014). Unlike studies of other disorders, the studies of ADHD also can specifically point to the potential for NF for children, giving reason to suspect that NF can be an effective treatment modality. Readers interested in thorough reviews of NF for ADHD fill find extensive and high-quality reviews from Cortese et al. (2016) and Micoulaud-Franchi et al. (2014). These reviews point to a general approach relying on EEG NF specifically theta/beta and SMR band NF. The reviews specifically challenge the field to improve in methods, with few studies embracing sham controls while showing notable and lasting effects for NF.

While studies of NF for ADHD are well captured in previous reviews, relevant to ER is the common symptom of impulsive behavior. Blushke et al. (2016) specifically pointed to the potential for NF to impact impulsivity-related structures within the brain. Other research points to how interventions can improve efficacy. For example, Fuchs et al. (2003) specifically isolated SMR/low beta ratio for children with high hyperactivity, on the hypothesis that the SMR band reflects inhibition of cortical-limbic activity (Sterman et al., 1969). The group used a separate NF paradigm, beta/theta, for children with the inattentive subtype ADHD.

Anxiety

Éismont et al. (2008) showed that teenagers with anxiety have lower SMR, beta1, and alpha power compared to less anxious teenagers, demonstrating that there are measurable differences in neural functioning in children with symptoms of anxiety. Scheinost et al. (2013) provided RTfMRI NF to adults with contamination anxiety, a common symptom of obsessive-compulsive disorder (OCD). Not only did the NF lead to measurable connectivity changes, but improved symptoms measured several days after NF. Participants received psychosocial support to help identify strategies, and the NF explicitly served to reinforce those top-down strategies. This design is important and unusual in NF studies, demonstrating that the effects of the training can last for at least a moderate amount of time post-NF.

Other Disorders

EEG NF has shown promise for reducing binge eating behavior compared to waitlist controls—subjects who participated in three NF sessions experienced reduced symptoms and connectivity changes that persisted for several days (Schmidt and Martin, 2016). Fielenbach et al. (2017) proposes a waitlist-controlled RCT of SMR EEG NF looking at impulsivity and substance abuse disorders, and should produce informative results once data becomes available. Depression, while likely having a different epigenesis than disorders that heavily implicate regulation circuitry, has seen promising preliminary work targeting the vlPFC (Linden et al., 2012). This is relevant as it targets the same goal-selection and language mechanisms seen in ER NF.

DISCUSSION

ER is a vital skill that develops across the lifetime. In healthy development, ER skills confer adaptive advantages that enable personal achievement and social participation. It is vital to nurture emotional competencies, including ER, from an early age. From developmental, educational, and clinical perspectives, ER deserves the same attention and care that traditional cognitive skills (like language or numeracy) receive. Clinically, emotional dysregulation symptomatically presents in many common pediatric mental health disorders, including ADHD and anxiety. Given the commonality and cross-cutting nature of dysregulation symptomatology, the diagnoses themselves may well be different behavioral projects of the same neurological pathology.

The neurological picture around emotional dysregulation has been developing since Phineas Gage had an unfortunate run-in with a railroad spike. Keyed around frontal-limbic connectivity, the "reptilian brain" in the amygdala provides initial responses to stimuli and drives early behaviors through the fight-or-flight reflex. The higher order "mammalian brain" provides a brake through top-down control where impulsive signals are regulated and alternative strategies are evaluated and potentially pursued. These skills grow throughout the lifespan, beginning with dendritic growth in infancy and culminating with myelination in very late adolescence.

NF is an emerging therapeutic tool for a range of mental health disorders. In a world where access to mental health professionals can be difficult, the promise of consistent, repeatable therapy provides hope not just for improved efficacy but also for improved access. With over 20 years of research in NF for ADHD, the field is still in its infancy in conducting well-designed and powered clinical trials. While the current state of therapeutic ER NF leaves a lot of room for growth, there's reason to continue. Because the neural mechanisms of ER are well understood, there is hope that specific targeting of these regions will allow for meaningful therapeutic growth for patients.

What we see in research now are the early efforts to enable this. In sum, the current literature points to a brain that changes with NF. These changes are meaningful shifts in BOLD response in the brain, sometimes lasting even as the NF is withdrawn. In analogous disorders, the effects of NF training seem to provide lasting effects, over both short and long term.

There are still limitations to the NF approach as well, which need to be addressed in future research. First is the theoretical orientation around reappraisal. Reappraisal is a tool for regulation that depends on strong PFC connectivity, which is vulnerable to "disconnection" due to chemical cascading during times of acute stress. Reappraisal is a top-down approach of regulation, and like other top-down approaches, it will fail if the higher order brain structures are not physically capable of providing the needed control. Future work must explore NF to enable bottom-up ER, either by focusing on muting the initial response or strengthening the frontal-limbic network to the point where minimal energy is needed to access its tools.

Methodologically, future studies will continue to improve. Current studies have predominantly focused on adults, though ER disorders typically develop in childhood. By the time adulthood sets in, pathways and behaviors are so ingrained that much more intervention is required to affect change. Early interventions are vital, and work in pediatric populations will no doubt begin to arise.

The other major methodological concerns are the use of control groups, in particular sham controls, and the need for a growing focus on symptomatic and functional outcomes. The current state of research reflects the nascent nature of ER research, where researchers are working to better understand how different regions of the brain can be impacted with NF. More robust control groups will allow for more detailed statements about the efficacy of NF. NF interventions are fairly obvious to participants, leading to the potential for placebo effects. Symptomatic and functional measures will allow us to make connections for how these interventions can improve patients' quality of life, which is the ultimate goal of any intervention. Building a deeper understanding of brain structures will undoubtedly pay dividends as we craft interventions. Ultimately, the shared goal is to use these interventions to improve quality of life for patients and globally find ways to promote healthy development. Researchers should strive to collect measures that allow us to tie impacts of interventions to behaviors, and the long-term impacts of those interventions

Technologically, real-time fMRI in its current form does not hold great promise for accessibility of NF. NIRS alleviates this concern, though studies of NIRS NF for ER are few. Future improvements in MRI technology may alleviate this concern as well. Because of the desire to target deep and specific brain structures, EEG has not become a major tool for ER NF, proving a strong contrast from pediatric ADHD.

Taken together, we are witnessing the very beginning of NF for ER. Given the primacy of ER in healthy development, and the therapeutic potential to strengthen specific neural structures and networks, we are at the beginning of a very promising technology.

References

Andrade, B.F., Waschbusch, D.A., Doucet, A., King, S., McGrath, P.J., Stewart, S.H., et al., 2012. Social information processing of positive and negative hypothetical events in children with ADHD and conduct problems and controls. J. Atten. Disord. 16 (6), 491–504.

Applehans, B.M., Luecken, L.J., 2006. Attentional processes, anxiety, and the regulation of cortisol reactivity. Anxiety Stress Coping 19 (1), 81–92.

Arns, M., de Ridder, S., Strehl, U., Breteler, M., Coenen, A., 2009. Efficacy of neurofeedback treatment in ADHD: the effects on inattention, impulsivity and hyperactivity: a meta-analysis. Clin. EEG Neurosci. 40 (3), 180–189. Available from: https://doi.org/10.1177/155005940904000311.

Arns, M., Drinkenburg, W., Kenemans, J.L., 2012. The effects of QEEG-informed neurofeedback in ADHD: an open-label pilot study. Appl. Psychophysiol. Biofeedback 37 (3), 171–180.

Arns, M., Conners, C.K., Kraemer, H.C., 2013. A decade of EEG theta/beta ratio research in ADHD: a meta-analysis. J. Atten. Disord. 17 (5), 374–383.

Arnsten, A.F.T., 2009. Stress signalling pathways that impair prefrontal cortex structure and function. Nat. Rev. Neurosci. 10 (6), 410–422. Available from: https://doi.org/10.1038/nrn2648.

Arnsten, A.F.T., 2011. Catecholamine influences on dorsal prefrontal cortical networks. Biol. Psychiatry 69 (12), e89–e99. Available from: https://doi.org/10.1016/j.biopsych.2011.01.027.

Aron, E.N., Aron, A., 1997. Sensory-processing sensitivity and its relation to introversion and emotionality. J. Pers. Soc. Psychol. 73 (2), 345–368.

Attwell, D., Laughlin, S.B., 2001. An energy budget for signaling in the grey matter of the brain. J. Cereb. Blood Flow Metab. 21 (10), 1133–1145.

Banich, M.T., 2009. Executive function: the search for an integrated account. Curr. Dir. Psychol. Sci. 18 (2), 89–94. Available from: https://doi.org/10.1111/j.1467-8721.2009.01615.x.

Banks, S.J., Eddy, K.T., Angstadt, M., Nathan, P.J., Luan Phan, K., 2007. Amygdala-frontal connectivity during emotion regulation. Soc. Cogn. Affect. Neurosci. 2 (4), 303–312. Available from: https://doi.org/10.1093/scan/nsm029.

Beesdo, K., Knappe, S., Pine, D.S., 2009. Anxiety and anxiety disorders in children and adolescents: developmental issues and implications for DSM-V. Psychiatr. Clin. North Am. 32 (3), 483–524.

Berking, M., Margraf, M., Ebert, D., Wupperman, P., Hofmann, S.G., Junghanns, K., 2011. Deficits in emotion-regulation skills predict alcohol use during and after cognitive–behavioral therapy for alcohol dependence. J. Consult. Clin. Psychol. 79 (3), 307–318. Available from: https://doi.org/10.1037/a0023421.

Berntson, G.G., Cacioppo, J.T., Quigley, K.S., 1993. Respiratory sinus arrhythmia: autonomic origins, physiological mechanisms, and psychophysiological implications. Psychophysiology 30 (2), 183–196.

Bianchi, V., Brambilla, P., Garzitto, M., Colombo, P., Fornasari, L., Bellina, M., et al., 2017. Latent classes of emotional and behavioural problems in epidemiological and referred samples and their relations to DSM-IV diagnoses. Eur. Child Adolesc. Psychiatry 26 (5), 549–557. Available from: https://doi.org/10.1007/s00787-016-0918-2.

Bissonette, G.B., Powell, E.M., Roesch, M.R., 2013. Neural structures underlying set-shifting: roles of medial prefrontal cortex and anterior cingulate cortex. Behav. Brain Res. 250, 91–101. Available from: https://doi.org/10.1016/j.bbr.2013.04.037.

Bjureberg, J., Ljotsson, B., Tull, M.T., Hedman, E., Sahlin, H., Lundh, L., et al., 2016. Development and validation of a brief version of the difficulties in emotion regulation scale: the DERS-16. J. Psychopathol. Behav. Assess. 2, 284. Available from: https://doi.org/10.1007/s10862-015-9514-x.

Blakemore, S.J., Choudhury, S., 2006. Development of the adolescent brain: implications for executive function and social cognition. J. Child Psychol. Psychiatry 47 (3-4), 296–312.

Bluschke, A., Broschwitz, F., Kohl, S., Roessner, V., Beste, C., 2016. The neuronal mechanisms underlying improvement of impulsivity in ADHD by theta/beta neurofeedback. Sci. Rep. 6, 31178. Available from: https://doi.org/10.1038/srep31178.

Bradley, B., Westen, D., Mercer, K.B., Binder, E.B., Jovanovic, T., Crain, D., et al., 2011. Association between childhood maltreatment and adult emotional dysregulation in a low-income, urban, African American sample: moderation by oxytocin receptor gene. Dev. Psychopathol. 23 (2), 439–452. Available from: https://doi.org/10.1017/S0954579411000162.

Brans, K., Koval, P., Verduyn, P., Lim, Y.L., Kuppens, P., 2013. The regulation of negative and positive affect in daily life. Emotion 13 (5), 926–939. Available from: https://doi.org/10.1037/a0032400.

Brasell, H., 1987. The effect of real-time laboratory graphing on learning graphic representations of distance and velocity. J. Res. Sci. Teach. 24 (4), 385–395.

Brotman, M., Schmajuk, M., Rich, B., Dickstein, D.P., Guyer, A.E., Costello, E.J., et al., 2006. Prevalence, clinical correlates, and longitudinal course of severe mood dysregulation in children. Biol. Psychiatry 60 (9), 991–997. Available from: https://doi.org/10.1016/j.biopsych.2006.08.042.

Brühl, A.B., Scherpiet, S., Sulzer, J., Stämpfli, P., Seifritz, E., Herwig, U., 2014. Real-time neurofeedback using functional MRI could improve down-regulation of amygdala activity during emotional stimulation: a proof-of-concept study. Brain Topogr. 27 (1), 138–148. Available from: https://doi.org/10.1007/s10548-013-0331-9.

Bush, G., Luu, P., Posner, M., 2000. Cognitive and emotional influences in anterior cingulate cortex. Trends Cogn. Sci. 4 (6), 215–222. Available from: https://doi.org/10.1016/S1364-6613(00)01483-2.

Campbell, S.B., Shaw, D.S., Gilliom, M., 2000. Early externalizing behavior problems: toddlers and preschoolers at risk for later maladjustmentDev. Psychopathol. 12 (3), 467–488 Retrieved from . Available from: http://www.ncbi.nlm.nih.gov/pubmed/11014748.

Caria, A., Sitaram, R., Veit, R., Begliomini, C., Birbaumer, N., 2010. Volitional control of anterior insula activity modulates the response to aversive stimuli. A real-time functional magnetic resonance imaging study. Biol. Psychiatry 68 (5), 425–432. Available from: https://doi.org/10.1016/j.biopsych.2010.04.020.

Clarke, A.R., Barry, R.J., Dupuy, F.E., Heckel, L.D., McCarthy, R., Selikowitz, M., et al., 2011. Behavioural differences between EEG-defined subgroups of children with attention-deficit/hyperactivity disorder. Clin. Neurophysiol. 22 (7), 1333–1341.

Coccaro, E.F., McCloskey, M.S., Fitzgerald, D.A., Phan, K.L., 2007. Amygdala and orbitofrontal reactivity to social threat in individuals with impulsive aggression. Biol. Psychiatry 62 (2), 168–178. Available from: https://doi.org/10.1016/j.biopsych.2006.08.024.

Coghill, D., Seth, S., Matthews, K., 2014. A comprehensive assessment of memory, delay aversion, timing, inhibition, decision making and variability in attention deficit hyperactivity disorder: advancing beyond the three-pathway models. Psychol. Med. 44 (9), 1989–2001.

Correll, C.U., Carlson, H.E., 2006. Endocrine and metabolic adverse effects of psychotropic medications in children and adolescents. J. Am. Acad. Child Adolesc. Psychiatry 45 (7), 771–791. Available from: https://doi.org/10.1097/01.chi.0000220851.94392.30.

Correll, C.U., Manu, P., Olshanskiy, V., Napolitano, B., Kane, J.M., Malhotra, A.K., 2009. Cardiometabolic risk of second-generation antipsychotic medications during first-time use in children and adolescents. JAMA 302, 1765–1773.

Cortese, S., Ferrin, M., Brandeis, D., Holtmann, M., Aggensteiner, P., Daley, D., et al., 2016. Neurofeedback for attention-deficit/hyperactivity disorder: meta-analysis of clinical and neuropsychological outcomes from randomized controlled trials. J. Am. Acad. Child Adolesc. Psychiatry 55 (6), 444–455. Available from: https://doi.org/10.1016/j.jaac.2016.03.007.

Cramer, S.C., Sur, M., Dobkin, B.H., O'Brien, C., Sanger, T.D., Trojanowski, J.Q., et al., 2011. Harnessing neuroplasticity for clinical applications. Brain 134 (Pt 6), 1591–1609. Available from: https://doi.org/10.1093/brain/awr039.

Crick, N.R., Dodge, K.A., 1994. A review and reformulation of social information-processing mechanisms in children's social adjustment. Psychol. Bull. 115 (1), 74.

Davidson, R.J., 2000. Dysfunction in the neural circuitry of emotion regulation—a possible prelude to violence. Science 289 (5479), 591–594. Available from: https://doi.org/10.1126/science.289.5479.591.

Davidson, R.J., Jackson, D.C., Kalin, N.H., 2000. Emotion, plasticity, context, and regulation: perspectives from affective neuroscience. Psychol. Bull. 126 (6), 890.

Delgado, M.R., Nearing, K.I., LeDoux, J.E., Phelps, E.A., 2008. Neural circuitry underlying the regulation of conditioned fear and its relation to extinction. Neuron 59 (5), 829–838. Available from: https://doi.org/10.1016/j.neuron.2008.06.029.

Denham, S.A., 2007. Dealing with feelings: How children negotiate the worlds of emotions and social relationships. Cogn. Brain, Behav. 11 (1), 1.

Dennis, T.A., Chen, C., 2007. Emotional face processing and attention performance in three domains: neurophysiological mechanisms and moderating effects of trait anxiety. Int. J. Psychophysiol. 65 (1), 10–19. Available from: https://doi.org/10.1016/j.ijpsycho.2007.02.006.

Diamond, A., 2002. Normal development of prefrontal cortex from birth to young adulthood: cognitive functions, anatomy, and biochemistry. Principles of Frontal Lobe Function. Oxford University Press, pp. 466–503. Available from: http://doi.org/10.1093/acprof:oso/9780195134971.003.0029.

Diamond, A., Barnett, W.S., Thomas, J., Munro, S., 2007. Preschool program improves cognitive control. Science (New York, NY) 318 (5855), 1387.

Diamond, A., 2013. Executive functions. Annu. Rev. Psychol. 64 (1), 135–168. Available from: https://doi.org/10.1146/annurev-psych-113011-143750.

Dunn, W., Myles, B.S., Orr, S., 2002. Sensory processing issues associated with Asperger syndrome: a preliminary investigation. Am. J. Occup. Ther. 56 (1), 97–102. Available from: https://doi.org/10.5014/ajot.56.1.97.

Egner, T., Gruzelier, J.H., 2004. EEG biofeedback of low beta band components: frequency-specific effects on variables of attention and event-related brain potentials. Clin. Neurophysiol. 115, 131–139.

Eisenberg, N., Fabes, R.A., Guthrie, I.K., Reiser, M., 2000. Dispositional emotionality and regulation: their role in predicting quality of social functioningJ. Pers. Soc. Psychol. 78 (1), 136–157Retrieved from . Available from: http://www.ncbi.nlm.nih.gov/pubmed/10653511.

Éismont, E.V., Aliyeva, T.A., Lutsyuk, N.V., Pavlenko, V.B., 2008. EEG correlates of different types of anxiety in 14- to 15-year-old teenagers. Neurophysiology 40 (5–6), 377–384. Available from: https://doi.org/10.1007/s11062-009-9063-6.

Elkins, R.M., McHugh, R.K., Santucci, L.C., Barlow, D.H., 2011. Improving the transportability of CBT for internalizing disorders in children. Clin. Child Fam. Psychol. Rev. 14 (2), 161–173.

Etkin, A., Büchel, C., Gross, J.J., 2015. The neural bases of emotion regulation. Nat. Rev. Neurosci. 16 (11), 693–700. Available from: https://doi.org/10.1038/nrn4044.

Ferrer, E., Helm, J.L., 2013. Dynamical systems modeling of physiological coregulation in dyadic interactions. Int. J. Psychophysiol. 88, 296–308. Available from: https://doi.org/10.1016/j.ijpsycho.2012.10.013.

Fielenbach, S., Donkers, F.C., Spreen, M., Bogaerts, S., 2017. Neurofeedback as a treatment for impulsivity in a forensic psychiatric population with substance use disorder: study protocol of a randomized controlled trial combined with an N-of-1 clinical trial. JMIR Res. Protoc. 6 (1), e13. Available from: https://doi.org/10.2196/resprot.6907.

Fuchs, T., Birbaumer, N., Lutzenberger, W., Gruzelier, J.H., Kaiser, J., 2003. Neurofeedback treatment for attention-deficit/hyperactivity disorder in children: a comparison with methylphenidate. Appl. Psychophysiol. Biofeedback 28 (1), 1–12. Available from: https://doi.org/10.1023/A:1022353731579.

Fuster, J.M., 2002. Frontal lobe and cognitive development. J. Neurocytol. 31 (3-5), 373–385.

Garon, N., Bryson, S.E., Smith, I.M., 2008. Executive function in preschoolers: a review using an integrative framework. Psychol. Bull. 134 (1), 31–60. Available from: https://doi.org/10.1037/0033-2909.134.1.31.

Gee, D.G., Humphreys, K.L., Flannery, J., Goff, B., Telzer, E.H., Shapiro, M., et al., 2013. A developmental shift from positive to negative connectivity in human amygdala–prefrontal circuitry. J. Neurosci. 33 (10), 4584–4593.

Gevensleben, H., Moll, G.H., Rothenberger, A., Heinrich, H., 2014. Neurofeedback in attention-deficit/hyperactivity disorder—different models, different ways of application. Front. Hum. Neurosci. 8, 846. Available from: https://doi.org/10.3389/fnhum.2014.00846.

Goel, V., Dolan, R.J., 2003. Reciprocal neural response within lateral and ventral medial prefrontal cortex during hot and cold reasoning. NeuroImage 20 (4), 2314–2321. Available from: https://doi.org/10.1016/j.neuroimage.2003.07.027.

Goldin, P.R., McRae, K., Ramel, W., Gross, J.J., 2008. The neural bases of emotion regulation: reappraisal and suppression of negative emotion. Biol. Psychiatry 63 (6), 577–586.

Gorgolewski, K.J., Alfaro-Almagro, F., Auer, T., Bellec, P., Capotă, M., Chakravarty, M.M., et al., 2017. BIDS apps: improving ease of use, accessibility, and reproducibility of neuroimaging data analysis methods. PLoS Comput. Biol. 13 (3), e1005209.

Gospic, K., Mohlin, E., Fransson, P., Petrovic, P., Johannesson, M., Ingvar, M., 2011. Limbic justice—amygdala involvement in immediate rejection in the Ultimatum Game. PLoS Biol. 9 (5), e1001054. Available from: https://doi.org/10.1371/journal.pbio.1001054.

Grant, K., McMeekin, E., Jamieson, R., Fairfull, A., Miller, C., White, J., 2012. Individual therapy attrition rates in a low-intensity service: a comparison of cognitive behavioural and person-centred therapies and the impact of deprivation. Behav. Cogn. Psychother. 40 (2), 245–249.

Gratz, K.L., Roemer, L., 2004. Multidimensional assessment of emotion regulation: development, factor structure, and initial validation of the Difficulties in Emotion Regulation Scale. J. Psychopathol. Behav. Assess. 26 (1), 41–54. Available from: https://doi.org/10.1007/s10862-008-9102-4.

Gross, J., 2002. Emotion regulation (reappraisal and suppression): affective, cognitive, and social consequences. Psychophysiology 39 (3), 281–291. Available from: https://doi.org/10.1017/S0048577201393198.

Gross, J.J., 1998. Antecedent- and response-focused emotion regulation: divergent consequences for experience, expression, and physiology. J. Pers. Soc. Psychol. 74, 234–237.

Gruzelier, J., 2009. A theory of alpha/theta neurofeedback, creative performance enhancement, long distance functional connectivity and psychological integration. Cogn. Process. 10, 101–109.

Gruzelier, J., 2014. Differential effects on mood of 12–15 (SMR) and 15–18 (beta1) Hz neurofeedback. Int. J. Psychophysiol. 93 (1), 112–115.

Gullone, Eleonora, et al., 2010. The normative development of emotion regulation strategy use in children and adolescents: a 2-year follow-up study. J. Child Psychol. Psychiatry 51 (5), 567–574.

Hamilton, J.P., Glover, G.H., Hsu, J.J., Johnson, R.F., Gotlib, I.H., 2011. Modulation of subgenual anterior cingulate cortex activity with real-time neurofeedback. Hum. Brain Mapp. 32 (1), 22–31.

Hammond, D.C., 2007. What is neurofeedback? J. Neurotherapy 10 (4), 25–36.

Hanslmayer, S., Sauseng, P., Doppelmayr, M., Schabus, M., Klimesch, W., 2006. Increasing individual upper alpha by neurofeedback improves cognitive performance in human subjects. Appl. Psychophysiol. Biofeedback 30, 1–10.

Herman, K.C., Borden, L.A., Reinke, W.M., Webster-Stratton, C., 2011. The impact of the Incredible years parent, child, and teacher training programs on children's co-occurring internalizing symptoms. Sch. Psychol. Q. 26 (3), 189.

Herschell, A.D., Kolko, D.J., Baumann, B.L., Davis, A.C., 2010. The role of therapist training in the implementation of psychosocial treatments: a review and critique with recommendations. Clin. Psychol. Rev. 30 (4), 448–466. Available from: https://doi.org/10.1016/j.cpr.2010.02.005.

Hinshaw, S.P., 1992. Externalizing behavior problems and academic underachievement in childhood and adolescence: causal relationships and underlying mechanisms. Psychol. Bull. 111 (1), 127–155. Available from: https://doi.org/10.1037//0033-2909.111.1.127.

Holtmann, Sonuga-Barke, Cortese, Brandeis, 2014. Neurofeedback for ADHD: a review of current evidence. Child Adolesc. Psychiatr. Clin. N. Am. 23 (4), 789–806.

Hudak, J., Blume, F., Dresler, T., Haeussinger, F.B., Renner, T.J., Fallgatter, A.J., et al., 2017. Near-infrared spectroscopy-based frontal lobe neurofeedback integrated in virtual reality modulates brain and behavior in highly impulsive adults. Front. Hum. Neurosci. 11, 1–13. Available from: https://doi.org/10.3389/fnhum.2017.00425 (September).

Insel, T., Cuthbert, B., Garvey, M., Heinssen, R., Pine, D.S., Quinn, K., et al., 2010. Research domain criteria (RDoC): toward a new classification framework for research on mental disorders. Am. J. Psychiatry 167 (7), 748–751.

Janak, P.H., Tye, K.M., 2015. From circuits to behaviour in the amygdala. Nature 517 (7534), 284–292. Available from: https://doi.org/10.1038/nature14188.

Johnston, S.J., Boehm, S.G., Healy, D., Goebel, R., Linden, D.E.J., 2010. Neurofeedback: a promising tool for the self-regulation of emotion networks. NeuroImage 49 (1), 1066–1072. Available from: https://doi.org/10.1016/j.neuroimage.2009.07.056.

Johnstone, J., 2008. A three-stage neuropsychological model of neurofeedback: historical perspectives. Biofeedback 36 (4), 142–147.

Johnstone, T., van Reekum, C.M., Urry, H.L., Kalin, N.H., Davidson, R.J., 2007. Failure to regulate: counterproductive recruitment of top-down prefrontal-subcortical circuitry in major depression. J. Neurosci. 27 (33), 8877–8884.

Kadosh, K.C., Luo, Q., de Burca, C., Sokunbi, M.O., Feng, J., Linden, D.E., et al., 2016. Using real-time fMRI to influence effective connectivity in the developing emotion regulation network. NeuroImage 125, 616–626.

Keizer, A.W., Verschoor, M., Verment, R.S., Hommel, B., 2010a. The effect of gamma enhancing neurofeedback on the control of feature bindings and intelligence measures. Int. J. Psychophysiol. 75, 25–32.

Keizer, Verment, Hommel, 2010b. Enhancing cognitive control through neurofeedback: a role of gamma-band activity in managing episodic retrieval. NeuroImage 49 (4), 3404–3413.

Kim, M.J., Gee, D.G., Loucks, R.A., Davis, F.C., Whalen, P.J., 2011a. Anxiety dissociates dorsal and ventral medial prefrontal cortex functional connectivity with the amygdala at rest. Cereb. Cortex 21 (7), 1667–1673. Available from: https://doi.org/10.1093/cercor/bhq237.

Kim, M.J., Loucks, R.A., Palmer, A.L., Brown, A.C., Solomon, K.M., Marchante, A.N., et al., 2011b. The structural and functional connectivity of the amygdala: from normal emotion to pathological anxiety. Behav. Brain Res. 223 (2), 403–410. Available from: https://doi.org/10.1016/j.bbr.2011.04.025.

Kim-Spoon, J., Cicchetti, D., Rogosch, F.A., 2013. A longitudinal study of emotion regulation, emotion lability-negativity, and internalizing symptomatology in maltreated and nonmaltreated children. Child Dev. 84 (2), 512–527.

Knox, M., Lentini, J., Cummings, T.S., McGrady, A., Whearty, K., Sancrant, L., 2011. Game-based biofeedback for paediatric anxiety and depression. Ment. Health Fam. Med. 8 (3), 195–203.

Koenderink, M.T., Uylings, H.B.M., Mrzljak, L., 1994. Postnatal maturation of the layer III pyramidal neurons in the human prefrontal cortex: a quantitative Golgi analysis. Brain Res. 653 (1), 173–182.

Kohn, N., Eickhoff, S.B., Scheller, M., Laird, A.R., Fox, P.T., Habel, U., 2014. NeuroImage neural network of cognitive emotion regulation—an ALE meta-analysis and MACM analysis. NeuroImage 87, 345–355. Available from: https://doi.org/10.1016/j.neuroimage.2013.11.001.

Krain, A.L., Castellanos, F.X., 2006. Brain development and ADHD. Clin. Psychol. Rev. 26 (4), 433–444.

Larkin, P., Jahoda, A., MacMahon, K., 2013. The social information processing model as a framework for explaining frequent aggression in adults with mild to moderate intellectual disabilities: a systematic review of the evidence. J. Appl. Res. Intellect. Disabil. 26 (5), 447–465. Available from: https://doi.org/10.1111/jar.12031.

Lévesque, J., Beauregard, M., Mensour, B., 2006. Effect of neurofeedback training on the neural substrates of selective attention in children with attention-deficit/ hyperactivity disorder: a functional magnetic resonance imaging study. Neurosci. Lett. 394 (3), 216–221.

Linden, D.E.J., Habes, I., Johnston, S.J., Linden, S., Tatineni, R., Subramanian, L., et al., 2012. Real-time self-regulation of emotion networks in patients with depression. PLoS ONE 7 (6). Available from: https://doi.org/10.1371/journal.pone.0038115.

Lubar, J.F., 1991. Discourse on the development of EEG diagnostics and biofeedback for attention-deficit/hyperactivity disorders. Biofeedback Self Regul. 16 (3), 201–225.

Lubar, J.F., Swartwood, M.O., Swartwood, J.N., O'Donnell, P.H., 1995a. Evaluation of the effectiveness of EEG neurofeedback training for ADHD in a clinical setting as measured by changes in T.O.V.A. scores, behavioral ratings, and WISC-R performance. Biofeedback Self Regul. 20 (1), 83–99. Available from: https://doi.org/10.1007/BF01712768.

Lubar, J.F., Swartwood, M.O., Swartwood, J.N., Timmermann, D.L., 1995b. Quantitative EEG and auditory event-related potentials in the evaluation of attention-deficit/hyperactivity disorder: effects of methylphenidate and implications for neurofeedback training. J. Psychoeducational Assess. 34, 143–160.

Marxen, M.J., Jacob, M.K., Müller, D.N., Hellrung, L., Riedel, P., Epple, R., et al., 2016. Amygdala regulation following fMRI-neurofeedback without instructed strategies. Front. Hum. Neurosci. 10, 1–14.

Masterpasqua, F., Healey, K.N., 2003. Neurofeedback in psychological practice. Prof. Psychol.: Res. Pract. 34 (6), 652.

Metcalfe, J., Mischel, W., 1999. A hot/cool-system analysis of delay of gratification: dynamics of willpower. Psychol. Rev. 106 (1), 3.

Micoulaud Franchi, J.-A., Geoffroy, P.A., Efond, G., Elopez, R., Ebioulac, S., Ephilip, P., 2014. EEG Neurofeedback treatments in children with ADHD: an updated meta-analysis of randomized controlled trials. Front. Hum. Neurosci. 13 (8), 906.

Miller, A.L., Gouley, K.K., Seifer, R., Dickstein, S., Shields, A., 2004. Emotions and behaviors in the head start classroom: associations among observed dysregulation, social competence, and preschool adjustment. Early Educ. Dev. 15 (2), 147–165. Available from: https://doi.org/10.1207/s15566935eed1502_2.

Miller, A.L., Rathus, J.H., Linehan, M., 2007. Dialectical Behavior Therapy With Suicidal Adolescents. Guilford Press, New York, NY, ©2007.

Monk, C.S., Telzer, E.H., Mogg, K., Bradley, B.P., Mai, X., Louro, H.M.C., et al., 2008. Amygdala and ventrolateral prefrontal cortex activation to masked angry faces in children and adolescents with generalized anxiety disorder. Arch. Gen. Psychiatry 65 (5), 568. Available from: https://doi.org/10.1001/archpsyc.65.5.568.

Monti, R.P., Lorenz, R., Braga, R.M., Anagnostopoulos, C., Leech, R., Montana, G., 2017. Real-time estimation of dynamic functional connectivity networks. Hum. Brain Mapp. 38 (1), 202–220.

Murphy, K.J., Brunberg, J.A., 1997. Adult claustrophobia, anxiety and sedation in MRI. Magn. Reson. Imaging 15 (1), 51–54.

Nicholson, A.A., Rabellino, D., Densmore, M., Frewen, P.A., Paret, C., Kluetsch, R., et al., 2017. The neurobiology of emotion regulation in posttraumatic stress disorder: amygdala downregulation via real-time fMRI neurofeedback. Hum. Brain Mapp. 38 (1), 541–560. Available from: https://doi.org/10.1002/hbm.23402.

Niv, S., 2013. Clinical efficacy and potential mechanisms of neurofeedback. Pers. Individ. Differences . Available from: https://doi.org/10.1016/j.paid.2012.11.037.

Nocavo, R.W., 2002. Anger control therapy. In: Hersen, M. (Ed.), Encyclopedia of Psychotherapy. University of South Florida, Tampa, FL, pp. 41–48.

Ochsner, K.N., Gross, J.J., 2005. The cognitive control of emotionTrends Cogn. Sci. 9 (5), 242–249 Retrieved from Available from: http://www.ncbi.nlm.nih.gov/pubmed/15866151.

Ochsner, K.N., Bunge, S.A., Gross, J.J., Gabrieli, J.D.E., 2002. Rethinking feelings: an fMRI study of the cognitive regulation of emotion. J. Cogn. Neurosci. 14 (8), 1215–1229. Available from: https://doi.org/10.1162/089892902760807212.

Ochsner, K.N., Silvers, J.A., Buhle, J.T., 2012. Functional imaging studies of emotion regulation: a synthetic review and evolving model of the cognitive control of emotion. Ann. N. Y. Acad. Sci. 1251 (1), E1–E24. Available from: https://doi.org/10.1111/j.1749-6632.2012.06751.x.

Olfson, M., Druss, B.G., Marcus, S.C., 2015. Trends in mental health care among children and adolescents. N. Engl. J. Med. 372 (21), 2029–2038. Available from: https://doi.org/10.1056/NEJMsa1413512.

Olson, E.A., Collins, P.F., Hooper, C.J., Muetzel, R., Lim, K.O., Luciana, M., 2009. White matter integrity predicts delay discounting behavior in 9- to 23-year-olds: a diffusion tensor imaging study. J. Cogn. Neurosci. 21 (7), 1406–1421.

Oram, R., Ryan, J., Rogers, M., Heath, N., 2017. Emotion regulation and academic perceptions in adolescence. Emotional Behav. Difficulties 22 (2), 162–173. Available from: https://doi.org/10.1080/13632752.2017.1290896.

Oswald, D.P., Mazefsky, C.A., 2006. Empirically supported psychotherapy interventions for internalizing disorders. Psychol. Sch. 43 (4), 439–449.

Panksepp, J., 2004. Affective Neuroscience: The Foundations of Human and Animal Emotions. Oxford University Press, New York.

Paret, C., Ruf, M., Gerchen, M.F., Kluetsch, R., Demirakca, T., Jungkunz, M., et al., 2016. FMRI neurofeedback of amygdala response to aversive stimuli enhances prefrontal-limbic brain connectivity. NeuroImage. Available from: https://doi.org/10.1016/j.neuroimage.2015.10.027.

Phelps, E.A., Delgado, M.R., Nearing, K.I., Ledoux, J.E., 2004. Extinction learning in humans: role of the amygdala and vmPFC. Neuron 43 (6), 897–905. Available from: https://doi.org/10.1016/j.neuron.2004.08.042.

Phillips, M.L., Ladouceur, C.D., Drevets, W.C., 2008. A neural model of voluntary and automatic emotion regulation: implications for understanding the pathophysiology and neurodevelopment of bipolar disorder. Mol. Psychiatry 13 (9), 833–857. Available from: https://doi.org/10.1038/mp.2008.65.

Piaget, J., 1954. The Construction of Reality in the Child. Basic Books, New York, NY.

Pina, A., Silverman, W., Weems, C., Kurtines, W., Goldman, M., Peterson, Lizette, et al., 2003. A comparison of completers and noncompleters of exposure-based cognitive and behavioral treatment for phobic and anxiety disorders in youth. J. Consult. Clin. Psychol. 71 (4), 701–705.

Porges, S.W., 1995. Orienting in a defensive world. Psychophysiology 32 (4), 301–318.

Porges, S.W., 2001. The polyvagal theory: phylogenetic substrates of a social nervous system. Int. J. Psychophysiol. 42, 123–146. Available from: https://doi.org/10.1016/S0167-8760(01)00162-3.

Rawana, J.S., Flett, G.L., McPhie, M.L., Nguyen, H.T., Norwood, S.J., 2014. Developmental trends in emotion regulation: A systematic review with implications for community mental health. Can. J. Commun. Ment. Health 33 (1), 31–44.

Rogoff, B., Mistry, J., Göncü, A., Mosier, C., Chavajay, P., Heath, S.B., et al., 1993. Guided participation in cultural activity by toddlers and caregivers. Monogr. Soc. Res. Child Dev. 58 (8), . Available from: https://doi.org/10.2307/1166109i.

Roy, A.K., Fudge, J.L., Kelly, C., Perry, J.S.A., Daniele, T., Carlisi, C., et al., 2013. Intrinsic functional connectivity of amygdala-based networks in adolescent generalized anxiety disorder. J. Am. Acad. Child Adolesc. Psychiatry 52 (3), 290–299. Available from: https://doi.org/10.1016/j.jaac.2012.12.010. e2.

Safaie, J., Grebe, R., Moghaddam, H.A., Wallois, F., 2013. Toward a fully integrated wireless wearable EEG-NIRS bimodal acquisition system. J. Neural. Eng. 10 (5), 056001.

Satpute, Lieberman, 2006. Integrating automatic and controlled processes into neurocognitive models of social cognition. Brain Res. 1079 (1), 86–97.

Scarpa, A., 2015. Physiological arousal and its dysregulation in child maladjustment. Curr. Dir. Psychol. Sci. 24 (5), 345–351. Available from: https://doi.org/10.1177/0963721415588920.

Schäfer, J.Ö., Naumann, E., Holmes, E.A., Tuschen-Caffier, B., Samson, A.C., 2017. Emotion regulation strategies in depressive and anxiety symptoms in youth: a meta-analytic review. J. Youth Adolescence 46 (2), 261–276. Available from: https://doi.org/10.1007/s10964-016-0585-0.

Schatz, D.B., Rostain, A.L., 2006. ADHD with comorbid anxiety: a review of the current literature. J. Atten. Disord. 10 (2), 141–149.

Scheinost, D., Stoica, T., Saksa, J., Papademetris, X., Constable, R.T., Pittenger, C., et al., 2013. Orbitofrontal cortex neurofeedback produces lasting changes in contamination anxiety and resting-state connectivity. Transl. Psychiatry 3, e250. Available from: https://doi.org/10.1038/tp.2013.24.

Schmidt, J., Martin, A., 2016. Neurofeedback against binge eating: a randomized controlled trial in a female subclinical threshold sample. Eur. Eat. Disord. Rev. 24 (5), 406–416. Available from: https://doi.org/10.1002/erv.2453.

Schoenberg, P.L.A., David, A.S., 2014. Biofeedback for psychiatric disorders: a systematic review. Appl. Psychophysiol. Biofeedback 39 (2), 109–135. Available from: https://doi.org/10.1007/s10484-014-9246-9.

Schultz, D., Shaw, D.S., 2003. Boys' maladaptive social information processing, family emotional climate, and pathways to early conduct problems. Soc. Dev. 12 (3), 440. Available from: https://doi.org/10.1111/1467-9507.00242.

Shweder, R. a, Haidt, J., Horton, R., Joseph, C., 2008. The cultural psychology of the emotions: ancient and renewed. Handb. Emotions . Available from: https://doi.org/10.2307/2076468.

Sitaram, R., Ros, T., Stoeckel, L., Haller, S., Scharnowski, F., Lewis-Peacock, J., et al., 2017. Closed-loop brain training: the science of neurofeedback. Nat. Rev. Neurosci. 18 (2), 86–100. Available from: https://doi.org/10.1038/nrn.2016.164.

Smith, L.B., Thelen, E., Titzer, R., McLin, D., 1999. Knowing in the context of acting: the task dynamics of the A-not-B error. Psychol. Rev. 106 (2), 235–260. Available from: https://doi.org/10.1037/0033-295X.106.2.235.

Sterman, M.B., 1996. Physiological origins and functional correlates of EEG rhythmic activities: implications for self-regulation. Biofeedback Self Regul. 21, 3–33. Available from: https://doi.org/10.1007/BF02214147.

Sterman, M.B., Wyrwicka, W., Howe, R., 1969. Behavioral and neurophysiological studies of the sensorimotor rhythm in the cat. Electroencephalogr. Clin. Neurophysiol. 27, 678–679.

Stormont, M., 2002. Externalizing behavior problems in young children: contributing factors and early intervention. Psychol. Sch. 39 (2), 127–138. Available from: https://doi.org/10.1002/pits.10025.

Strehl, U., Leins, U., Goth, G., Klinger, C., Hinterberger, T., Birbaumer, N., 2006. Self regulation of slow cortical potentials: a new treatment for children with attention deficit/hyperactivity disorder. Pediatrics 118, e1530–e1540.

Studer, P.H., Kratz, O., Moll, G., Heinrich, H., Gevensleben, H., Rothenberger, A., et al., 2014. Slow cortical potential and theta/beta neurofeedback training in adults: effects on attentional processes and motor system excitability. Front. Hum. Neurosci. 8, (July).

Sulzer, J., Sitaram, R., Blefari, M.L., Kollias, S., Birbaumer, N., Stephan, K., et al., 2013. Neurofeedback-mediated self-regulation of the dopaminergic midbrain. NeuroImage 83, 817–825.

Teplan, M., 2002. Fundamentals of EEG measurement. Meas. Sci. Rev. 2 (2), 1–11. Available from: https://doi.org/10.1021/pr070350l.

Thomas, J., 2016. Psychotherapy techniques for somatization: cognitive-behavioral therapy, hypnosis, and biofeedback. J. Am. Acad. Child Adolesc. Psychiatry 55 (10), S34–S35.

Uecker, M., Zhang, S., Frahm, J., 2010. Nonlinear inverse reconstruction for real-time MRI of the human heart using undersampled radial FLASH. Magn. Reson. Med. 63 (6), 1456–1462.

Vohs, K.D., Heatherton, T.F., 2000. Self-regulatory failure: A resource-depletion approach. Psychol. Sci. 11 (3), 249–254.

Vygotsky, L.S., 1980. Mind in Society: The Development of Higher Psychological Processes. Harvard University Press, Cambridge, MA.

Weiskopf, N., 2012. Real-time fMRI and its application to neurofeedback. NeuroImage 62 (2), 682–692.

Weissman, M.M., Wolk, S., Wickramaratne, P., Goldstein, R.B., Adams, P., Greenwald, S., et al., 1999. Children with prepubertal-onset major depressive disorder and anxiety grown up. Arch. Gen. Psychiatry 56 (9), 794–801.

Wenck, L.S., Leu, P.W., D'Amato, R.C., 1996. Evaluating the efficacy of a biofeedback intervention to reduce children's anxiety. J. Clin. Psychol. 52 (4), 469–473. Available from: http://doi.org/10.1002/(SICI)1097-4679(199607)52:4 < 469::AID-JCLP13 > 3.0.CO;2-E.

Wheeler Maedgen, J., Carlson, C.L., 2000. Social functioning and emotional regulation in the attention deficit hyperactivity disorder subtypes. J. Clin. Child Psychol. 29 (1), 30–42.

Winstanley, C.A., 2004. Contrasting roles of basolateral amygdala and orbitofrontal cortex in impulsive choice. J. Neurosci. 24 (20), 4718–4722. Available from: https://doi.org/10.1523/JNEUROSCI.5606-03.2004.

Yang, Y., Raine, A., 2009. Prefrontal structural and functional brain imaging findings in antisocial, violent, and psychopathic individuals: a meta-analysis. Psychiatry Res.: Neuroimaging 174 (2), 81–88.

Zeman, J., Cassano, M., Perry-Parrish, C., Stegall, S., 2006. Emotion regulation in children and adolescents. J. Dev. Behav. Pediatr. 27 (2), 155–168.

Zotev, V., Krueger, F., Phillips, R., Alvarez, R.P., Simmons, W.K., Bellgowan, P., et al., 2011. Self-regulation of amygdala activation using real-time FMRI neurofeedback. PLoS ONE 6 (9). Available from: https://doi.org/10.1371/journal.pone.0024522.

Zotev, V., Phillips, R., Young, K.D., Drevets, W.C., Bodurka, J., 2013. Prefrontal control of the amygdala during real-time fMRI neurofeedback training of emotion regulation. PLoS ONE 8 (11), e79184. Available from: https://doi.org/10.1371/journal.pone.0079184.

Zotev, V., Phillips, R., Yuan, H., Misaki, M., Bodurka, J., 2014. Self-regulation of human brain activity using simultaneous real-time fMRI and EEG neurofeedback. NeuroImage. Available from: https://doi.org/10.1016/j.neuroimage.2013.04.126.

Further Reading

Ekman, P., 2009. Darwin's contributions to our understanding of emotional expressions. Philos. Trans. R. Soc. Lond. B: Biol. Sci. (1535), 3449. Available from: https://doi.org/10.1098/rstb.2009.0189.

Kahneman, D., 2011. Thinking, Fast and Slow. Farrar, Straus and Giroux, New York, NY.

Lazarus, R., Folkman, S., 1984. Stress, Appraisal, and Coping. Springer Pub, New York, NY.

Linehan, M., 1993. Cognitive-Behavioral Treatment of Borderline Personality Disorder. Guilford Press, New York, NY.

Thompson, R.A., 1994. Emotion regulation: a theme in search of definition. Monogr. Soc. Res. Child Dev. 59, 25–52.

Thompson, R.A., 2001. Development in the first years of life. Future Child. 11 (1), 21–33. Available from: https://doi.org/10.2307/1602807.

Wamboldt, M.Z., Wamboldt, F.S., 2000. Role of the family in the onset and outcome of childhood disorders: selected research findings. J. Am. Acad. Child Adolesc. Psychiatry 39 (10), 1212–1219. Available from: https://doi.org/10.1097/00004583-200010000-00006.

Ziv, Y., Hadad, B.S., Khateeb, Y., 2014. Social information processing in preschool children diagnosed with autism spectrum disorder. J. Autism Dev. Disord. 44 (4), 846. Available from: https://doi.org/10.1007/s10803-013-1935-3.

CHAPTER

13

Chronotherapy for Adolescent Major Depression

Inken Kirschbaum-Lesch, Martin Holtmann and Tanja Legenbauer

LWL-University Hospital for Child and Adolescent Psychiatry and Psychotherapy Hamm, Ruhr University Bochum, Germany

OUTLINE

Neurotechnological Pediatric Neuropsych
DOI: https://doi.org/10.1016/B978-0-12-812777-3.00013-1

DEPRESSION IN CHILDREN AND ADOLESCENTS

Epidemiology and Clinical Picture

Depressive disorders are among the most prominent health problems in youth. Approximately 11% of children and adolescents are affected (Ravens-Sieberer et al., 2007), with a lifetime prevalence rate of 20% by late adolescence (Lewinsohn et al., 1998). In addition, around 25% of adolescents are affected by subthreshold depression, that is, depressive symptoms that fall short of the diagnostic criteria for major depression (Klein et al., 2009). The prevalence of depression is twice as high in girls as it is in boys, and youth with a high number of psychosocial risks show a higher symptom load (KiGGS survey, Bettge et al., 2008). Once affected, the risk of recurrence of adolescent depression within 5 years is as high as 40% (Lewinsohn et al., 1994). In the long term, childhood depression constitutes a key risk factor for drug abuse, suicidal behavior, and relapses in adulthood (Measelle et al., 2006; Patton et al., 2014).

Clinical depression in adults is marked by a depressed mood, loss of interest in almost all activities (anhedonia), fatigue or loss of energy, impaired concentration, feelings of worthlessness or guilt, insomnia or hypersomnia, suicidal ideation, and loss of appetite or increased appetite (Diagnostic and Statistical Manual of Mental Disorders, Text Revision: American Psychiatric Association, 2000). Childhood depression, by contrast, is characterized—in addition to marked sadness and social isolation—by rather somatic symptoms like sleep problems and loss of appetite (Mehler-Wex and Kölch, 2008). In adolescents, clinical depression is characterized by social isolation, anhedonia, and fear for the future, and additionally by increased irritability, low frustration tolerance, aggressive behavior, and sleep problems (Mehler-Wex and Kölch, 2008). Insomnia, defined as difficulties in initiating/maintaining sleep and/or nonrestorative sleep accompanied by decreased daytime functioning (Luca, Luca and Calandra, 2013), is a particularly common feature of depression in adolescents (Chellappa et al., 2009).

Of currently depressed adolescents, 75% report comorbid sleep problems including later sleep onset and offset as well as low sleep efficiency (Ivanenko and Johnson, 2008; Robillard et al., 2015). Recent evidence emphasizes the role of comorbid sleep disturbances in adolescent depression: the presence of sleep problems seems to be associated with a stronger impairment in psychosocial domains and a more serious clinical picture with marked depressive symptoms and a longer duration of the depressive episode (O'Brien et al., 2011; Sunderajan et al., 2010).

STATE-OF-THE-ART TREATMENTS FOR DEPRESSED CHILDREN AND ADOLESCENTS

The treatment of depression in children and adolescents is multimodal and includes psychotherapy (mostly cognitive-behavioral and interpersonal approaches), psychosocial approaches, and medication (Mehler-Wex and Kölch, 2008). Despite extensive research efforts, treatment outcomes for depression in children and adolescents remain unsatisfactory (Emslie et al., 2012). Treatment with selective serotonin reuptake inhibitors in depressed youth may worsen existing sleep problems (Winokur et al., 2001) and thus impair full recovery from the illness (Emslie et al., 2012; Luca et al., 2013). Moreover, residual symptoms of fatigue, mood and sleep disturbances, and concentration difficulties often still exist after treatment with antidepressants (Kennard et al., 2006).

Despite available treatments, remission rates for moderate-to-severe depression in youth are below 40% after 3 and 6 months of combined medication and psychotherapy, even in carefully designed gold standard studies (Kennard et al., 2006; Emslie et al., 2010). In a 5-year follow-up study, Curry et al. (2011) found that 42.9% of the partial and full responders suffered from a recurrence. The German clinical practice guideline for the treatment of depression in adolescents concluded that due to "the glaring lack of clinical studies (...) there is a pressing need for intervention research" (Dolle and Schulte-Körner, 2013). Given that depression in childhood and adolescence constitutes a key risk factor for drug abuse, suicidal behavior, and relapses in adulthood (Measelle et al., 2006; Patton et al., 2014), further research on alternative and especially additional treatment forms is urgently needed. Consequently, in the last decade, there has been a focus on chronotherapeutic approaches as an additive and noninvasive treatment component.

INNOVATIVE TREATMENT APPROACHES: CHRONOTHERAPEUTICS

Chronotherapeutic treatments such as bright light therapy (BLT), wake therapy (WT), and sleep phase advance (SPA) are noninvasive and beneficial treatments with high compliance and almost no side effects (Moscovici and Kotler, 2009). In particular, BLT has been emphasized both as mono- and adjunctive therapy for depression in adults and consequently, BLT has been included in several international recommendations and guidelines (e.g., American Psychiatric Association, 2000; Bauer et al., 2013; Ravindran et al., 2016). However, there is also uncertainty concerning the efficacy of BLT on nonseasonal depression: both the British NICE recommendations (National Collaborating Centre for Mental Health, 2009) and the World Federation of Societies of Biological Psychiatry are more hesitant with their recommendations (Bauer et al., 2013). The German clinical practice guideline for the treatment of unipolar depression in adults emphasizes the application of WT as an add-on therapy to psychological and pharmacological approaches in order to achieve fast responses and increase standard treatment effects (DGPPN et al., 2012), whereas the combination of WT with SPA, BLT, or antidepressants is suggested in order to sustain its rapid effects. Importantly, there is only preliminary evidence for children and adolescents and most studies do not include children younger than 12 years. Hence, due to lack of empirical evidence, the application of BLT and other chronotherapeutic approaches is not recommended for children and adolescents in any guidelines. However, preliminary findings of studies applying BLT to adolescents are promising. In Box 13.1, a short overview about the main chronotherapeutic approaches is given. Further details for the application in youth are given in the second part of the chapter.

The Circadian Rhythm

Circadian rhythm describes the natural sleep–wake rhythm throughout the night (darkness) and day (light) that regulates our life. Two interacting processes control the circadian rhythm: the circadian process, which is also defined as process C ("Two process model" by Borbely, 1982) and the homeostatic process (process S). Process C comprises the "circadian pacemaker," which is located in the suprachiasmatic nuclei (SCN) of the hypothalamus and regulates, among other things, body temperature, cortisol, and melatonin release. Process S describes a homeostatic factor of sleep pressure, which indicates the need for sleep. It increases during the day and decreases during sleep. Irregularities in these processes have been observed in patients with sleep disturbances. For example, disturbances in process S can lead to a reduced or deficient level of sleep pressure and disturbances in process C can lead to a shift to a later sleep onset (Borbely, 1982; Wirz-Justice et al., 2013). Irregularities in the sleep–wake processes such as

BOX 13.1

CHRONOTHERAPEUTIC APPROACHES

Bright light therapy: During BLT, patients are exposed to ultraviolet (UV)-filtered broad-spectrum white light (mostly with a light intensity of 10,000 lx for 30 minutes). BLT in the morning leads to earlier melatonin secretion in the evening so that the patient becomes tired earlier.

Wake therapy: WT is known as the fastest acting antidepressant and helps regulate disturbed sleep–wake phases. WT means that the person has to stay awake the whole night and does not sleep until 5:00 p.m. the next day. The mood-enhancing effect arises immediately, but there is a high risk of relapse after restorative sleep. Therefore, WT may be repeated after a few days and supported by a combination with other chronotherapeutics such as BLT or SPA.

Sleep phase advance: SPA moves the time of sleep onset and awakening forward and is used in combination with WT to prevent relapse. After one night of total sleep deprivation (WT), patients are instructed to go to bed earlier (e.g., at 5:00 or 6:00 p.m.). They will be woken around midnight/1:00 p.m. and then have to stay awake the second half of the night.

decreased sleep quality or reduced sleep were found to be associated with depressed mood as well as mood swings, poor functioning, and increased risk of psychiatric illnesses (Wirz-Justice et al., 2009b).

Adjusting the Internal Clock by Chronotherapeutics

According to the two process model, irregularities in the circadian rhythm can be re-synchronized by chronotherapeutics. WT increases the reduced level of sleep pressure (process S) to a normal level (Wirz-Justice et al., 2009a). The underlying mechanisms of the antidepressant effect of WT are not yet completely understood. It has been hypothesized that WT, similar to antidepressant drugs, acts on monoaminergic systems (Wirz-Justice et al., 2009a) and changes the activity of the neurotransmitter systems of serotonin (Salomon et al., 1994), noradrenalin (Mueller et al., 1993), and dopamine (Ebert and Berger, 1998). Furthermore, it has been suggested that WT increases the levels of thyroid hormones and interacts with glycogen synthase kinase 3β and glutamate (Benedetti and Colombo, 2011).

Despite a fast response to WT, many patients relapse after recovering sleep. Therefore, WT is often combined with BLT and/or SPA (Wirz-Justice et al., 2009a). Light, as an "external zeitgeber," influences process C through the secretion of the sleep hormone melatonin. Light falls onto the retina and is transferred to the SCN via retinal ganglion cells. In the SCN, the light suppresses the secretion of melatonin and these changes in the melatonin level influence the sleep–wake phases (Lewy, 2007). As a result, wakefulness is supported and a delayed sleep phase can be normalized. However, longer darkness leads to an increased secretion of melatonin (Maywood et al., 2007; Lewy, 2007). Thus, the effects of BLT and melatonin are contrarious. BLT in the morning leads to advanced melatonin secretion in the evening, whereas the melatonin secretion is delayed when persons are exposed to light in the evening. In contrast, melatonin taken as a sleeping pill leads to an increased melatonin level, meaning that the person becomes tired immediately (Wirz-Justice et al., 2009a). In addition to these indirect effects of BLT, a direct effect on mood is also assumed. Recent data suggest that BLT applied over a period of at least 2 weeks positively influences serotonergic and noradrenergic systems, with more light leading to higher levels of serotonin (e.g., Stephenson et al., 2012). However, research that enhances the understanding of how BLT enfolds its effect is still at its beginning and the underlying mechanisms are not clear.

In summary, chronotherapeutic approaches synchronize a disturbed sleep–wake rhythm, enabling patients to establish a healthier rhythm. Several neurotransmitter and hormonal systems seem to be addressed and altered by chronotherapeutic interventions; however, clear evidence that provides an empirical basis for the assumed underlying mechanisms is lacking.

EMPIRICAL EVIDENCE FOR THE EFFECTIVENESS OF CHRONOTHERAPEUTIC TREATMENTS

Bright Light Therapy

BLT was initially developed for the treatment of seasonal affective disorders. A large body of evidence [e.g., several meta-analyses and randomized controlled trials (RCTs)] strongly supports the effectiveness of BLT in patients with seasonal depression (e.g., Golden et al., 2005). In the past few years, meta-analyses and RCTs have also suggested positive effects of BLT in *adults* with nonseasonal depression. In a Cochrane review on BLT in nonseasonal depression, studies with a higher methodological quality rating showed unequivocal superiority of BLT over control treatment (effect size 0.90, CI 0.31–1.50; Tuunainen et al., 2004).

In a recent RCT, both BLT monotherapy and its combination with fluoxetine were superior to placebo, while fluoxetine treatment alone was not (Lam et al., 2016). Recent meta-analyses support these findings and emphasize a significant reduction of depressive symptoms in nonseasonal depression (e.g., Al-Karawi and Jubair, 2016). In particular, BLT appears to be efficacious when administered for 2–5 weeks (Al-Karawi and Jubair, 2016). However, the evidence is still inconclusive regarding optimal duration and intensity of the light exposure as well as the ideal time point and recommended duration of single treatment sessions.

Studies examining the effectiveness of chronotherapeutics in *children and adolescents* are scarce. However, first evidence suggests that BLT is also effective in children and adolescents. Swedo et al. (1997) conducted the first RCT with BLT in children and adolescents suffering from seasonal depression. Participants (aged 7–17 years) received either 60 minutes of BLT plus 2 hours dawn simulation (active treatment) or 60 minutes of wearing clear glasses plus 5 minutes of low-intensity dawn simulation (placebo) for 1 week. Children younger than 9 years obtain a lower light intensity level (2500 vs 10,000 lx) because of concerns about the exposure of high light intensity levels to children's eyes (Swedo et al., 1997). The results showed a significant reduction in depressive symptoms in the active treatment group compared to placebo. However, age-related effects have not been addressed by the authors and hence, no clear recommendations based on this study can be deducted for younger children.

Another randomized cross-over design study investigated the effect of BLT compared to dim light placebo in 27 adolescents (aged 14–17 years) (Niederhofer and von Klitzing, 2011). A significant reduction in depressive symptoms was found in both groups, but BLT was superior with respect to reduced Beck Depression Inventory (BDI) scores, increased melatonin levels in the evening and decreased melatonin levels in the morning. Bogen et al. (2015) performed an RCT in an inpatient setting to compare BLT (active treatment) with dim light placebo in 58 children and adolescents. The authors found no direct effect of BLT on depression but emphasized phase-shifting effects and a positive influence of BLT on sleep parameters and chronotype: depressive symptoms improved in both groups, but a stable improvement in sleep parameters only emerged in the BLT group. Moreover, again only in the BLT group, the results showed a shift in circadian preference toward morningness (phase advance) measured with a chronotype questionnaire. Interestingly, improvements in sleep quality and the phase advance predicted the reduction in depressive symptoms.

The described studies used BLT boxes with different light intensities. Patients were instructed to sit in front of a light therapy lamp and to look into the light at least once per minute (e.g., Bogen et al., 2015).

The distance between the eyes of the children and the light therapy box had to be 65 cm to reach a light intensity level of 10,000 lx (Gest et al., 2015). Due to the limitations of this method (i.e., inflexibility, inconvenience), other recent studies used light glasses instead of light boxes (Kirschbaum-Lesch et al., 2018; Viola et al., 2014). The results of these studies suggest that light glasses seem to be as efficient as light boxes. An overview of the studies investigating the effectiveness of BLT in adolescents is displayed in Table 13.1.

Research on circadian preference supports the outlined preliminary findings concerning the effectiveness of BLT in adolescents. Due to changes in circadian preference during adolescence, the SPAs following BLT seem to be fruitful to treat sleep disturbances in children and adolescents. The circadian preference, known as the "chronotype," describes whether a person generally prefers earlier bedtimes and waking-up times (morning type), later bedtimes and waking-up times (evening type), or in between (intermediate type). The chronotype depends on environmental (light, work, and school schedules) and genetic factors as well as age, with the chronotype changing over the lifetime. Children are morning types and adolescents evening types, reaching a maximum in their "lateness" at around the age of 20. With increasing age, they become earlier again, moving toward an intermediate type (advancing: Roenneberg et al., 2004). Thus, in adolescence, there is a sleep phase delay (toward an evening type), which in turn is associated with sleep problems, sleepiness during the day, depressive symptoms, and lower well-being (Keller et al., 2016; Randler, 2011; Russo et al., 2017). It is assumed that BLT may work against this shift toward eveningness in adolescents, thus stabilizing the sleep–wake phases on a more "morningness" level consequently improving depressive symptoms.

Summing up the evidence, BLT can be recommended as an add-on treatment in combination with psychotherapeutic and psychopharmacological treatment for adult patients with major depression and comorbid sleep disturbances, whereas clear evidence for children and adolescents is lacking. However, the kind of sleep problems is not characterized well and it may be that BLT is not helpful for those who sleep more compared to those who show less sleep. It has to be stated that up to now, studies suffered from methodological issues such as heterogeneity in various aspects (e.g., light intensity, the duration of exposure to light, treatment duration, diagnostic categories included, control conditions, etc.) and small sample sizes. Hence, it is difficult to compare and generalize the results. Replication studies are therefore needed with larger samples and comparable methodology in order to clarify for whom BLT is helpful and evaluate the best way to apply it. Nevertheless, the outlined preliminary findings are promising and provide first evidence that BLT is applicable in children and adolescents who are 12 years or older with very good

TABLE 13.1 Studies Investigating the Effectiveness of BLT in Adolescents With Depression

Study	Treatment	Duration of trial	Illuminance (lx)	Exposure time	Disorder	Age	*N*	Major findings
Bogen et al. (2015)	BLT and dim light placebo	2 weeks	10,000 vs 100–150	45 min	Major depression, moderate or severe episode	12–18	57	Significant reduction of depressive symptoms in both groups. BLT-only leads to improvements in sleep quality compared to placebo
Gest et al. (2015)	BLT and a combination of BLT plus WT	2 weeks BLT and 1 night WT	10,000	45 min	Major depression, moderate or severe episode	13–18	62	Significant reduction of depressive symptoms and improvement of sleep quality in both groups, but in the long run, BLT was superior to the combined trial
Kirschbaum-Lesch et al. (2018)	BLT with light glasses	4 weeks	10,000	30 min	Major depression, moderate or severe episode	12–18	39	BLT with light glasses showed similarly effects on depressive symptoms and sleep quality as light boxes
Niederhofer and von Klitzing (2011)	BLT and dim light placebo	2 weeks	2500 vs 50	60 min	Mild depressive disorder	14/17	2	BLT was superior to placebo
Niederhofer and von Klitzing (2011)	BLT and dim light placebo	2 weeks	2500 vs 50	60 min	Mild depressive disorder	14–17	28	Significant reduction of depressive symptoms in both groups, but BLT was superior
Papatheodorou and Kutcher (1994)	BLT	1 week	10,000	45–60 min	Bipolar disorder, depressive episode	16–22	7	Significant reduction of depressive symptoms
Saha et al. (2000)	BLT	2 weeks	10,000	3–4 h	Seasonal affective disorder	4	1	Reduction of depressive symptoms within 3–4 days
Swedo et al. (1997)	BLT, dust simulator, dark glasses, and dim light placebo	1 week	2500 vs 10,000 vs 250	60 min	Seasonal affective disorder	7–17	28	BLT was superior to placebo

WT, Wake therapy; *BLT*, Bright light therapy; *N*, Sample size.
Adapted from Gest, S., Legenbauer, T., Bogen, S., Schulz, C., Pniewski, B., Holtmann, M., 2014. Chronotherapeutics: an alternative treatment of juvenile depression. Med. Hypotheses 82, 346–349.

effects on sleep problems and more indirect effects on mood problems. For children under the age of 12 years, evidence is lacking.

Wake Therapy

Since BLT needs some time to reach a certain level of effectiveness (Martiny et al., 2015), it can be combined with WT, which can be repeated during treatment with BLT. WT, also known as total or partial sleep deprivation, was shown to be one of the fastest acting antidepressants, with antidepressant effects already emerging within hours (Moscovici and Kotler, 2009; Wirz-Justice et al., 2005). As most antidepressants need up to 2 weeks to be effective, WT is an effective and fast intervention in severe major depression. Since WT disturbs sleep–wake phases, it is scheduled at the beginning of a chronotherapeutic treatment (Wirz-Justice et al., 2013). Changes occurring after WT are similar to those occurring after pharmacotherapy (Wirz-Justice et al., 2013). Studies in adults comparing total sleep deprivation with partial sleep deprivation have found that being awake in the second half of the night is superior to being awake in the first half (Wirz-Justice and Van den Hoofdakker, 1999). Thus, the second half of the night seems to be crucial for the antidepressant effect of WT.

Even fewer studies have investigated the effectiveness of WT in children and adolescents with depression. One study examined the antidepressant effect of one night of WT in patients with severe depression compared to patients with other psychiatric disorders (Naylor et al., 1993). The results showed an antidepressant effect only for depressed patients. In contrast to WT in adults, in whom depressive symptoms increased again after only one night of recovery sleep (Wirz-Justice et al., 2009a), the antidepressant effect of one night of WT in depressed adolescents (aged 12–17 years) lasted for 3–5 days.

To the best of our knowledge, only one previous study, which was conducted by our research group, has offered a combined treatment of BLT plus WT in children and adolescents (Gest et al., 2015). Sixty-two depressed inpatients (aged 13–18 years) were randomized to either a group receiving one night of WT followed by 2 weeks of morning BLT or a BLT-only group. Based on previous findings in adults and in minors, it was expected that both treatments would reduce depressive symptoms and sleep problems and that the combined treatment would yield faster and greater improvements compared to BLT-only. Both groups reported similar improvements in sleep quality and depressive symptoms after 2 weeks of intervention. Moreover, sleep parameters improved markedly and predicted the reduction in depressive symptoms, as was previously shown by Bogen et al. (2015). Contrary to expectation, the combined treatment of BLT plus WT seemed to have no additional benefit in adolescents. However, one major limitation was the application of only one night of WT, which might be one reason for the lack of impact of WT on the depressive symptoms.

To further understand this outcome and elaborate on the effects of WT in depressed adolescents, a secondary analysis of the data was performed addressing prompt and longer term effects of WT and BLT by means of actigraphy and sleep diary data (Kirschbaum et al., 2017). It was shown that sleep parameters (e.g., total sleep time and sleep efficiency) improved after one night of total sleep deprivation. This result is in line with the homeostatic rebound effect that has been found in adults after one night of sleep deprivation (Wirz-Justice et al., 2009a). However, these effects declined in the second week of intervention. Moreover, the BLT-only group showed a sleep phase advance in the second week of intervention. An overview of studies investigating the effectiveness of WT in adolescents with depression is given in Table 13.2. Taken together,

TABLE 13.2 Studies Investigating the Effectiveness of WT in Adolescents With Depression

Study	Treatment	Disorder	Age	*N*	Major findings
Detrinis et al. (1990)	WT	Major depression	Adolescents	4	The patients experienced improved mood and psychomotor activity
Gest et al. (2015)	WT and BLT	Major depression, moderate or severe episode	13–18	62	Significant reduction of depressive symptoms and improvement of sleep quality in both groups, but in the long run, BLT was superior to the combined trial
King et al. (1987)	Partial sleep deprivation, medication	Major depression	12	1	The patient reported mild beneficial effects on mood
Naylor et al. (1993)	WT	Major depression, major depression in remission, and psychiatric controls	12–17	17	The antidepressant effect after WT was shown for 3–5 days in responders

WT, Wake therapy; *BLT*, Bright light therapy; *N*, Sample size.
Adapted from Gest, S., Legenbauer, T., Bogen, S., Schulz, C., Pniewski, B., Holtmann, M., 2014. Chronotherapeutics: an alternative treatment of juvenile depression. Med. Hypotheses 82, 346–349.

further research is needed to determine whether or not WT should be used in children and adolescents.

Sleep Phase Advance

Combinations of chronotherapeutics in addition to psychotherapeutic and psychopharmacological treatment are used to stabilize and maintain effects on depressive symptoms and sleep problems. For instance, Voderholzer et al. (2003) combined one night of WT with 3 days of SPA. After one night of total sleep deprivation, patients were asked to go to bed at 5:00 p.m. and to get up at 12:00 a.m. On the second night, the sleep phase was shifted to 7:00 p.m.–2:00 a.m. and on the third night to 9:00 p.m.–4:00 a.m. SPA in addition to WT was able to prevent relapse for at least 6 days (Voderholzer et al., 2003). Chronotherapeutics were also shown to reduce the number of days of hospitalization required in depressed adults (Wirz-Justice et al., 2009a). As to our knowledge, no studies applied SPA to children and adolescents. Major Points of BLT, WT and SPA are displayed in Box 13.2.

BOX 13.2

MAJOR POINTS

- First evidence suggests that BLT is effective in children and adolescents.
- Due to methodological differences, studies on BLT are not easy to compare. In addition, they suffer from small sample sizes.
- First studies investigating the effectiveness of WT in children and adolescents showed prompt effects of one night of WT, but these effects decline after a few days.
- One study showed that BLT alone led to greater effects than the combination of BLT and WT.
- Studies using SPA to sustain effects of BLT and WT are still lacking in children and adolescents.
- Further research in children and adolescents is needed to determine whether WT should be used in children and adolescents and whether combinations of chronotherapeutics are able to prevent relapse and increase effect sizes.

APPLICATION OF CHRONOTHERAPEUTICS IN CLINICAL PRACTICE—WHAT DO WE NEED TO KNOW AND WHAT DO WE HAVE TO CONSIDER?

How to Apply Bright Light Therapy

During BLT, patients are exposed to UV-filtered broad-spectrum white light. In most controlled clinical trials, white light of about 10,000 lx is used, which is comparable to skylight 40 minutes after sunrise (Wirz-Justice et al., 2013). This intensity requires an exposure time of 30 minutes (Wirz-Justice et al., 2013). When using lower light intensity, longer exposure times are needed. BLT can be applied either with light therapy boxes or with light glasses. Fig. 13.1 illustrates a light box and light glasses, which were used in pilot studies by our research group. When using light therapy boxes, the distance between the box and the eyes of the adolescent has to be considered to reach the desired light intensity level.

In addition to the already mentioned points of (1) light intensity, (2) duration of exposure, and (3) time of day of exposure, a further factor that has to be addressed is the treatment duration. Empirical evidence suggests that treatment durations of 2–4 weeks might be sufficient; however, it seems that longer durations are superior to shorter treatment durations in adults (Al-Karawi and Jubair, 2016). So far, the optimal treatment duration of BLT for adolescents has not been identified; the limited evidence suggests that at least 2 weeks of BLT should be completed. A possible approach in the treatment with BLT is illustrated in Fig. 13.2.

How to Apply Wake Therapy/Sleep Phase Advance

During WT, patients are required to stay awake the entire night and the following day until approximately 5:00 p.m. Short naps are also not

FIGURE 13.1 A bright light therapy Box (e.g., PhysioLight LD1100; DAVITA) and a light glasses (e.g., luminette; Lucimed, SA) which were used in pilot trials (Bogen et al., 2015; Kirschbaum-Lesch et al., 2018).

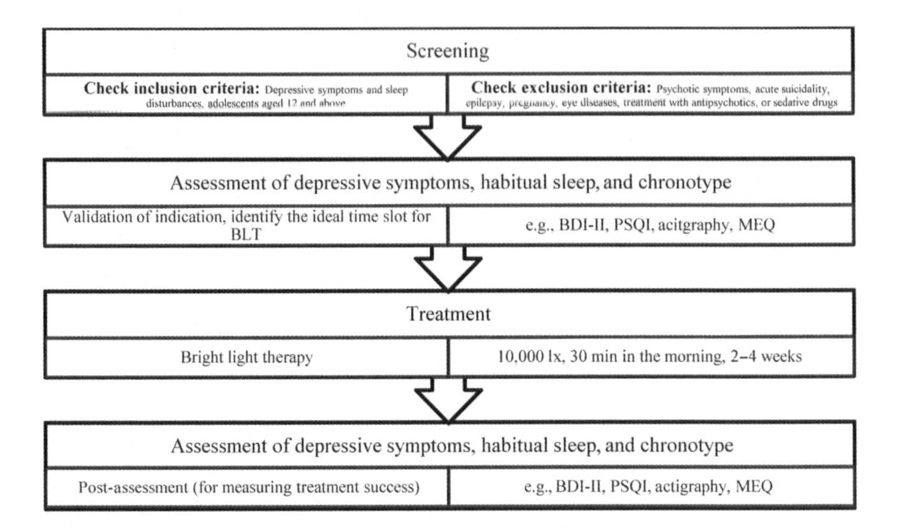

FIGURE 13.2 Flowchart. *BDI-II*, Beck depression inventory-II; *PSQI*, Pittsburgh sleep quality index; *MEQ*, Morningness–eveningness questionnaire.

allowed in this time frame, because even a short nap can induce a depressive relapse (Wirz-Justice et al., 2009a). Patients should be monitored the entire night to ensure compliance, to avoid short naps, to motivate them to stay awake, and to monitor mood fluctuations. In particular, patients should be kept occupied between 3:00 and 4:00 a.m., because this is the time frame with the highest level of sleepiness (Wirz-Justice et al., 2009a). No sedative medication should be given before WT. Stimulants like caffeine are allowed and should be ideally consumed in small portions through the night. To maintain wakefulness, the following activities are recommended: playing group games or computer games, watching TV, internet or DVD movies, cooking, or doing stretching exercises (Gest et al., 2015). Furthermore, it is important to provide one warm light meal and snacks whenever the patient is hungry. Good ways to fight sleepiness are short walks indoors and particularly outdoors, and splashing cold water on the face. It might be easier for patients to stay awake and perform the activities in a group rather than alone.

The following day, breakfast should start earlier and again, until recovering sleep, short naps are not allowed. Outdoor activities should be avoided in the afternoon and the recovery sleep can start in the early evening and should be protected from noise (Wirz-Justice et al., 2009a). When WT is combined with SPA, after one night of total sleep deprivation, patients are instructed to go to bed at 5:00 p.m. and to get up at 12:00 a.m. and then stay awake. The following nights, the sleep onset is shifted 1 or 2 hours later (e.g., Voderholzer et al., 2003). Since evidence on the effects of SPA is only available for adult samples, no guidance can be given for application in adolescents.

Possible Side Effects

Based on our clinical experience, side effects are rare. Side effects which were reported by depressed adolescent inpatients, especially in the treatment with BLT, are headache, dizziness, nausea, eye irritability, and mild visual complaints (e.g., seeing spots in front of the eyes). These adverse events are infrequent and mild and usually occur after the beginning of treatment and subside after a few days.

What Else Has to Be Considered

Chronotherapeutics, especially BLT, are indicated to be embedded in the therapy of children and adolescents with affective disorders and comorbid sleep disturbances. Previous studies investigated the effects of chronotherapeutics in children and adolescents from the age of 12 years. Whether BLT is applicable in children under the age of 12 years remains unclear, because no studies investigated the effect of BLT in this young age group due to ethical considerations regarding possible side effects on eye development. BLT is considered as alternative treatment with only a small number of side effects such as headaches and nausea; however, the mood-enhancing effect and its possible impact on the serotonergic system lead to an exclusion of patients with acute suicidality. Also, chronotherapeutics should be avoided in patients with psychotic symptoms (due to reports on a worsening in delusions) or mania. Further contraindications are epilepsy (for WT, since prolonged wakefulness is a known trigger for seizures; Wirz-Justice et al., 2013), pregnancy, hypersensitivity to light, and eye diseases (for BLT; Gest et al., 2015). Combined treatment with antipsychotics (due to photosensitizing effects of some neuroleptics, e.g., phenothiazine) or sedative drugs (counteracting the effects of WT) is also not recommended, but treatment with antidepressant drugs can be continued (Wirz-Justice et al., 2013).

As studies applying chronotherapeutics in depressed adolescents in an outpatient setting are preliminary in nature, the following recommendations refer to an inpatient setting. First, appropriate diagnostic procedures (questionnaires and methods measuring depressive symptoms, sleep, and chronotype) will be described, and practical guidance for the application of chronotherapeutics will then be provided.

Measurements

Questionnaires measuring depressive symptoms and sleep problems are recommended in order to confirm an indication for BLT and review the success of treatment. For the assessment of depressive mood during the past 2 weeks, the revised BDI (BDI-II; Beck et al., 1996) has shown good validity and internal consistency within samples of adolescents from the age of 13 onward (Besier et al., 2008; Kühner et al., 2007). The BDI-II is often used in clinical studies to measure changes in depressive symptoms. A cutoff score of >13 indicates a clinically relevant depression.

To measure clinician-rated depressive symptoms, the Children's Depression Rating Scale-Revised (CDRS-R; Keller et al., 2011; Poznanski & Mokros, 1996) can be used. It is a widely used semi-structured interview and assesses the changes in depressive symptoms. The CDRS-R is suitable for children and adolescents between 6 and 17 years of age and reveals good psychometric properties (Cronbach's alpha = 0.85; interrater-reliability = 0.92).

Self-reported sleep quality can be assessed using the Pittsburgh Sleep Quality Index (PSQI; Buysse et al., 1989), which is the most frequently used measure in clinical and research practice (Mollayeva et al., 2016). The PSQI consists of 19 items and assesses sleep over the period of 1 month. Higher scores indicate worse sleep, and a cutoff score of five differentiates between good and bad sleepers (sensitivity 89.6%, specificity 86.5%; Buysse et al., 1989). The questionnaire has been developed for adults in the first place; however, some studies also apply it in adolescent samples. An alternative assessment tool specifically created for the use in children and adolescents is the Sleep Disturbance Scale for Children (Bruni et al., 1996).

As studies have shown characteristic changes in sleep architecture in depression, for example, reductions in nonrapid eye movement sleep production, disruptions of sleep continuity, a decrease in rapid eye movement (REM) sleep latency, and an increase in REM sleep duration (e.g., Wang et al., 2015), more objective assessments such as Electroencephalography (EEG) are recommended. Also, a well-established objective measure of sleep is actigraphy, which records sleep behavior objectively over longer time periods and, especially in combination with sleep diaries, is more reliable than self-report measures (Ancoli-Israel et al., 2003). Sleep variables of interest that can be assessed using actigraphy and sleep diary are sleep onset and offset, time in bed, sleep duration, sleep efficiency, and nighttime awakening (for more information, see Kirschbaum et al., 2017).

The Morningness–Eveningness Questionnaire (MEQ; Horne and Östberg, 1976) measures an individual's chronotype and circadian phase shift. The MEQ consists of 16 items and assesses sleep habits and individual circadian preferences. It classifies respondents into one of five groups: "definitely morning type," "moderate morning type," "definitely evening type," "moderate evening type," and "neutral type." The MEQ is suitable for adolescent and adult samples. An online version of the MEQ can be downloaded as a PDF file in several languages (Danish, English, French, German, Italian, Japanese, Portuguese, Russian, and Spanish) at www.cet.org. The MEQ was developed in adult samples; however, it has been applied by our work group in adolescent samples (age range 12–18; Bogen et al., 2015; Gest et al., 2015).

Timing of BLT

Studies reveal a superior effect of BLT compared to placebo at all times of the day, but BLT in the morning shows far better responses than BLT in the evening (Wirz-Justice et al., 2013). Since the individual timing of BLT is crucial for treatment response, it is recommended to identify the ideal time slot for BLT (Wirz-Justice et al., 2009a) depending on the individual circadian rhythm. The dim light melatonin onset (DLMO), representing the onset of the melatonin secretion under dim light conditions, can be regarded as gold standard for characterizing the individual circadian rhythm (Pandi-Perumal et a1., 2007; Wirz-Justice et al., 2009a).

Due to the low feasibility of measuring melatonin concentration in clinical practice, the DLMO can be roughly estimated using the MEQ score, as the two have been found to correlate ($r = -0.73$;

TABLE 13.3 Timing of BLT, 8.5 h After Estimated Melatonin Onset (DLMO) Based on MEQ Score

MEQ score	Time of day (a.m.)	MEQ score	Time of day (a.m.)
23–26	8:15–8:45	50–53	6:30–7:00
27–30	8:00–8:30	54–57	6:15–6:45
31–34	7:45–8:15	58–61	6:00–6:30
35–38	7:30–8:00	62–65	5:45–6:15
39–41	7:15–7:45	66–68	5:30–6:00
42–45	7:00–7:30	69–72	5:15–5:45
46–49	6:45–7:15	73–76	5:00–5:30

Note: Persons with MEQ scores below 23 and above 76 should choose the closest time, because these scores have not yet been clinically evaluated (Wirz-Justice et al., 2009a).
Adapted from Wirz-Justice, A., Benedetti, F., Terman, M., 2009a. Chronotherapeutics for Affective Disorder: A Clinicians Manual for Light and Wake Therapy. Karger, Basel.

Wirz-Justice et al., 2009a). Thus, an algorithm was developed that permits the time of melatonin onset to be estimated from the MEQ score. The algorithm represents the time approximately 8.5 hours after melatonin onset, which seems to be the ideal time slot to facilitate antidepressant responses and circadian phase advances (Wirz-Justice et al., 2009a). Table 13.3 represents the ideal time frame of BLT based on MEQ scores (adapted from Wirz-Justice et al., 2009a). Beginning BLT in a window of 7.5–9.5 hours after melatonin onset is recommended.

CONCLUSION

Chronotherapeutics are noninvasive and beneficial treatments with high compliance and almost no side effects. Especially for BLT, first evidence suggests an improvement of sleep quality and depressive symptoms as well as earlier sleep times (circadian phase shift toward morningness) after treatment in depressed adolescents. Against the background of a biologically determined delay of the circadian rhythm in adolescence, chronotherapeutic approaches seem to be a promising add-on treatment in combination with psychotherapeutic and psychopharmacological approaches for children and adolescents with major depression and comorbid sleep disturbances. The empirical evidence gathered to date suffers from methodological issues. Replication studies with larger sample sizes and state-of-the-art methodology are needed in order to clarify how BLT should best be applied and for whom it is helpful. Furthermore, additional research investigating the effectiveness of WT and SPA in children and adolescents with depression is necessary.

References

Al-Karawi, D., Jubair, L., 2016. Bright light therapy for nonseasonal depression: meta-analysis of clinical trials. J. Affect. Disord. 198, 64–71.

American Psychiatric Association, 2000. Diagnostic and Statistical Manual of Mental Disorders, Text Revision: DSM-IV-TR, Fourth ed American Psychiatric Association (APA), Washington, DC.

Ancoli-Israel, S., Cole, R., Alessi, C., Chambers, M., Moorcroft, W., Pollak, S., 2003. The role of actigraphy in the study of sleep and circadian rhythms. Sleep 1, 342–392.

Bauer, M., Pfennig, A., Severus, E., et al., 2013. World Federation of Societies of Biological Psychiatry (WFSBP) guidelines for biological treatment of unipolar depressive disorders, part 1: update 2013 on the acute and continuation treatment of unipolar depressive disorders. World J. Biol. Psychiatry 14, 334–385.

Beck, A.T., Steer, R.A., Brown, G.K., 1996. Beck Depression Inventory Manual. The Psychological Corporation, San Antonio, TX.

Benedetti, F., Colombo, C., 2011. Sleep deprivation in mood disorders. Neuropsychobiology 64, 141–151.

Besier, T., Goldbeck, L., Keller, F., 2008. Psychometric properties of the Beck depression inventory-II (BDI-II) among adolescent psychiatric patients. Psychother. Psychosom. Med. Psychol. 58 (2), 63–68.

Bettge, S., Wille, N., Barkmann, C., Schulte-Markwort, M., Ravens-Sieberer, U., 2008. Depressive symptoms of children and adolescents in a German representative sample: results of the BELLA study. Eur. Child Adolesc. Psychiatry 17 (1), 71–81.

Bogen, S., Legenbauer, T., Gest, S., Holtmann, M., 2015. Lighting the mood of depressed youth: feasibility and efficacy of a 2 week-placebo controlled bright light treatment for juvenile inpatients. J. Affect. Disord. 190, 450–456.

Borbely, A.A.A., 1982. A two process model of sleep regulation. Hum. Neurobiol. 1, 195–204.

Bruni, O., Ottaviano, S., Guidetti, V., Romoli, M., Innocenzi, M., Cortesi, F., et al., 1996. The Sleep Disturbance Scale for Children (SDSC) Construct ion and validation of an instrument to evaluate sleep disturbances in childhood and adolescence. J. Sleep Res. 5 (4), 251–261.

Buysse, D.J., Reynolds III, C.F., Monk, T.H., Berman, S.R., Kupfer, D.J., 1989. The Pittsburgh sleep quality index: a new instrument for psychiatric practice and research. Psychiatry Res. 28 (2), 193–213.

Chellappa, S.L., Schroder, C., Cajochen, C., 2009. Chronobiology, excessive daytime sleepiness and depression: is there a link? Sleep Med. 10, 505–514.

Curry, J., Silva, S., Rohde, P., Ginsburg, G., Kratochvil, C., Simons, A., et al., 2011. Recovery and recurrence following treatment for adolescent major depression. Arch. Gen. Psychiatry 68 (3), 263–269.

Detrinis, R., Harris, J., Allen, R., et al., 1990. Effects of partial sleep deprivation in children with major depression and attention deficit hyperactivity disorder (ADHD). Sleep Res. 19, 322.

DGPPN, 2012. S3-Guidelines for the Diagnostic and Treatment of Unipolar Depression.

Dolle, K., Schulte-Körner, G., 2013. The treatment of depressive disorders in children and adolescents. Dtsch Ärzteblatt Int. 110 (50), 854–860.

Ebert, D., Berger, M., 1998. Neurobiological similarities in antidepressant sleep deprivation and psychostimulant use: a psychostimulant theory of antidepressant sleep deprivation. Psychopharmacology 140, 1–10.

Emslie, G.J., Mayes, T., Porta, G., Vitiello, B., Clarke, G., Wagner, K.D., et al., 2010. Treatment of resistant depression in adolescents (TORDIA): week 24 outcomes. Am. J. Psychiatry 167 (7), 782–791.

Emslie, G.J., Kennard, B.D., Mayes, T.L., Nakonezny, P.A., Zhu, L., Tao, R., et al., 2012. Insomnia moderates outcome of serotonin-selective reuptake inhibitor treatment in depressed youth. J. Child Adolesc. Psychopharmacol. 22, 21–28.

Gest, S., Legenbauer, T., Bogen, S., Schulz, C., Pniewski, B., Holtmann, M., 2015. Chronotherapeutics: an alternative treatment of juvenile depression. Med. Hypotheses 82, 346–349.

Gest, S., Holtmann, M., Bogen, S., Schulz, C., Pniewski, B., Legenbauer, T., 2015. Chronotherapeutic treatments for depression in youth. Eur. Child. Adolesc. Psychiatry 25 (2), 151–161.

Golden, R.N., Gaynes, B.N., Ekstrom, R.D., Hamer, R.M., Jacobsen, F.M., Suppes, T., et al., 2005. The efficacy of light therapy in the treatment of mood disorders: a review and meta-analysis of the evidence. Am. J. Psychiatry 162, 656–662.

Horne, J.A., Östberg, O., 1976. A self-assessment questionnaire to determine morningness–eveningness in human circadian rhythms. Int. J. Chronobiol. 4, 97–110.

Ivanenko, A., Johnson, K., 2008. Sleep disturbances in children with psychiatric disorders. Semin. Pediatr. Neurol. 15, 70–78.

Keller, F., Grieb, J., Ernst, M., Sproeber, N., Fegert, J.M., Koelch, M., 2011. Children's Depression Rating Scale – Revised (CDRS-R): Development of a German version and psychometric properties in a clinical sample. Z. Kinder Jugendpsychiatr. Psychother. 39, 179–185.

Keller, L.K., Zöschg, S., Grünewald, B., Roenneberg, T., Schulte-Körne, G., 2016. Chronotyp und Depression bei Jugendlichen—ein Review. Z. Kinder Jugendpsychiatr. Psychother. 44 (2), 113–126.

Kennard, B.D., Silva, S., Vitiello, B., Kratochvil, C., Simons, A., 2006. Remission and residual symptoms after shortterm treatment in the treatment of adolescents with depression study (TADS). J. Am. Acad. Child Adolesc. Psychiatry 45, 1401–1411.

King, B.H., Baxter Jr, L.R., Stuber, M., Fish, B., 1987. Therapeutic sleep deprivation for depression in children. J. Am. Acad. Child Adolesc. Psychiatry 26, 928–931.

Kirschbaum, I., Straub, J., Gest, S., Holtmann, M., Legenbauer, T., 2017. Short-term effects of wake–and bright light therapy on sleep in depressed youth. Int. J. Chronobiol. 35 (1), 101–110.

Kirschbaum-Lesch, I., Gest, S., Holtmann, M., Legenbauer, T., 2018. Feasibility and efficacy of four weeks of morning bright light therapy with light glasses in depressed adolescent inpatients. Z. Kinder Jugendpsychiatr. Psychother. 46 (5), 423–429.

Klein, D.N., Shankman, S.A., Lewinsohn, P.M., Seeley, J.R., 2009. Subthreshold depressive disorder in adolescents: predictors of escalation to full-syndrome depressive disorders. J. Am. Acad. Child Adolesc. Psychiatry 48 (7), 703–710.

Kühner, C., Bürger, C., Keller, F., Hautzinger, M., 2007. Reliability and validity of the revised Beck Depression Inventory (BDI-II). Results from the German samples. Nervenarzt 78, 651–656.

Lam, R.W., Levitt, A.J., Levitan, R.D., Michalak, E.E., Cheung, A.H., Morehouse, R., et al., 2016. Efficacy of bright light treatment, fluoxetine, and the combination in patients with nonseasonal major depressive disorder: a randomized clinical trial. JAMA Psychiatry 73, 56–63.

Lewinsohn, P.M., Clarke, G.N., Seeley, J.R., Rohde, P., 1994. Major depression in community adolescents: age at onset, episode duration, and time to recurrence. J. Am. Acad. Child Adolesc. Psychiatry 33, 809–818.

Lewinsohn, P.M., Rohde, P., Seeley, J.R., 1998. Major depressive disorder in older adolescents: prevalence, risk factors, and clinical implications. Clin. Psychol. Rev. 18, 765–794.

Lewy, A.J., 2007. Melatonin and human chronobiology. Cold Spring Harb. Symp. Quant. Biol. 72, 623–636.

Luca, A., Luca, M., Calandra, C., 2013. Sleep disorders and depression: brief review of the literature, case report, and nonpharmacologic interventions for depression. Clin. Intervent. Aging 8, 1033–1039.

Martiny, K., Refsgaard, E., Lund, V., Lunde, M., Thoughaard, B., Lindberg, L., et al., 2015. Maintained superiority of chronotherapeutics vs. exercise in a 20-week randomized follow-up trial in major depression. Acta Psychiatr. Scand. 131, 446–457.

Maywood, E.S., O'Neill, J.S., Reddy, A.B., Chesham, J.E., Prosser, H.M., Kyriacou, C.P., et al., 2007. Genetic and molecular analysis of the central and peripheral circadian clockwork of mice. Cold Spring Harb. Symp. Quant. Biol. 72, 85–94.

Measelle, J.R., Stice, E., Hogansen, J., 2006. Developmental trajectories of co-occurring depressive, eating, antisocial, and substance abuse problems in adolescent girls. J. Abnorm. Psychol. 115 (3), 524–538.

Mehler-Wex, C., Kölch, M., 2008. Depression in children and adolescents. Dtsch. Ärzteblatt Int. 105, 149–155.

Mollayeva, T., Thurairajah, P., Burton, K., Mollayeva, S., Shapiro, C.M., Colantonio, A., 2016. The Pittsburgh sleep quality index as a screening tool for sleep dysfunction in clinical and non-clinical samples: a systematic review and meta-analysis. Sleep Med. Rev. 25, 52–73.

Moscovici, L., Kotler, M., 2009. A multistage chronobiologic intervention for the treatment of depression: a pilot study. J. Affect. Disord. 116, 201–2017.

Mueller, H.-U., Riemann, D., Berger, M., Mueller, W.E., 1993. The influence of total sleep deprivation on urinary excretion of catecholamine metabolites in major depression. Acta Psychiatr. Scand. 88, 16–20.

Naylor, M.W., King, C.A., Landsay, K.A., Evans, T., Armelagos, J., Shain, B.N., et al., 1993. Sleep deprivation in depressed adolescents and psychiatric controls. J. Am. Acad. Child Adolesc. Psychiatry 32 (4), 753–759.

Niederhofer, H., von Klitzing, K., 2011. Bright light treatment as add-on therapy for depression in 28 adolescents: a randomized trial. Prim. Care Companion CNS Disord. 13 (6).

O'Brien, E.M., Chelminski, I., Young, D., Dalrymple, K., Hrabosky, J., Zimmerman, M., 2011. Severe insomnia is associated with more severe presentation and greater functional deficits in depression. J. Psychiatr. Res. 45 (8), 1101–1105.

Pandi-Perumal, S.R., Smits, M., Spence, W., Srinivasan, V., Cardinali, D.P., Lowe, A.D., et al., 2007. Dim light melatonin onset (DLMO): a tool for the analysis of circadian phase in human sleep and chronobiological disorders. Prog. Neuropsychopharmacol. Biol. Psychiatry 31 (1), 1–11.

Papatheodorou, G., Kutcher, S., 1994. The effect of adjunctive light therapy on ameliorating breakthrough depressive symptoms in adolescent-onset bipolar disorder. J. Psychiatry Neurosci. 20, 226–232.

Patton, G.C., Coffey, C., Romaniuk, H., Mackinnon, A., Carlin, J.B., Degenhardt, L., et al., 2014. The prognosis of common mental disorders in adolescents: a 14-year prospective cohort study. Lancet 383, 1404–1411.

Poznanski, E.O., Mokros, H.B., 1996. Children's Depression Rating Scale, Revised (CDRS-R) Manual. Western Psychological Services, Los Angeles, CA.

Randler, C., 2011. Association between morningness-eveningness and mental health in adolescents. Psychol. Health Med. 16 (1), 29–38.

Ravens-Sieberer, U., Wille, N., Bettge, S., Erhart, M., 2007. Psychische Gesundheit von Kindern und Jugendlichen in Deutschland. Bundesgesundheitsblatt 5, 871–878.

Ravindran, A.V., Balneaves, L.G., Faulkner, G., et al. Canadian Network for Mood and Anxiety Treatments (CANMAT, 2016). Clinical Guidelines for the Management of Adults with Major Depressive Disorder: Section 5. Complementary and Alternative Medicine Treatments. Canadian journal of psychiatry Revue canadienne de psychiatrie, 61(9), 576–587.

Robillard, R., Hermens, D.F., Naismith, S.L., White, D., Rogers, N.L., Ip, T.K., et al., 2015. Ambulatory sleep-wake patterns and variability in young people with emerging mental disorders. J. Psychiatry Neurosci. 40, 28–37.

Roenneberg, T., Kuehnle, T., Pramstaller, P.P., Ricken, J., Havel, M., Guth, A., et al., 2004. A marker for the end of adolescence. Curr. Biol. 14.

Russo, P., Biasi, V., Cipolli, C., Mallia, L., Caponera, E., 2017. Sleep habits, circadian preference, and school performance in early adolescents. Sleep Med 29, 20–22.

Saha, S., Pariante, C.M., McArdle, T.F., Fombonne, E., 2000. Very early onset seasonal affective disorder: a case study. Eur. Child Adolesc. Psychiatry 9, 135–138.

Salomon, R.M., Delgado, P.L., Licinio, J., Krystal, J.H., Heninger, G.R., Charney, D.S., 1994. Effects of sleep deprivation on serotonin function in depression. Soc. Biol. Psychiatry 36, 840–846.

Stephenson, K.M., Schroder, C.M., Bertschy, G., Bourgin, P., 2012. Complex interaction of circadian and non-circadian effects of light on mood: shedding new light on an old story. Sleep Med. Rev. 16, 445–454.

Sunderajan, P., Gaynes, B.N., Wisniewski, S.R., Miyahara, S., Fava, M., Akingbala, F., et al., 2010. Insomnia in patients with depression: a STAR*D report. CNS Spectr. 15 (6), 394–404.

Swedo, S.E., Allen, A.J., Glod, C.A., Clark, C.H., Teicher, M.H., Richter, D., et al., 1997. A controlled trial of light therapy for the treatment of pediatric seasonal affective disorder. J. Am. Acad. Child Adolesc. Psychiatry 36 (6), 816–821.

Tuunainen, A., Kripke, D.F., Endo, T., 2004. Light therapy for nonseasonal depression. Cochrane Database Syst. Rev., CD004050.

Viola, A., Hubbard, J., Comtet, H., Hubbard, I., Delloye, E., Ruppert, E., & Bourgin, P. (2014). Benefi cial effect of morning light after one night of sleep deprivation: Light glasses versus light box administration (Poster). Society for Light Treatment and Biological Rhythms Annual Meeting, Vienna, Austria.

Voderholzer, U., Valerius, G., Schaerer, L., Riemann, D., Giedke, H., Schwärzler, F., et al., 2003. Is the antidepressive effect of sleep deprivation stabilized by a three day phase advance of the sleep period? A pilot study. Eur. Arch. Psychiatry Clin. Neurosci. 253, 68–72.

Wang, Y.Q., Li, R., Zhang, M.Q., Zhang, Z., Qu, W.M., Huang, Z.L., 2015. The neurobiological mechanisms and treatments of REM sleep disturbances in depression. Curr. Neuropharmacol. 13 (4), 543–553.

Winokur, A., Gary, K.A., Rodner, S., Rae-Red, C., Fernando, A.T., Szuba, M.P., 2001. Depression, sleep physiology, and antidepressant drugs. Depress. Anxiety 14, 19–28.

Wirz-Justice, A., Benedetti, F., Berger, M., 2005. Chronotherapeutics (light and wake therapy) in affective disorders. Psychol. Med. 35, 939–944.

Wirz-Justice, A., Benedetti, F., Terman, M., 2009a. Chronotherapeutics for Affective Disorder: A Clinicians Manual for Light and Wake Therapy. Karger, Basel.

Wirz-Justice, A., Bromundt, V., Cajochen, C., 2009b. Circadian disruption and psychiatric disorders: the importance of entrainment. Sleep Med. Clin. 4 (2), 273–284.

Wirz-Justice, A., Benedetti, F., Terman, M., 2013. Chronotherapeutics for Affective Disorders: A Clinician's Manual for Light and Wake Therapy, second ed Karger, Basel.

Wirz-Justice, A., Van den Hoofdakker, R.H., 1999. Sleep deprivation in depression: what do we know, where do we go? Biol. Psychiatry 46, 445–453.

Further Reading

Dubicka, B., Hadley, S., Roberts, C., 2006. Suicidal behaviour in youths with depression treated with new-generation antidepressants—meta-analysis. Br. J. Psychiatry 189, 393–398.

Gregory, A.M., OConnor, T.G., 2002. Sleep problems in childhood: a longitudinal study of developmental change and association with behavioral problems. J. Am. Acad. Child Adolesc. Psychiatry 41, 964–971.

Gregory, A.M., Sadeh, A., 2012. Sleep, emotional and behavioral difficulties in children and adolescents. Sleep Med. Rev. 16, 129–136.

Holtmann, M., Bolte, S., Poustka, F., 2006. Suicidality in depressive children and adolescents during treatment with selective serotonin reuptake inhibitors review and meta-analysis of the available randomised, placebo controlled trials. Nervenarzt 77, 1332–1337.

Neumeister, A., Goessler, R., Lucht, M., Kapitany, T., Bamas, C., Kasper, S., 1996. Bright light therapy stabilizes the antidepressant effect of partial sleep deprivation. Biol. Psychiatry 39, 16–21.

Whittington, C.J., Kendall, T., Fonagy, P., Cottrell, D., Boddington, E., 2004. Selective serotonin reuptake inhibitors in childhood depression: systematic review of published versus unpublished data. Lancet 363, 1341–1345.

CHAPTER

14

Conclusions and Future Directions for Neurotechnology and Brain Stimulation Treatments in Pediatric Psychiatric and Neurodevelopmental Disorders

Lindsay M. Oberman[1] *and Peter G. Enticott*[2]

[1]Center for Neuroscience and Regenerative Medicine, Henry M. Jackson Foundation for the Advancement of Military Medicine, Rockville, MD, United States [2]Cognitive Neuroscience Unit, School of Psychology, Deakin University, Geelong, VIC, Australia

OUTLINE

DOI: https://doi.org/10.1016/B978-0-12-812777-3.00014-3

NEUROTECHNOLOGY AND BRAIN STIMULATION IN PEDIATRIC PSYCHIATRIC AND NEURODEVELOPMENTAL DISORDERS

The chapters in this volume provide the most up-to-date findings in the area of device-based treatments for pediatric Psychiatric and Neurodevelopmental disorders. The tools and techniques reviewed included environmental stimulation (Chapter 3: Environmental Stimulation Modulating the Pathophysiology of Neurodevelopmental Disorders), transcranial magnetic stimulation (Chapters 5–8), transcranial direct current stimulation (Chapter 9: tDCS in Pediatric Neuropsychiatric Disorders), deep brain stimulation (Chapter 10: Deep Brain Stimulation for Pediatric Neuropsychiatric Disorders), neurofeedback (Chapters 11 and 12), and chronotherapy (Chapter 13: Chronotherapy for Adolescent Major Depression). Though each chapter explores the use of these tools and techniques in different Psychiatric and Neurodevelopmental disorders, there are several commonalities that are present across all chapters, including theoretical promise, supportive preliminary data, and small-scale studies with numerous limitations.

There is much reason to be optimistic. The past two decades have seen remarkable advancement in the understanding of brain networks involved in the pathophysiology of Psychiatric and Neurodevelopmental disorders fueled by advances in neurotechnology and basic neuroscience techniques. The brain stimulation and other device-based tools described in the preceding chapters have the capacity to engage and induce long-term modulation in targeted brain networks. In addition, with the exception of deep brain stimulation, the tools and techniques discussed herein are considered "minimal risk," even in pediatric populations. Certainly, all of the tools and techniques described in this volume have a more desirable side-effect profile than many more commonly prescribed pharmacological agents.

The degree to which brain target engagement leads to the desired behavioral improvements is still uncertain. The studies discussed in this volume almost exclusively were small-scale studies and many are open label. To date, there has not been any "Level 1" large-scale, placebo (sham)-controlled prospective clinical trial with unequivocal findings of any brain stimulation or device-based treatment approach in children with Psychiatric or Neurodevelopmental disorders. In fact, most of the evidence for efficacy in pediatric populations is extrapolated from the efficacy in adult studies. But, as reviewed in the initial chapters of this volume, children's brains and the underlying neuropathophysiology leading to pediatric disorders cannot and should not be assumed to be the same as adult's. Thus, we have a far way to go before brain stimulation and other neurotechnology should be offered as "Evidence-Based Practice."

OPEN QUESTIONS

Several questions remain unanswered when it comes to the use of neurotechnology in pediatric populations, including: (1) Dosing, (2) Treatment target, and (3) How and when to apply these interventions for optimal therapeutic effect.

Dosing

It is unclear whether the dosage used for adult protocols are appropriate or safe for children. As discussed in Chapter 3, Environmental Stimulation Modulating the Pathophysiology of Neurodevelopmental Disorders, from birth until puberty, the gross structure of the head and scalp is changing and growing; thus, the trajectory and strength of the magnetic and resultant electrical field will also change accordingly. In addition, once the field enters the brain, how it affects the immature neurons and networks also varies across the age span. Though most of neurogenesis is completed prenatally, as discussed in Chapter 2, The Developing Brain—Relevance to Pediatric Neurotechnology, there are many molecular and cellular changes across childhood affecting excitability, plasticity, and development of neurocircuits from birth until puberty. These varying levels of neurotransmitters, synaptogenesis, and establishment of functional circuits create a dynamic system, and it is very hard to predict what may be the best protocol to apply at any given age or developmental stage. There is also a paucity of data on the effects of repetitive transcranial magnetic stimulation (rTMS) on typically developing children. The ethical reasons for such a lack of data is understandable, as many parents are likely resistant stimulating the brains of their typically developing children, and IRBs may not consider this approach reflects an acceptable risk-to-benefit ratio. However, this lack of normative data makes it difficult to predict the effects of rTMS protocols on the developing human brain.

In addition, it is well known that neural circuitry (especially in the prefrontal cortex) does not achieve adult levels until age 25 (Casey et al., 2008). Longitudinal magnetic resonance imaging (MRI) studies show that just prior to puberty, there is a surge of neuronal proliferation and a thickening of gray matter (Baird et al., 1999; Giedd et al., 1999). Then, from puberty to approximately 25 years old, the brain undergoes dendritic pruning, eliminating unused synapses and myelination to increase the speed of impulse conduction across the brain's functional circuits (Arain et al., 2013). Thus, what is the appropriate intensity to apply brain stimulation at the various ages or developmental stages? Should children receive higher or lower intensities of stimulation? Intuitively, we may think lower (erring on the side of caution and safety), however, as discussed in Chapter 2, The Developing Brain—Relevance to Pediatric Neurotechnology, motor thresholds (the established basis for intensity in adult transcranial magnetic stimulation (TMS) research) are higher in children as compared to adults, so perhaps higher intensities of stimulation are required in children to induce similar levels of target engagement.

Beyond intensity, we also must consider other dosing parameters, including frequency, number of sessions, intersession interval, etc. Related to frequency, we know that the pediatric and pubertal brain are especially susceptible to seizure induction (Rakhade and Jensen, 2009; Hameed et al., 2017). In addition, the risk of TMS-induced seizures increases proportionate to increases in both intensity, frequency, and inter-train interval. Thus, high-frequency and small inter-train intervals may be more likely to induce seizures in children as compared to adults. Finally, increased plastic (and perhaps metaplastic) state of the pediatric brain may also influence the number of sessions necessary for clinical efficacy and the appropriate time between sessions (typically therapeutic protocols are applied daily for 4–6 weeks or more in adult applications). All of these factors will need to be considered when designing pivotal trials for pediatric stimulation.

Treatment Target

Another unanswered question is what brain region, or network, or frequency band should be targeted. Historically, the despite the fact that both Neurological and Psychiatric disorders stem from pathology in the brain, the main difference lies in the relative lack of reliably identifiable structural or chemical etiology. Thus, what brain structure or network should one stimulate, or what frequency band should one target that will lead to behavioral benefit? By analogy, if a patient complains to the doctor that he can't walk (behavioral symptom), and by all objective accounts based on standardized behavioral evaluations, the doctor confirms the patient's deficits. The appropriate treatment may be a splint or a cast (if he has a broken bone) or may be physical therapy (if his problem is muscular) or may involve anti-inflammatory medication or another treatment depending on the underlying cause of the behavioral symptom. Thus, before treatment, the doctor would need to do further testing before he can prescribe a given treatment. The same can be said for Psychiatric or Neurodevelopmental disorders. Certainly, with advances in brain imaging and a focus on identifying neural mechanisms of Psychiatric disorders, we can make educated hypotheses about potential targets.

Once a potential target is identified, and supported by the literature, there are a few additional steps that need to be evaluated prior to determining if this is the optimal target for a given individual (at a specific age or developmental stage) with this behaviorally defined disorder. First, does the literature that you are basing your target on come from individuals of approximately the same age or developmental stage? Many studies, due to difficulty with recruitment and in an effort to generalize the findings, enroll participants across broad age ranges. However, as discussed above, one cannot assume that a given target is stable across the age span. What may be associated with a behavioral symptom at one age may not be at a different age. Second, even if the majority of individuals with a given disorder show a given dysfunction in the targeted brain region (or network or frequency band,) one needs to confirm it on an individual level. Most "biomarkers" for Psychiatric and Neurodevelopmental disorders are based on group means, rather than individual data. It is unreasonable to expect that everyone with a given Psychiatric or Neurodevelopmental disorder will show a single neural pathology. If they did, the disorder would likely be classified as Neurological rather than Psychiatric and be diagnosed biologically rather than behaviorally. So one needs to confirm that the target of the stimulation or device-based treatment is in fact impaired in the given individual.

Finally, one needs to decide upon the ultimate primary outcome. If the disorder is defined as a complex set of symptom domains [such as autism spectrum disorder (ASD), where the core deficits are both impairments in social communication as well as the presence of restricted and repetitive behaviors, or schizophrenia, where there are both positive and negative symptoms], a given intervention is likely not going to treat every aspect of that disorder as each symptom domain may be the result of dysfunction in different brain regions/networks. Thus, one's behavioral target should be confined to the behavioral domain that is predicted to be subserved by the targeted brain region/network. It then follows that if the goal is a therapeutic intervention, then the primary outcome measure should be behavioral and clinically meaningful. If a given intervention targets a specific brain region/network and the result is a modulation of that brain region/network, the study has shown brain target engagement but has not shown clinical efficacy. In the end, patients do not come to doctors to treat "impairments in functional connectivity" or an "imbalance in beta:theta ratio."

It reasonable to believe, but not a given, that if a neural mechanism underlies a given behavioral impairment then modulating it will lead to behavioral improvements. It is possible that the neural mechanism does in fact underlie the behavioral symptoms, but as a result of brain plasticity, that mechanism is no longer being used for its intended purpose. For example, in a study conducted by Cohen et al. (1997), they showed that disrupting the function of primary visual cortex in blind participants who used their visual cortex to read Braille leads to errors on the task, but the same stimulation had no effect on tactile performance in normal-sighted subjects. The authors concluded that blindness from an early age leads to plastic changes whereby visual cortex is now recruited for somatosensory processing. Similarly, in individuals with Neurodevelopmental disorders, it is possible that they use a different part of the brain to process, for example, social stimuli, and thus, stimulating regions implicated in "typical" processing of social stimuli may not result in the expected improvements because that function has been "remapped" to another region.

How and When to Apply an Intervention for Optimal Therapeutic Effects

Once a brain target is identified and dysfunction in that brain target has been reasonably associated with the behavioral symptom of interest, there still remains the question of how to modulate that network for optimal treatment outcomes. If a brain region is underactive, would increased activity lead to behavioral improvements? Similarly, if a region is overactive would suppressing its activity be therapeutic? One could imagine three options. First, the straightforward and intuitive option. Yes, "normalizing" the activity in the brain will lead to a therapeutic effect. But what if, as in the example above, this region is underactive because it isn't used for the expected function? In the second case, "normalizing" activity may result in no effect on the targeted behavior, because in this individual, he may not be using this network for the expected function. Finally, what if pathologically increased activity in a given network is actually compensatory? Perhaps, the child needs the increased activity to function at the level that they do. Thus, suppressing activity may lead to a decrement in performance in a given domain. As it is difficult to prospectively predict which of these options may be correct, it is critical to conduct proof of mechanism and target-engagement studies before engaging in a randomized clinical trial.

Also unclear is when is the optimal time for modulating a given brain network. It is likely that dysfunction of brain circuitry precedes behavioral symptom presentation. For example, in some cases, differences can be identified in infants who go on to develop ASD (Varcin and Nelson, 2016). Thus, should we intervene at that point as these abnormalities predict later deficits? If so, could the development of the behavioral symptoms be avoided? If this were the case, this may be quite desirable. However, what are the long-term results of such a modulation? Could we be setting the child's brain on a different trajectory that may have negative consequences later in life? In addition, as discussed above, the association between network "dysfunction" and behavioral symptom is not 1:1. If we intervene before a behavioral symptom manifests, it is possible that the behavioral symptom may never have developed. Thus, in the end, if the ultimate target is a disabling symptom, then perhaps we should wait until said symptom manifests and becomes disabling rather than trying to predict who may or may not benefit from a preemptive brain modulation.

CONCLUSIONS AND FUTURE DIRECTIONS

With all of this said, where do we go from here? What is the future of neurotechnology and brain stimulation approaches for pediatric Psychiatric and Neurodevelopmental disorders? Do the challenges and ethical concerns outweigh the potential? We would like to think not. With an estimated 241 million youth worldwide affected by Psychiatric or Neurodevelopmental disorders, and inadequate existing interventional options, we must continue to develop innovative solutions. As with much of science, multidisciplinary collaboration is critical. Clinical Neuroscientists must continue to stay informed, and better collaborate with Basic Neuroscientists to design protocols that are informed by brain development. Those working with preclinical animal models of pediatric Psychiatric and Neurodevelopmental disorders must work to translate and validate their findings into human clinical trials. We must advocate for funding for properly powered, properly controlled research studies capable of answering some of the unanswered questions posed above. We need to start conducting "Level 1" large-scale, placebo (sham)-controlled prospective clinical trials to establish empirical evidence of efficacy.

We need to recognize that a "one size fits all" treatment across the age span and across individuals will not be effective in treating the complex behavioral sequelae that affect children and adolescents with Psychiatric and Neurodevelopmental disorders. We need to take a personalized approach to determine which behavior to target, who is most likely to respond to which treatment approach based on moderating and mediating factors, and how and when to intervene. We need to stratify our sample based on physiological biomarkers, age/developmental stage, and clinical presentation. We need to continue to validate biomarkers that not only have diagnostic and prognostic potential, but also those that predict response or track with clinical change. We need to use adaptive designs, similar to those being used in large-scale cancer trials, to efficiently hone in on the most optimal parameters (see Alexander et al. (2013 for an example of an adaptive trial design that could be applied to brain stimulation protocols for use in pediatric Psychiatric and Neurodevelopmental disorders).

We recognize that what we are proposing is ambitious and not without challenges. We recognize that large-scale trials such as those required to answer these questions take time and require a huge amount of operational and regulatory oversight. However, our children and their families deserve better, and the existing options are simply not good enough. If effective, these innovative, non-pharmacological, and targeted interventions will markedly alter the way we treat children with Psychiatric and Neurodevelopmental disorders. The benefits of such treatments will impact not only the child, but the adult they will become, their family, caretakers, and community. We must push forward; the mental health and future well-being of our children depends on it.

References

Alexander, B.M., Wen, P.Y., Trippa, L., Reardon, D.A., Yung, W.K., Parmigiani, G., et al., 2013. Biomarker-based adaptive trials for patients with glioblastoma—lessons from I-SPY 2. Neuro Oncol. 15 (8), 972–978.

Arain, M., Haque, M., Johal, L., Mathur, P., Nel, W., Rais, A., et al., 2013. Maturation of the adolescent brain. Neuropsychiatr. Dis. Treat. 9, 449–461.

Baird, A.A., Gruber, S.A., Fein, D.A., Maas, L.C., Steingard, R.J., Renshaw, P.F., et al., 1999. Functional magnetic resonance imaging of facial affect recognition in children and adolescents. J. Am. Acad. Child Adolesc. Psychiatry 38 (2), 195–199.

Casey, B.J., Jones, R.M., Hare, T.A., 2008. The adolescent brain. Ann. N.Y. Acad. Sci. 1124, 111–126.

Cohen, L.G., Celnik, P., Pascual-Leone, A., Corwell, B., Falz, L., Dambrosia, J., et al., 1997. Functional relevance of cross-modal plasticity in blind humans. Nature 389 (6647), 180–183.

Giedd, J.N., Blumenthal, J., Jeffries, N.O., Castellanos, F.X., Liu, H., Zijdenbos, A., et al., 1999. Brain development during childhood and adolescence: a longitudinal MRI study. Nat. Neurosci. 2 (10), 861–863.

Hameed, M.Q., Dhamne, S.C., Gersner, R., Kaye, H.L., Oberman, L.M., Pascual-Leone, A., et al., 2017. Transcranial magnetic and direct current stimulation in children. Curr. Neurol. Neurosci. Rep. 17 (2), 11.

Rakhade, S.N., Jensen, F.E., 2009. Epileptogenesis in the immature brain: emerging mechanisms. Nat. Rev. Neurol. 5 (7), 380–391.

Varcin, K.J., Nelson III, C.A., 2016. A developmental neuroscience approach to the search for biomarkers in autism spectrum disorder. Curr. Opin. Neurol. 29 (2), 123–129.

Index

Note: Page numbers followed by "*f*," "*t*," and "*b*" refer to figures, tables, and boxes, respectively.

F

G

H

I

N

V

W

Z

Printed in the United States
By Bookmasters